Charles J. Byrne

Lunar Orbiter Photographic Atlas of the Near Side of the Moon

With 619 Illustrations and a CD-ROM

Charles J. Byrne
Image Again
Middletown, NJ
USA

Cover illustration: Earth-based photograph of the full Moon from the "Consolidated Lunar Atlas" on the Website of the Lunar and Planetary Institute.

British Library Cataloging-in-Publication Data
Byrne, Charles J., 1935–
Lunar Orbiter photographic atlas of the near side of the Moon
1. Lunar Orbiter (Artificial satellite) 2. Moon–Maps 3. Moon–Photographs from space
I. Title
523.3′0223
ISBN 1852338865

Library of Congress Cataloging-in-Publication Data
Byrne, Charles J., 1935–
Lunar Orbiter photographic atlas of the near side of the Moon : with 619 figures /
Charles J. Byrne.
p. cm.
Includes bibliographical references and index.
ISBN 1-85233-886-5 (acid-free paper)
1. Moon–Maps. 2. Moon–Photographs from space. 3. Moon–Remote-sensing images.
4. Lunar Orbiter (Artificial satellite) I. Title.
G1000.3.B9 2005
523.3′022′3–dc22 2004045006

ISBN 1-85233-886-5 Printed on acid-free paper.

Printed in China. (EXP/EVB)

9 8 7 6 5 4 3 2 1 SPIN 10978726

Springer Science+Business Media
springeronline.com

Preface

The Moon is Earth's nearest neighbor. Since the dawn of intelligence, our eyes have seen the Moon, puzzling over its shady figures, its phases, its motions in the sky, and its relation to tides. Even the smallest telescopes resolve the shadows into a heavily cratered surface, stimulating the imagination. Each advance of the astronomer's art has revealed new insights into the nature of the lunar surface, until curiosity and competition led the American and Russian space programs to send orbital cameras, robotic landers and rovers, and the Apollo astronaut exploration teams to the Moon.

The Lunar Orbiter program, a series of five photographic spacecraft launched in 1966 and 1967, was motivated by the need to find and certify safe and interesting landing sites for the Apollo spacecraft. When the Lunar Orbiter program was started (1964), no spacecraft had landed on the Moon, but the Apollo program was committed to safely land the Lunar Module, with two astronauts on board. At the time, I was working in the lunar environment group of Bellcomm, Inc. AT&T established Bellcomm at the request of the National Aeronautics and Space Administration (NASA) to support the Apollo project headquarters group. Our responsibility included the challenging assignment of finding a safe landing site for a vehicle about the size of a helicopter, with a half-meter (0.5-m, 20-inch) ground clearance and limited ability to land on a slope. Of course, we had very little information about the lunar surface at such scales. There was some information from lunar photometry and radar scatter measurements, but there were strong uncertainties about what aspects of the surface were being measured; in particular, the soil strength was an unknown. Speculation raised possibilities of dust floated by static electricity or fragile glasslike lava.

The requirements for Lunar Orbiter were established to achieve the best possible resolution within the state of the art and to obtain imagery of that resolution over a significant percentage of the area available for Apollo landings. Targeted Apollo landing sites had to be as smooth as possible over a large enough area to accommodate the down-range and cross-range navigation errors, determined by the tracking, and control uncertainties associated with factors such as the largely unknown gravity anomalies.

NASA's Langley Research Center was chosen to manage the Lunar Orbiter program. I had the pleasure of drafting the specifications and participating in the selection of contractors. The resulting spacecraft and camera designs of Boeing Aircraft and Eastman Kodak (respectively) were capable of enormous data collection capacity, even in today's terms. All together, about 1000 pairs of medium- and high-resolution exposures were made during the five missions. The negatives were developed in orbit, scanned, and transmitted to photographic and magnetic tape recorders in the three stations of the Deep Space Network operated by the Jet Propulsion Laboratory (JPL) in California, Spain, and Australia. Each exposure results in one medium-resolution frame and one long high-resolution frame, usually presented as three subframes.

Although five missions were planned to compensate for possible failures, either of spacecraft or rejection of initial target sites, the survey for early Apollo landing sites was completed in the first three missions. As a result, the fourth mission was used for a comprehensive survey of the near side of the Moon; these are the photographs that are the primary contents of this book. The fifth mission examined many scientific sites at very high resolution, surveyed a few additional landing sites for later Apollo missions, and improved coverage of the far side of the Moon. Since the Lunar Orbiter missions returned their extensive photographic coverage of Earth's Moon, the pictures have been the basic reference for high-resolution topographic information. The most often referenced images are the comprehensive set of about 600 images selected by Bowker (1971) and the near side set of about 450 images selected by Whitaker (1970).

Following the thorough coverage of the Lunar Orbiter program, the Apollo Command and Service Module, in orbit during landing missions, provided additional coverage of the equatorial regions with its mapping and panoramic cameras. The Clementine mission provided a comprehensive survey of altitude, albedo (intrinsic brightness), and multispectral data in 1994. Lunar Prospector provided gamma ray spectroscopy in 1998. The data from these spacecraft has added insight into the mineral composition of nearly all the lunar surface, extending surveys of the equatorial region by Apollo. The interpretation of the remote sensing data has been supported by ground truth from analysis of lunar rocks and soil returned by Apollo and Luna missions. Despite the advances of these later missions, the Lunar Orbiter photographs, taken at a low sun angle, remain the primary source of topographic images and are used extensively in current scientific presentations, papers, and books.

At the time of Lunar Orbiter's design, image scanning technology was much less advanced than it is today. The methods used resulted in artifacts in the images that distract a viewer. Scanning and transmission limitations required the subframes to be reassembled from 20 to 30 framelets of 35-millimeter film. There are fine bright lines running across each subframe between the framelets, and there are brightness variations from the spacecraft's scanner that appear as streaks within the framelets. The artifacts are particularly distracting when the images are printed at high contrast to show subtle topographic features and brightness variations. Lunar scientists have become used to these artifacts, but they detract from their value to students and casual observers.

Since the first photos were received, I have wanted to clean up the scanning artifacts, but at the time it would have been very expensive. The priorities were to examine the photos and start using them, rather than to improve their visual quality. Advances in the art of computation and the capacity of modern computers have enabled processing of the photos to remove nearly all the scanning artifacts, resulting in clear images that are much easier to view. Drawing on an understanding of the nature of the artifacts, I have written programs that measure and compensate for the systematic artifacts and in addition apply filtering techniques similar to those published by Lisa Gaddis of the United States Geological Survey (USGS) (Gaddis, 2001).

Lunar Orbiter photography has been archived as hard copy photographs, each about 60 centimeters (cm) (about 2 feet) wide, at each NASA Regional Planetary Image Facility, including one at the Lunar and Planetary Institute (LPI) in Houston, Texas. LPI has digitized this important archival source and published the images in the Digital Lunar Orbiter Photographic Atlas of the Moon on the LPI web site (www.lpi.usra.edu/research/lunar_orbiter/). Further, the LPI staff has added annotations to the photos, clearly outlining many of the features and labeling them with their internationally recognized names.

A team led by Jeff Gillis carried out this important work; Jeff was supported by Washington University at St. Louis and LPI. He is currently with the University of Hawaii. LPI technical and administrative support was provided by Michael S. O'Dell, Debra Rueb, Mary Ann Hager, and James A. Cowan with assistance from Sandra Cherry, Mary Cloud, Renee Dotson, Kin Leung, Jackie Lyon, Mary Noel, Barbara Parnell, and Heather Scott. The selection of photos on the LPI Digital Archive is that selected for *Lunar Orbiter Photographic Atlas of the Moon* by Bowker and Hughes (see the reference section for details).

The annotated photos in this atlas provide full coverage of the near side, including nearly all the features whose names have been approved by the International Astronomical Union (IAU). All the high- and medium-resolution photos of Lunar Orbiter 4 (except for a few that were found unacceptable for Bowker, 1971) have been cleaned and are in the enclosed compact disc. Most of these, selected for nonredundancy and interest of features, are printed along with the annotated photos. All the photos (before processing by the author) are courtesy of NASA and LPI.

Photos with annotated overlays label the major features within each of the photos, including the landing sites of manned and unmanned spacecraft. These overlays were extracted digitally from those published by LPI. I added additional annotations to provide latitude, longitude, and scale information and also to bring the set of features up to date with the Gazetteer of Planetary Nomenclature, the list maintained for the IAU by the United States Geological Survey (USGS) Astrogeology Program. Notes with each photo point out salient aspects of the features. The combination of cleaned photos, labeled features, and notes are intended to serve as powerful aids to learning the geography and geology of the near side of the Moon as well as valuable reference material.

Throughout the project of cleaning the photos and writing this book, helpful suggestions and comments were made by Jeff Gillis, Mary Ann Hager, Paul Spudis, Lisa Gaddis, Debbie Martin, Michael Martin, Ewen Whitaker, and Brad Jolliff.

My dear wife Mary worked long and patiently as system administrator, picture processor, and reviewer.

Special gratitude goes to Don Wilhelms, USGS Astrogeology (retired), whose books have been major sources and who was kind enough to review this book and set me straight on lunar geology.

Charles J. Byrne

Table of Contents

Chapter 1

Overview of the Atlas

1.1. Content

The atlas presents full coverage of the nearside of the Moon with a series of photos all taken from orbit by Lunar Orbiter 4, with a nearly vertical viewpoint and sunlight at approximately 20 degrees (°) from the horizontal. Extensive computer processing has been used to improve the quality of these photos. Features whose names have been recognized by the International Astronomical Union (IAU) have been identified on an overlay and listed in the index of features. Notes on each page discuss the geologic processes that formed the features and controlled their interactions.

The high-resolution photos of this book typically cover about 200 by 300 kilometers (km). Approximately 350 of these photos cover the nearside of the Moon (with some overlap). The photos are grouped into chapters that each present one of the regions of the Moon. Six of these regions have been selected to focus on basins of the Moon, the largest coherent lunar features. Chapters on the North and South Polar Regions complete the coverage of the nearside. Figure 1.1 identifies these regions against the background of a photograph of the full Moon. Figure 1.2 identifies the regions of the central latitudes against a Mercator projection map, Figure 1.3 shows the North Polar Region, and Figure 1.4 shows the South Polar Region.

An introduction to each chapter surveys the relevant region, using wide coverage photos. An overview of the geology of each region and its features introduces the specialized vocabulary needed to describe lunar processes. Each chapter provides a guide to the high-resolution photos as an aid to relating them to the wide-angle views and to the Moon as a whole. Each high-resolution photo is marked with the latitude and longitude as a further guide to locating features on the maps of Figures 1.2 to 1.4.

An enclosed compact disc (CD) contains all the cleaned-up photographs of Lunar Orbiter Mission 4, with an index listing all the officially named lunar features shown in this book. These photos, when viewed on a monitor with appropriate magnification, show more detail than can be seen in the printed pictures.

1.2. How to Use This Atlas

A telescopic observer or photographer of the Moon, working at high resolution, views a small area under lighting that varies both with time and with position. The angle of view is nearly vertical at the center of the Moon as we see it, but severely foreshortened near the limb (edge of the visible Moon). One might ask: What features are in the area? How do they relate to the rest of the Moon? How would the feature look relative to others if seen at similar lighting conditions and viewing angles? What is the current understanding of the processes that formed these features?

The following is a summary of how an interested observer might answer these questions, with the aid of this atlas:

1. Relate a feature viewed in a telescope, photo, or electronic image to its position on the full Moon photo of Figure 1.1.
2. Note the region in which it appears; consult the Table of Contents to find the appropriate chapter.
3. Determine the approximate latitude and longitude of the feature from the maps in Figures 1.2 and 1.3. These coordinates can be used with the chart in the beginning of the relevant chapter to narrow the search for the corresponding photo.
4. Once the photo or photos that cover the area of interest are located, annotations will identify the name of the feature (if it has a formal name) and the names of nearby features.
5. The notes on the page or pages describe something of the processes that took place to form the feature or its surroundings.
6. The reference section refers to more detailed descriptions in the literature for many of the photos.

The process described above can be followed in reverse. If an observer is interested in a feature found by browsing the book, the latitude and longitude on the annotated photo can be used to locate the feature relative to larger features on the maps of Figures 1.2 to 1.4. Then an observer can feature-walk a telescopic field of view to the appropriate part of the Moon, just as one star-walks a field of view from one star group to another to bring dim deep-sky objects into sight.

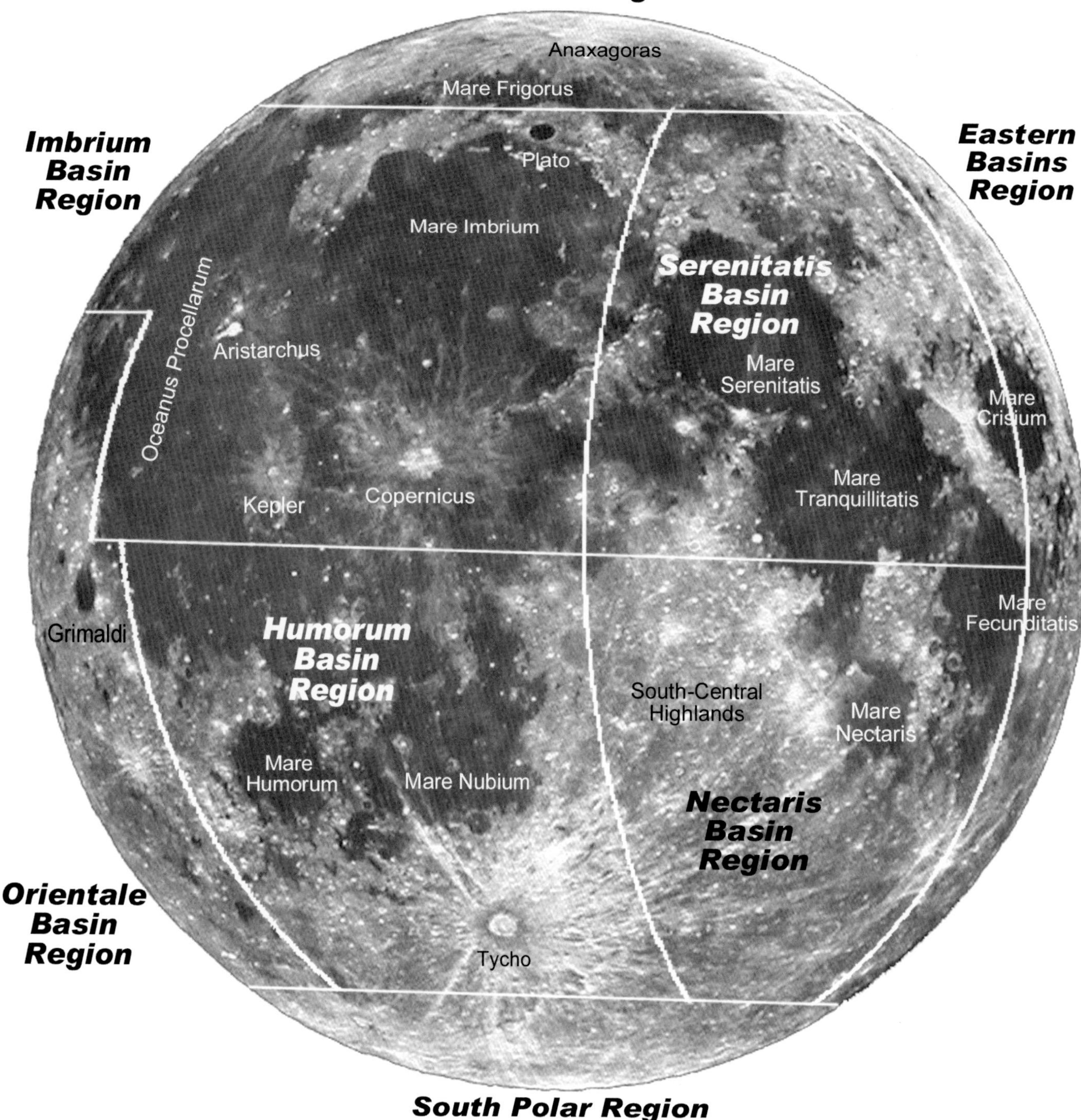

Figure 1.1. These regions of the Moon establish the scope of the corresponding chapters of this atlas. The base photo is from the Consolidated Lunar Atlas (LPI Web site).

Of course, these are only examples of ways this atlas can be useful. The overview chapters and the introductions to the regional chapters can be read as a book summarizing the largest features on the Moon and their relationships. Another way to use the atlas is to browse it. The notes need not be read sequentially. The atlas can also be used as a reference book. The feature index in the CD lists near side features with IAU names and directs the reader to a relevant page. The Source Notes section of the enclosed CD often identifies discussion of features in books of comprehensive coverage of the Moon, which refer in turn to additional research publications.

1.3. Large-Scale Maps

Figure 1.1 is the full Moon, photographed from Earth. A few outstanding features are identified as an aid to orientation when looking at the Moon. The dark maria (low-lying areas flooded by lava) form the patterns variously known as the man in the Moon, the rabbit, and so on.

The map in Figure 1.2 is based on data from the Clementine spacecraft. Clementine scanned the Moon from a polar orbit whose plane was aligned with the sun, so that the lunar surface shows its inherent brightness (called albedo),

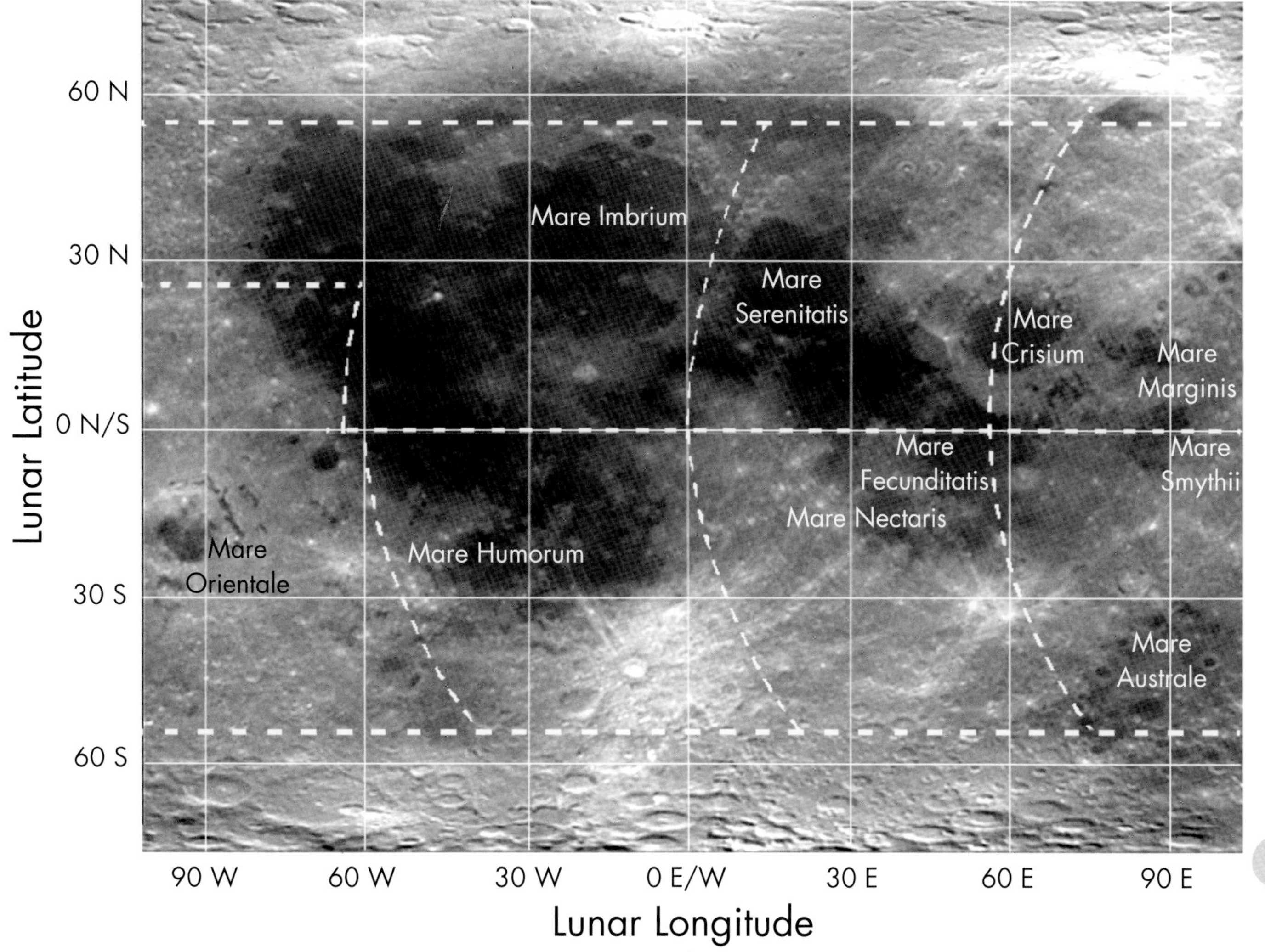

Figure 1.2. Map of the near side of the Moon, based on Clementine brightness data (NRL Web site). Region names are the same as in Figure 1.1; the north-south regional boundaries are curved to follow the Lunar Orbiter 4 photos, planned for uniformity of the sun angle. Mare Orientale and the features on the eastern limb are visible from Earth only at times of favorable libration and then are much foreshortened. Note that features in the polar regions are stretched horizontally in this Mercator projection.

as it appears from Earth when the Moon is full. This map shows features of the eastern and western limbs more clearly than they can be seen from Earth.

Figures 1.3 and 1.4 are polar projections, mosaics of Clementine scans of the polar regions. Clementine's brightness sensor continued to look almost directly down at the surface as the spacecraft passed nearly over the poles, but these maps show the topography of the features rather than the albedo because the sun is always low near the poles. Under such low illumination, crater walls throw shadows.

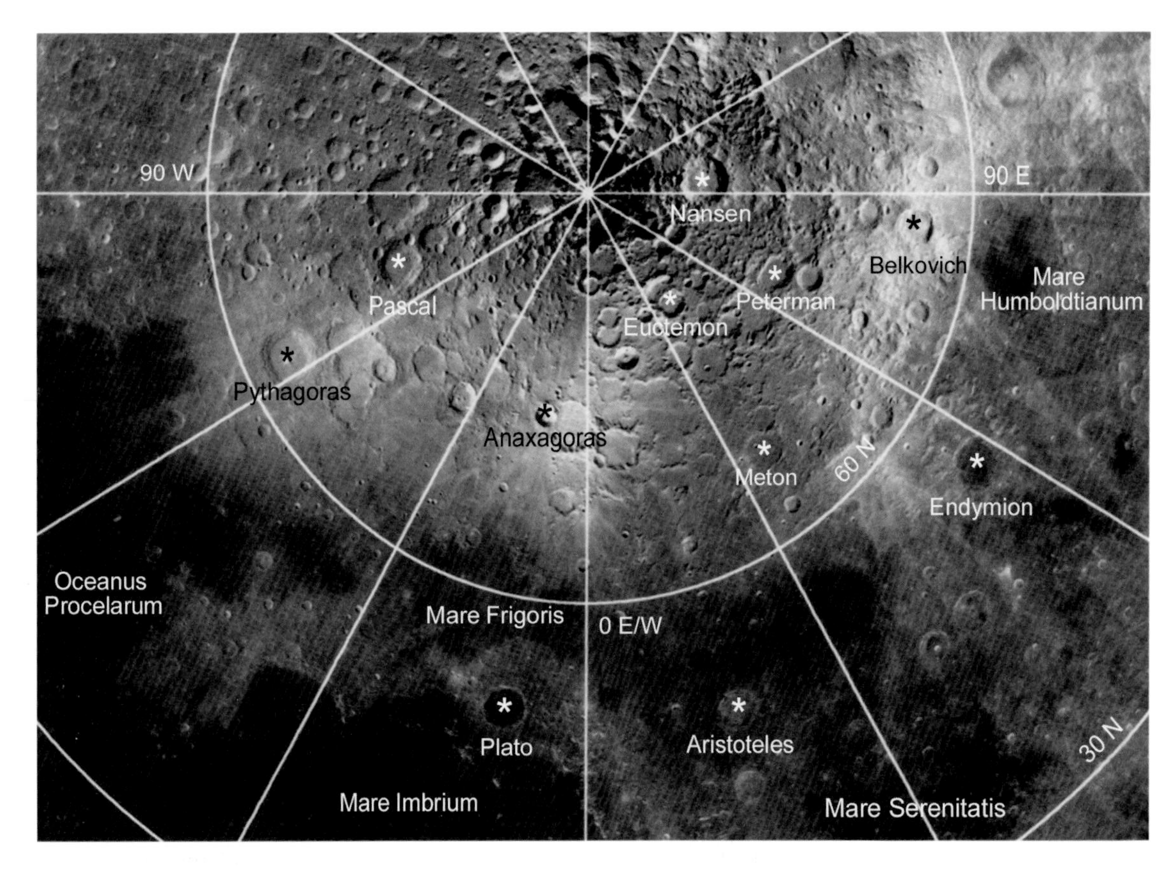

Figure 1.3. North Polar Region; near side is at bottom. This circumpolar projection is based on Clementine brightness data (USGS Web site). The boundary of the North Polar Region (not shown) is at 55° north latitude.

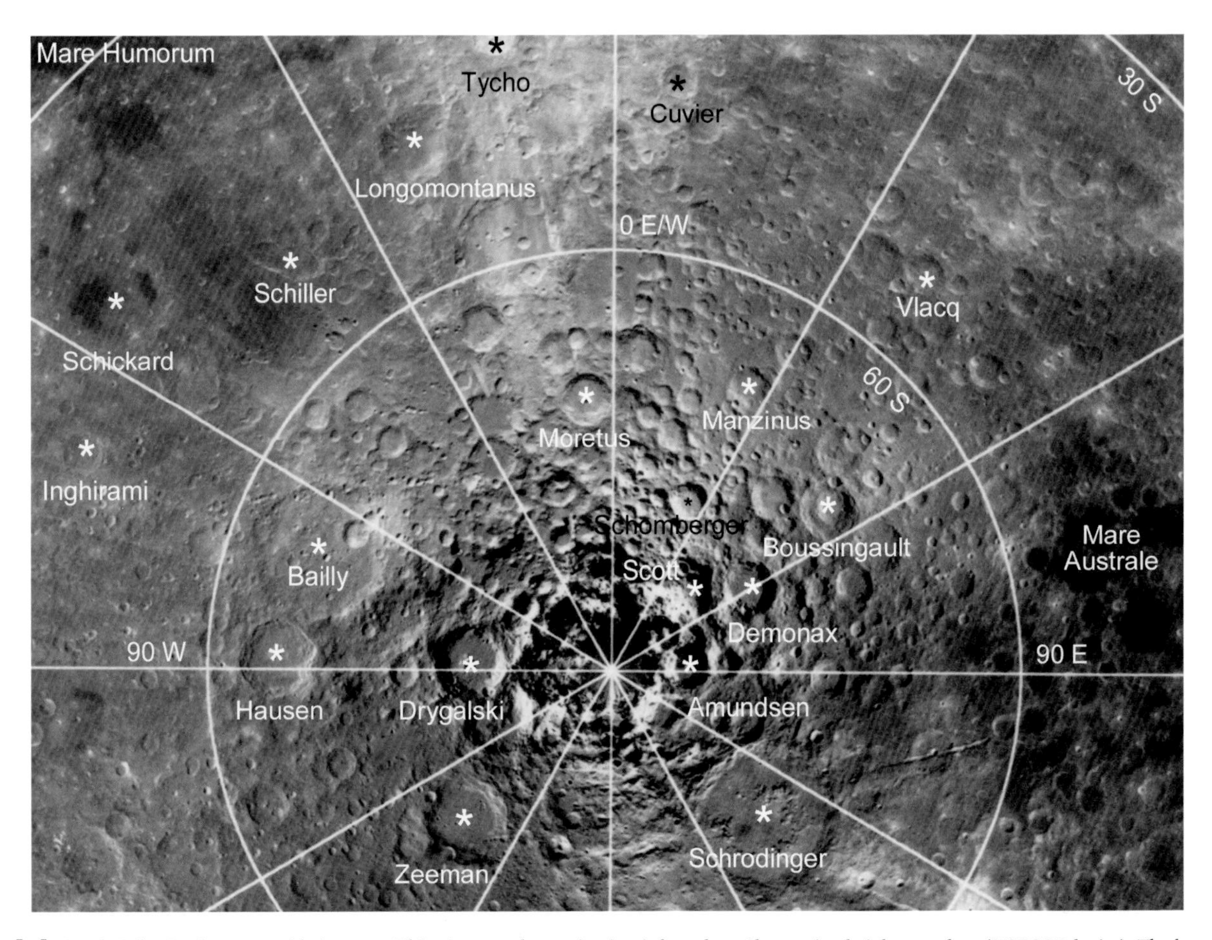

Figure 1.4. South Polar Region; near side is at top. This circumpolar projection is based on Clementine brightness data (USGS Web site). The boundary of the South Polar Region (not shown) is at 55° south latitude.

Chapter 2

Lunar Orbiter Mission 4

2.1. The Mission

Lunar Orbiter Mission 4 produced most of the photos in this book. This scientific mission was planned to systematically cover the nearside of the Moon. Extensive 1-meter (1-m) photography had been taken of the candidate Lunar Orbiter landing sites by missions 1 through 3, and the geologists wanted moderate-resolution photography (about 50 m) of so many large features on the nearside that a systematic survey mission was appropriate.

2.2. Mission Design

The layout of exposures from orbit depends on many aspects of the mission design. The spacecraft was placed in a high-inclination (nearly polar) orbit to provide full coverage of the nearside of the Moon. The orbit was such that the incidence angle of sunlight was about 20° (east to west; morning sun) throughout the photographic mission. While the 2-week photographic mission continued, the Moon revolved under the orbit with approximately uniform sun angle from the east limb to the west limb, presenting a new surface under the spacecraft at each orbital pass. The period of the orbit, the altitude of the spacecraft, and the width of the high-resolution image were coordinated so that high-resolution images taken on successive orbits overlapped by about 10% at the lunar equator, with more overlap at high latitudes. Successive high-resolution photographs taken on the same orbit were timed to overlap between about 10% as well. The overlap of photos avoids gaps in coverage, assists the construction of mosaics, and provides partial stereo coverage.

Most of the photos were taken vertically, except for those near the poles, where this could not be done because of the nature of the chosen orbit. The vertical photography minimizes foreshortening of the features to aid interpretation of the images.

Although all five of the Lunar Orbiters successfully provided extensive new photographic coverage, there were often problems, and Mission 4 was no exception. In this case, the problem involved erratic operation of the thermal door, a lens cap intended to protect the cameras from the cold of deep space when they were not actually making an exposure. Early in the mission, the optics chilled to the dew point of the moisture within the cameras and became fogged. The Boeing operations team and the science team designed a work-around. The thermal door was left open, but the spacecraft was oriented toward the sunlit part of the Moon so that the optics would be kept warm as much as possible.

Many photos of the eastern nearside were lost while the problem was analyzed and resolved, but additional exposures were made at the end of the mission to replace most of the missed coverage. These photos were taken on the "back side" of the orbit, with the spacecraft near apolune. As a consequence they are at lower resolution than the other photos and lighting is reversed, coming from west to east.

A schematic of the layout of the high-resolution images across the nearside of the Moon is shown in Table 2.1.

2.3. The Cameras

The Lunar Orbiter photography system consisted of two cameras, one with a short focal length and wide field of view (called the "medium-resolution camera") and one with a long focal length and a narrower field of view (called the "high-resolution camera"). The high-resolution field of view was centered within the field of view of the medium-resolution camera and was more than three times longer than it was wide.

The image motion compensation was provided by a feedback loop based on a circular scan of part of the high-resolution image. This mechanism, which had failed on Lunar Orbiter Mission 1 but worked flawlessly on all the later missions, moved the film platen in two dimensions to track the image.

Six fiducial marks ("sawteeth") for mapping were provided at the edge of the field of view. These marks have been trimmed from the images in this book.

2.4. The Film

The negative film was a fine-grained low-contrast 70-millimeter (70-mm) black-and-white standard Kodak product. The fine grain assured low graininess at high resolution. The low contrast (about 1 to 1) was selected to tolerate unknown variations in the brightness of the lunar surface at high resolution. A gray scale, resolution charts, and reticule marks (small crosses, barely visible on the images in this

Latitude Range	Photo Number																											
56 N–90 N				190	176		164		152		140	128		116		104		092		080	068				056				
27 N–56 N			189	183	175	170	163	158	151	145	139	134	127	122	115	110	103	098	091	086	079	074	067	062	055		177	165	024
0–27 N		196	168	182	174	169	162	157	150	144	138	133	126	121	114	109	102	097	090	085	078	073	066	061	054	181			018
0–27 S		195	187	181	173	168	161	156	149	143	137	132	125	120	113	108	101	096	089	084	077	072	065	060	053	046	039	027	
27 S–56 S		194	186	180	172	167	160	155	148	142	136	131	124	119	112	107	100	095	088	083	076	071	064	059	052	184, 045	038	178	009
56 S–90 S	193	179		166			154			130		118		106		094		082		070		058		044			005		
Longitude at Equator		95W	89W	82 W	76W	68W	62W	56W	49W	41W	35W	30W	23W	16W	10W	3W	4E	10E	16E	24E	30E	38E	43E	49E	57E	63E	70E	83E	90E

Table 2.1. Layout of the high-resolution photos from Lunar Orbiter Mission 4. Numbers are exposure numbers: the full title of each high-resolution subframe is LO4-XXXH1, -H2, or -H3, where XXX is the Photo Number shown above. See Chapter 4 for an explanation of the subframe suffixes H1, H2, and H3. Photos whose numbers are underlined were taken at the end of the mission, at higher altitude, and with reversed sunlight (from the west). Images shown next to each other in this table, either vertically or horizontally, overlap. All these photos are included in the enclosed CD, but some redundant photos have not been printed in this book. A table in each chapter shows the high-resolution photos for that region and indicates which ones are printed.

book) were exposed on the negative to assist in reconstruction and calibration of the images after scanning.

The film was developed in the spacecraft by the Kodak Bimat process. A second 70-mm film, its thick emulsion saturated with developing fluid, was pressed against the exposed negative by rollers. The two-film sandwich was stored in a buffer of rollers until development was complete and then they were separated.

Unfortunately, the Bimat process was susceptible to certain flaws. Small bubbles sometimes interposed between the negative and developer film, leaving various patterns on the negative. Also, rollers sometimes left bars across the film where the film was paused. These defects are easily distinguished from lunar features but unfortunately degrade a minority of the images.

2.5. Scanning and Reconstruction

The developed film was scanned by the available technology of the time; charge-controlled devices (CCDs) were still well in the future. Instead, a cathode ray tube (CRT) was used to provide a single scan line. Of course, this scan line could provide only the equivalent of about 800 pixels because of limitations on the phosphor grain and focus of the electron beam. This short (about 4 mm) scan line was aligned with the long dimension of the film. A mirror directed it across the 70-mm width of the film. A photocell detected variations in brightness as a measure of the density of the developed negative. After the width of the film was scanned, the film was advanced by a little less than the width of the scan line, providing overlap between scans to avoid gaps.

On the ground, the recorded signal was received at a Deep Space Network (DSN) site in California, Spain, or Australia, operated by the Jet Propulsion Laboratory (JPL). A CRT in the Kodak Ground Recording Equipment (GRE) transformed the signal into an image on 35-mm film. A segment of this film (called a framelet) represented a scan of the mirror across the 70-mm film in the spacecraft. There were two GREs, backed up by two Ampex video tape recorders (the original rotating head recorders) at each DSN site.

Once the 35-mm framelets were developed, they were trimmed by automatic machines and assembled on a light table. This was done at JPL (during the mission), Langley Research Center, and the Army Mapping Service. The assembled framelets were limited to about 60 centimeters (cm), 24 inches, in length; this was sufficient to fully reconstruct the medium-resolution frames, but the long high-resolution frames were reconstructed as a set of three subframes. Contact prints were made from these reconstructed frames and subframes; these are the primary records of the photos. Second-generation images at the same scale are stored at NASA Regional Planetary Image Facilities around the world.

The Army Mapping Service often used the tape recorders to adjust the contrast and brightness of the surface. Typical images are at a contrast of about 3 to 1 relative to the brightness variations of the lunar surface. This contrast exaggerates the brightness variation of topography, compressing the variations in the brightest and darkest parts of an image, but also amplifies the inherent brightness variation of rays and ejecta blankets from craters.

2.6. Scanning Artifacts

The complex deconstruction and reconstruction processes, involving multiple electrical, mechanical, optical, photographic, and manual processes, introduced several types of scanning artifacts. Thin bright lines, leaks from the light table between framelets, appear in most reconstructed frames and subframes. Sometimes negative framelets were assembled on the light tables, so these lines are dark on the positive prints. Variations in sensitivity of the phosphors on the two CRTs used in the process (one in the spacecraft and one on the ground) and possibly variations in the sensitivity of the rotating heads of the tape recorder cause systematic brightness variations in the series of framelets. This effect, which is quite distracting, is sometimes called the "venetian blind effect" because it appears as if the image has been projected on a set of venetian blinds.

2.7. Cleaning the Images

A computer program written by the author specifically for this purpose has cleaned the photographs in this book. Appendix A, in the enclosed CD, describes this process in detail. As a result, residual scanning artifacts are only occasionally visible in the printed photos. However, artifacts associated with the development process in the spacecraft remain.

Chapter 3
Overview of the Near Side of the Moon

3.1. Origin of the Moon

To understand these photos, one must understand the Moon, its geography, and the factors that have influenced its near side surface. The origin of the Moon was a matter of vigorous debate until about 12 years after the last Apollo mission. In 1984, after many years of analysis of the 300 kilograms (kg), 800 pounds, of rocks and soil returned by Apollo and Luna missions, a consensus was struck that has lasted for the subsequent 20 years. The following narrative retells this consensus origin story, without attempting to review the debate of alternative theories. The curious reader is encouraged to review the references.

As unlikely as it seems, another planet perhaps about the size of Mars and formed in the general orbital vicinity of Mars had its orbit perturbed by some unknown event and struck the early Earth a glancing blow. Just as a small car absorbs much of the energy of an impact with a large truck, the smaller planet, known as Theia, was partly vaporized, partly melted, and largely pulverized and thrown into space. Some of Theia escaped from Earth or fell to Earth, but most of it ended up in an orbit near Earth. Modeling suggests that much of the core of Theia, relatively dense, was pulled into Earth and joined Earth's core.

The pulverized and molten orbital components in near-Earth orbit (perhaps about four Earth radii according to the computer simulations) coalesced. As a result, the resulting body was heated by gravitational energy, supplemented by radioactive energy, and largely melted into a deep magma ocean. Part of Theia had become the Moon.

The strong tidal forces generated by the orbital proximity of these two bodies caused a loss of energy in the total system and a transfer of angular momentum between them. The Moon's rotation relative to its revolution about Earth stopped, so that the near side of the Moon always faces Earth (synchronism). The Moon's orbital radius increased, absorbing some of the angular momentum of the early Earth; that change slowed Earth's rotation to its current 24-hour period.

Most of these events occurred within the first 500 million years after the Earth had itself coalesced and took place in a relatively short interval, perhaps 100 million years (the coalescence of the Moon could have taken place within weeks after the impact). Of course, the chaotic impact would have thrown material into many orbits, some of which may have taken very much longer to impact the Moon. The heavily cratered surface reflects the arrival of many such delayed impactors as well as new arrivals from the solar system after the outer surface had cooled and become solid. The rate of arrival of these later impactors is a matter of intense interest. There is some evidence of a quiet period followed after a time by a brief period of a high rate of arrival, sometimes called the cataclysm to suggest a major chaotic event in the Solar System.

As a result of this history, our Moon is somewhat unusual in the solar system. It contains much the same chemical elements as the other rocky planets, but its crust contains a distinctly different quantitative distribution of those elements. The mineralogy is often familiar to geologists from their terrestrial experience, even to the point of the relevance of an extensive descriptive vocabulary. But the Moon is (to our current knowledge) unique in having experienced two distinct melting phases, one when Theia was formed as a planet and one when the Moon was formed from a part of Theia's debris. Each melting phase resulted in differentiation of the minerals, as heavy material sank and light material rose in two different gravity fields and two different thermal domains. A set of elements called KREEP that is concentrated when massive quantities of material are melted and solidified are unusually abundant in some maria. Erupting magma carried the doubly concentrated material to the surface.

3.2. The Near Side Versus the Far Side

The synchronization of the Moon's rotation with its revolution in the intense early tidal environment established that the near side would always face Earth and be the visible side (visible to humanity before the mid-twentieth century, that is).

Measurements of the Moon's gravitational field have established that the depth of the crust is less on the near side than on the far side. This implies that, for a given size of impactor or strength of internal processes, penetration of the crust is more likely on the near side.

3.3. Mare and Highlands

The Moon exhibits significant differences between the near side and near side. In particular, much more mare material is found on the near side. Mare material is the relatively dark material visible on the face of the Moon, contrasting with the more reflective highlands. It is made of relatively heavy minerals that have risen from the mantle of the Moon, below the crust. How can we explain this paradox of the rise of heavy materials through light materials? The answer lies in the decrease of density that accompanies heating. The material becomes lighter than crust because it is hotter. When it becomes exposed, it quickly cools and is locked in place by the rigidity of a surface exposed to the cold of deep space half the time. The association of most mare material with positive gravity anomalies called mascons (an abbreviation of "mass concentrations") also reveals their high density. Both remote sensing and analysis of lunar rock and soil samples returned to Earth have established that the materials of the maria are heavier than highland materials because of their higher proportion of iron and magnesium.

Historically, the distinction between mare materials and the more pristine highland materials (so called because they are in fact higher than mare, but also resemble terrestrial highlands in their relatively rugged topography) has been very important, because elementary observation distinguishes the dark, flat mare from the relatively bright, rugged, highlands.

Region	Latitude	Longitude
Orientale Basin Region (includes Grimaldi Basin)	19° S	95° W
Humorum Basin Region (includes Nubium Basin)	24° S	39.5° W
Imbrium Basin Region (includes Oceanus Procellarum)	35° N	17° W
Nectaris Basin Region (includes Tranquillitatis and Fecunditatis Basins and nearby highlands)	16° S	34° E
Serenitatis Basin Region (includes the Crisium Basin and nearby highlands)	17.5° N	58.5° E
Eastern Basins Region (the **Smythii Basin**, the Humboldtianum Basin, the Australe Basin, and intervening highlands)	2° S	87° E
North Polar Region (**North Pole**)	90° N	NA
South Polar Region (**South Pole**)	90° S	NA

Table 3.1. Landmarks for the regions of the Moon and their coordinates. Landmarks are in bold print. Note that the center of the Orientale Basin is actually around the western limb; it is included because much of the basin and its ejecta are on the near side.

3.4. Basins

As understanding grew, it no longer appeared that maria are fundamental to the structure of the lunar surface. Rather, it seems that mare regions are secondary to the formation of large basins by impactors that are large enough to compromise the integrity of the lunar crust. Such basins allow heavy, dark minerals to rise through the fractured crust from the hotter upper mantle below the crust. The basin associated with a mare region can be much larger and has a much more extensive effect on lunar topography. Typically, only the interior of the depression formed by the basin impactor floods (or partly floods) with mare material. However, deposits ejected from the basin can extend for a distance that is a multiple of the crater radius. Such basin ejecta blankets cover or partially cover much more area than the central depression.

Basins that have been flooded with lava to form a mare are named for the mare they contain (the Orientale Basin is named for Mare Orientale). Basins that lack mare fill are named for two craters that happen to be superposed on the basin (the Schiller-Zucchius Basin).

Because basins are more fundamental than their included maria, this introductory section emphasizes basins as landmarks for the geography of the Moon.

3.5. Landmarks for Geography

Strictly speaking, perhaps we should say selenography or lunography (writing about the Moon) instead of geography (writing about the Earth), but the term geography seems to be more evocative of what we mean. Geography is not geology: it refers to landforms and their relation, without emphasis on the mineralogy or the formation process of the landforms. It is difficult to grasp a good mental image of the major features of the Moon in terms of their global distribution and relationships. There is no equivalent of the organization of terrestrial geography in terms of continents and oceans that are so useful in establishing a mental and visual image of Earth.

This section introduces a high-level list of eight regions, centered on interesting, memorable focal points. They are chosen for broad distribution, to ensure roughly uniform coverage of the lunar near side. Each region is sufficiently small to support photographic and other images with reasonable distortion. The proposed regions, and the latitude and longitude of their focal points, are shown in Table 3.1.

Basins were chosen as many of the focal points because they are major modifiers of the surface geology, not only through their central rings and maria but also through their ejecta blankets. Specific basins were chosen as much for uniformity of spacing around the lunar globe as for size or interest. In addition to basins, the North and South Polar Regions were included because of the special significance of the shadowed craters in those regions and also because the photographic quality is quite different.

3.6. Descriptions of Landmark Regions

The following sections summarize the characteristics and prominent features of each region. Only enough description is included to differentiate the regions from each other; these descriptions are not intended to be comprehensive or to discuss the geology or stratigraphy of the regions in detail. No attempt is made to establish precise boundaries between the regions: that would be like attempting a precise boundary between the Atlantic and Arctic Oceans. These regions are proposed to serve as a method to organize imagery, an alternative to latitude and longitude, and no representation is made that they are of fundamental geologic or cartographic significance.

Orientale Basin Region

The Orientale Basin, with its central mare and multiple rings, has been called the archetype of basins because it is both large and relatively recent. Consequently, its structure is very clear. The region includes the smaller Grimaldi Basin and craters Schickard and Bailly.

Humorum Basin Region

The Humorum Basin region includes the Nubium Basin and craters Pitatus and Tycho. This region is particularly interesting for understanding the interactions of different types of features such as the manner in which mare floors encounter crater rims and highlands. The Fra Mauro Peninsula is an example of such a feature.

Imbrium Basin Region

The Imbrium Basin and its ejecta blanket dominate much of the near side. The region includes Oceanus Procellarum, and other neighboring maria: Vaporum, Serenitatis, and Frigoris. Craters in this region include Kepler, Copernicus, Archimedes, and Aristoteles. This entire region is rich in maria, ejecta blankets, and rays of craters. It also has a number of features (such as Vallis Schroteri) that are associated with the flooding of the mare floors.

Nectaris Basin Region

The northeastern part of the Nectaris region is rich in maria, including Fecunditatis and Tranquillitatis as well as Mare Nectaris. To the west and south, the region includes extensive highlands. To the west, parts of these highlands are covered with ejecta from the Imbrium Basin.

Serenitatis Basin Region

This region includes the Serenitatis and Tranquillitatis Basins, the northern part of the Fecunditatis Basin, and the western part of the Crisium Basin. The region shows the interplay of overlapping basins.

Eastern Basins Region

From Earth, this region covers the eastern limb (edge, as we see it) of the Moon. It includes the eastern part of the Crisium Basin, the Australe Basin, the Smythii Basin, and Mare Marginis.

North Polar Region

The heavy shadowing in both polar regions obscures both photography and passive spectral measurements, especially for the floors of basins and craters. The near side part of the region and the North Pole itself are largely covered with ejecta from the Imbrium Basin. Craters in this region include Nansen, Shackleton, and Anaxagoras. The region includes the Humboldtianum Basin.

South Polar Region

The South Polar Region is dominated by the rim of the South Pole–Aiken Basin and several smaller basins such as Schrodinger, Planck, and Bailly. Permanently shadowed crater floors in this region (as well as in the North Polar Region) are believed to harbor deposits of hydrogen or water ice. Permanently sunlit crater rims nearby are proposed as sites for solar power.

3.7. The Ages of the Lunar Features

An interesting attribute of a lunar feature is the time it was formed. The ages of different features on the Moon are inferred from several sources.

Estimating Ages

The most precise ages are determined from measurement of isotope ratios in our precious rock samples; association of the samples with specific features establishes their time of formation.

When rock samples are unavailable, geologists resort to less precise measures. For example, sharply defined features are inferred to be recent because meteorite bombardment softens the edges of features by a process called mass-wasting. Also, counts of crater densities can be used to estimate the time a surface has been exposed to bombardment.

Very often, even though absolute time cannot be determined, the sequence of formation can be determined. For example, ejecta from a basin may overlay a crater or lava from a mare may flow into a crater. Such layering and the branch of geology that studies it are called stratigraphy.

Named Age Ranges

Systematic study of stratigraphy, together with other clues, establishes a chain of evidence that leads to relative ages and estimated age ranges of most of the features of the Moon. These age ranges are named for associated archetype features. In order from the oldest to the youngest, the ranges are the Pre-Nectarian, Nectarian, Early Imbrian, Late Imbrian, Eratosthenian, and Copernican Periods or Epochs. Geologists use the term "epoch" as a subdivision of a period. The Early Imbrian and the Late Imbrian Epochs are subdivisions of the Imbrian Period. The boundary between the Pre-Nectarian and Nectarian Periods is the time of the impact that produced the Nectaris Basin. The Nectarian Period ends with the event that produced the Imbrium Basin. The Early Imbrian Epoch spans the time between the Imbrium and Orientale events.

Large basin-forming events are used to divide these time periods because such events spread their deposits so very far, as much as halfway around the Moon. This method allows us to relate many other features to large basin events by analyzing interactions with the widespread deposits. Because Orientale is the last big basin, however, the younger age ranges are based on large craters. The Eratosthenian Period is next, and then the Copernican Period (this interval is open ended; we live in the Copernican Period). These last two periods are not distinguished by bounding events, but on the degree of degradation of craters and their ejecta (especially the fading of rays) as they are exposed to further impacts and to the solar wind.

Chapter 4

Organization of the Photos

4.1. Grouping by Landmark Regions

Each chapter that follows displays images of a particular landmark region. A few medium-resolution frames are shown for each region to provide an overview. Only a few of the medium-resolution photos are presented because they have a high degree of overlap. The high-resolution photos are selected in rectangular blocks so that a reasonable degree of organization is imposed. The relation of the photos and their assignment to regions is shown in a table in each chapter.

4.2. Order: West to East and South to North

The regions are presented from the west limb around the nearside to the east limb; that is because Orientale, the newest and clearest basin and Imbrium, the largest nearside basin, are both toward the west and are should be viewed before the other regions. Keep in mind, however, that the mission and the rising sun move from east to west. To minimize confusion, the photos covering a particular longitude range are presented from south to north (in the order most of the photos were taken, and in order of the numbering convention for those photos). A few of the photos in the Eastern Basin Region were taken in reverse of the usual orbital direction, to compensate at the end of the mission for photos lost at the beginning of the mission.

4.3. Clean and Annotated Images

One page is provided for each high-resolution subframe or frame. The cleaned image is shown on the right side of each page and an overlay with the major features of the image labeled is on the left. A title line shows the photo number, lighting conditions, and spacecraft altitude at the time of the exposure. Latitude and longitude marks and a scale bar are shown in the annotated overlay. The scale bar is derived from the size of prominent surface features, as listed on the USGS Web site. Except for the apostrophe, diacritical marks in feature names have been omitted in both photos and text because they are not supported in the software used to annotate the photos. A list of names with diacritical marks as guides to pronunciation is in the enclosed CD.

4.4. High-Resolution Frames and Subframes

In most cases, subframes (designated H1, H2, and H3) are displayed. In some cases, the three subframes are reassembled as a full high-resolution frame (designated H). This is done either to show the relationship of the features across the full exposure or because the features displayed are deemed to be of relatively little interest, usually because of similarity to their surroundings. All subframes are presented at full detail in the enclosed CDs.

4.5. Discussion Notes

The photo pages contain notes that point out and discuss relevant aspects of the features and their relationships. These notes draw on an extensive lunar literature to include aspects of the stratigraphy, geology, and possible formation and modification processes of the features.

4.6. Enclosed Compact Disc (CD)

The enclosed CD contains:

- All high-resolution subframes from Lunar Orbiter Mission 4 (JPEG files)
- All medium-resolution frames from Lunar Orbiter Mission 4 (JPEG files)
- An index of nearside features whose names are recognized by the International Astronomical Union (IAU)
- A description of the process used to clean up scanning artifacts
- Source notes related to individual photos

The feature index includes all the near side features in the master IAU list maintained by the USGS that are larger than 6 km. Several smaller features and some far side features near the eastern limb, the western limb, the North Pole, or the South Pole are also included. Each of the features listed in the index can be found in the annotated photo (and the photo number) on the indicated page.

Chapter 5

Orientale Basin Region

5.1. Overview

Orientale, the Archetype Multi-Ringed Basin

Figure 5.1 shows the spectacular Orientale Basin, the newest large basin on the Moon. It is often called the archetype of basins because it reveals the structure of basins so clearly. Like typical large basins, it is multi-ringed; that is, there are a number of concentric rings both inside and outside of its major raised rim. Mare surfaces have been formed from lava seeping up from below its central floor. Additional lava has seeped up into the low troughs between the external rings.

Although the Orientale Basin has many features in common with all basins, it must be considered that there are significant differences between basins as well. The detailed structure of a basin depends on the nature of the target material. Orientale has formed in an area of thick, somewhat uniform crust. Although there are no major basins in its immediate vicinity, large craters have been identified that influenced its detailed structure.

Orientale Basin and its included Mare Orientale are located on the western edge of the Moon, the edge that rises and sets last. Terrestrial astronomers glimpse its eastern rings as the libration of the Moon turns that edge a bit toward us. Spacecraft Zond 3 photographed it in 1965 at essentially Earth-based resolution. However, the Lunar Orbiter 4 coverage showed the magnificence of its structure at a much higher resolution.

These photos created a major paradigm shift in minds of geologists who had been debating whether volcanism or impacts dominated the lunar surface. This newest basin had a fresh ejecta blanket, revealed in detail by the high-resolution photos. The unavoidable conclusion was that many geologic units whose origins were debated were in fact formed by ejecta, not only from Orientale but also from larger, older basins such as Imbrium. This insight was confirmed and strengthened by the pervasive discovery of impact breccia in the rocks returned by Apollo missions. Such rocks are formed by the hypervelocity shock of impacts welding preexisting rock fragments from diverse sources into larger rocks. They show a diverse set of shock effects, from changes of crystalline structure to partial or complete melting (Wilhelms, 1987).

Figure 5.1. LO4-187M. The Orientale Basin is the archetype of multi-ringed basins. The lava flow of Mare Orientale is in the dark center of the concentric rings. The dark mare surface to the upper right is Oceanus Procellarum, the largest deposit of mare on the Moon. The small dark circular feature is a mare deposit on the floor of the 440-km Grimaldi Basin. The bright rays in the upper right corner of the photo radiate from Glushko, a small crater near Oceanus Procellarum, not from Orientale.

A remarkable property of multi-ringed basins is the regularity of the radii of the concentric rings. Such rings are found not only on the Moon (where there is evidence of about 60 multi-ringed basins), but also on Earth, Mercury, Mars, the Jupiter moons Ganymede and Callisto, and the Saturn moons Tethys and Rhea (Spudis, 1993). Measurements of the rings of basins on all these bodies fit a specific relationship; successive rings, both internal and external to the topographic rim (the ring of highest elevation also called the main ring), have radii in the ratio of the square root of 2.

This relationship, equivalent to the doubling of the area of each successive ring, is very regular, but has no consensus theoretical basis at this time.

Crater Morphology as a Function of Size

Small impact craters do not have such concentric rings. The smallest craters have a transient cavity that is nearly hemispherical. The transient cavity is the zone of target material that was melted and pulverized by the hypervelocity impact. Some of this material is ejected, with most of it landing within one radius of the transient cavity from the rim of the transient cavity. This rim is a single topographic ring. Medium-sized craters have a distinctive central peak; the focused sound wave rebounding from the mass of pulverized and compressed material has lifted material nearly vertically, and it forms a mountain in the center of the crater. As the diameters of lunar craters exceed about 300 km (200 miles), the central peak breaks up into one or more internal rings, and external rings form as well. It is these very large craters that we call multi-ringed basins, or basins for short. The bottoms of large craters may penetrate near or through the boundary between the crust and the mantle.

Surroundings of the Orientale Basin

To the north and west, the ejecta blanket from Orientale has been deposited on highlands. To the northeast, the ejecta fell on the surface later covered by the mare material of Oceanus Procellarum. To the east and south lie highlands interrupted by other, smaller basins: Grimaldi and Humorum. The flank of Grimaldi shows fine striations radial to Orientale, so it must have formed earlier. Similarly, the Humorum Basin predates Orientale.

Figure 5.2 shows a mosaic of the eastern sector of the Orientale Basin Region, illustrating the structure of its rings.

Figure 5.2. Mosaic of Lunar Orbiter 4 high-resolution photos of the Orientale Basin, from the central mare at the left out past the concentric rings to the right. Dark mare material is found not only in the central part of the basin, but also in the troughs between the rings. Note the striations in the ejecta blanket radiating from the central portion of the basin. They have a stronger appearance to the northeast and southeast rather than directly toward the east. Because of the direction of solar illumination, ridges oriented in the east-west direction do not cast shadows. The smaller Grimaldi Basin in the upper right of the mosaic shows both a topographic ring that has contained most of the mare lava and a lower outer ring.

Subframes LO4-180H1 and LO4-167H1 are not printed in this chapter because they are redundant with the subframes to their east and west and show no additional features. The base photos are included in the enclosed CD.

5.2. High-Resolution Images

Table 5.1 shows the high-resolution images of the Orientale Basin region in schematic form.

The following pages show the high-resolution subframes from south to north and west to east; that is, they are in the order LO4-194H, LO4-195H1, LO4-195H2, LO4-195H3, LO4-186H1 . . . LO4-161H3. Note that the photos taken on a given orbit are presented in order; both frame and subframe numbers increase in the same order as the pages of this book. However, in moving from west to east to a new orbital sequence, the numbers decrease.

LO4-194H has been assembled as a complete frame. It shows the range of Orientale ejecta from the topographic rim to one radius away from that rim. In addition, it illustrates how subframes are contiguous parts of high-resolution frames. Similarly, LO4-169H and LO4-161H are displayed as full frames.

Latitude Range	Photo Number							
27 N–56 N			189	183	175	170	163	158
0–27 N			188	182	174	169	162	157
0–27 S		195	187	181	173	168	161	156
27 S–56 S		194	186	180	172	167	160	155
56 S–90 S	179			166			154	
Longitude at Equator		95 W	89 W	82 W	76 W	68 W	62 W	56 W

Table 5.1. The cells shown in white represent the high-resolution photos of the Orientale Basin Region (LO4-XXX H1, -H2, and -H3, where XXX is the Photo Number; LO4 means Lunar Orbiter Mission 4). The Imbrium Basin Region is to the northeast, the Humorum Basin Region is to the east, and the South Polar Region is to the south. The far side is to the west. The next number after LO4 is the exposure number, which increased as the mission progressed from east to west.

LO4-194H
Sun Elevation: 16.10°
Altitude: 3002.79 km

This full high-resolution frame shows the inner Orientale Basin at the top. Montes Cordillera is the topographic ring. Its scarp bounds the floor of the basin. The arc of Montes Rook shown here is the first inner ring of the Orientale Basin.

Striations show a heavy deposit of ejecta (the Hevelius Formation) from the inner Orientale Basin to the south. The character of the ejecta changes from the heavy, striated inner Hevelius Formation to the lighter, more uniform outer Hevelius Formation about 440 km away from the rim of the Montes Cordillera ring (930 km in diameter). Graff, Catalan, and Baade have been covered with a heavy deposit while Yakovin has received a lighter deposit. Several other craters in the 10- to 30-km range (probably secondaries from Orientale) left deposits of fine ejecta outside their rims.

The crater Hausen (out of the picture to the southwest) has more recently deposited chains of small secondary craters that can be seen on the floor of crater Pingre and outside its rim. Pingre may lie on the floor of an older basin called the Pingre-Hausen Basin. The large semicircular ridge east and north of Pingre is the boundary of that basin.

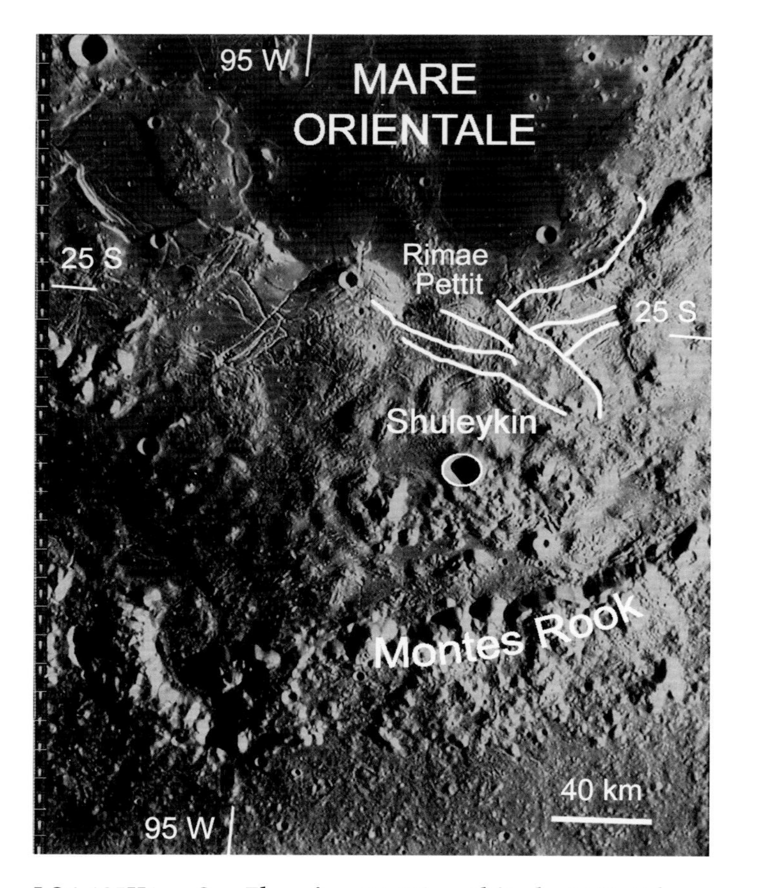

LO4-195H1 Sun Elevation: 14.50° Altitude: 2721.44 km

The inner Orientale Basin was flooded long after the impact by mare lava rising from below. Multiple shorelines appear in this photo, suggesting that the lava rose to a certain level and then receded, as is common on earth. A network of radial and circumferential fractures surrounds the border of the lava. The surface between the fractures is often very smooth. This area is called the Maunder Formation, named for the genetically unrelated Maunder crater north of Mare Orientale (see LO4-195H2). The Maunder Formation has been interpreted as the melt sheet of the basin. The fractures relieve stresses that could have been associated with cooling of the melt sheet, settling of underlying pulverized material, or flooding of the mare. Other basins may have had similar melt sheets that were completely hidden by a later covering of mare lava.

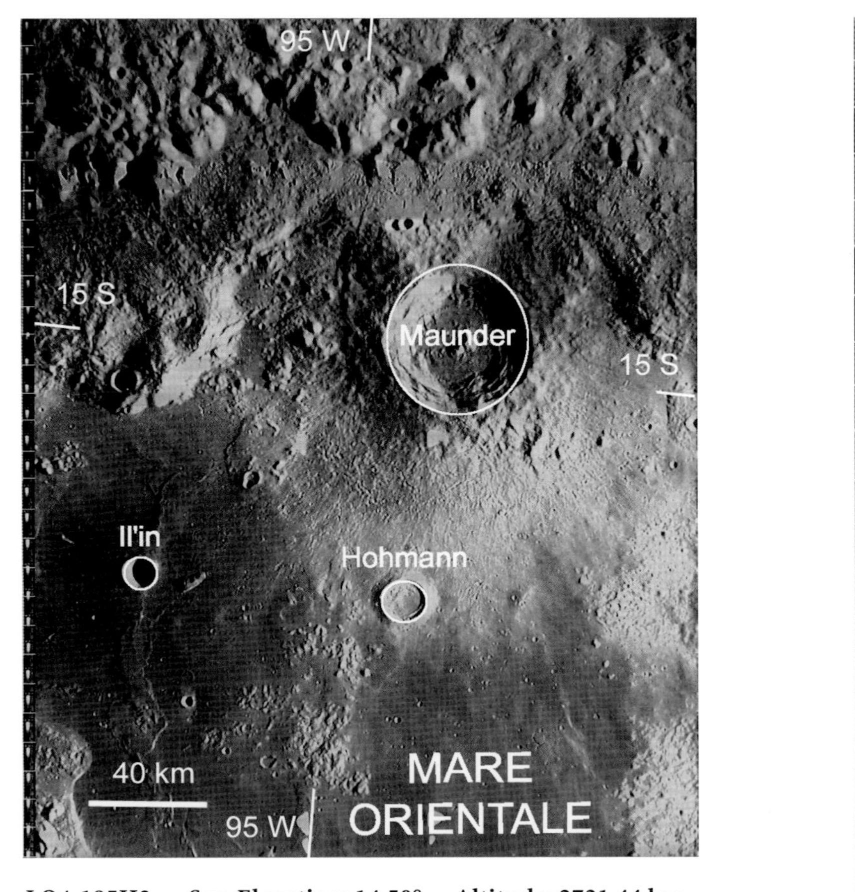

LO4-195H2 Sun Elevation: 14.50° Altitude: 2721.44 km

Maunder was obviously formed after the mare floor had solidified. It has a typical profile for a large (55-km) primary crater, with a well-formed central peak, terraced walls, an ejecta blanket with radial furrows, and a field of secondary craters. The melt sheet of the Maunder Formation (named after the crater) appears to have partially collapsed at either side of the Maunder crater, possibly due to the impact shock. The ejecta of some post-mare craters such as Il'in appear to have flow characteristics, as if they impacted while the mare was still semimolten or still hot enough to be melted by the impacts. See the note for Kopff with photo LO4-187H2.

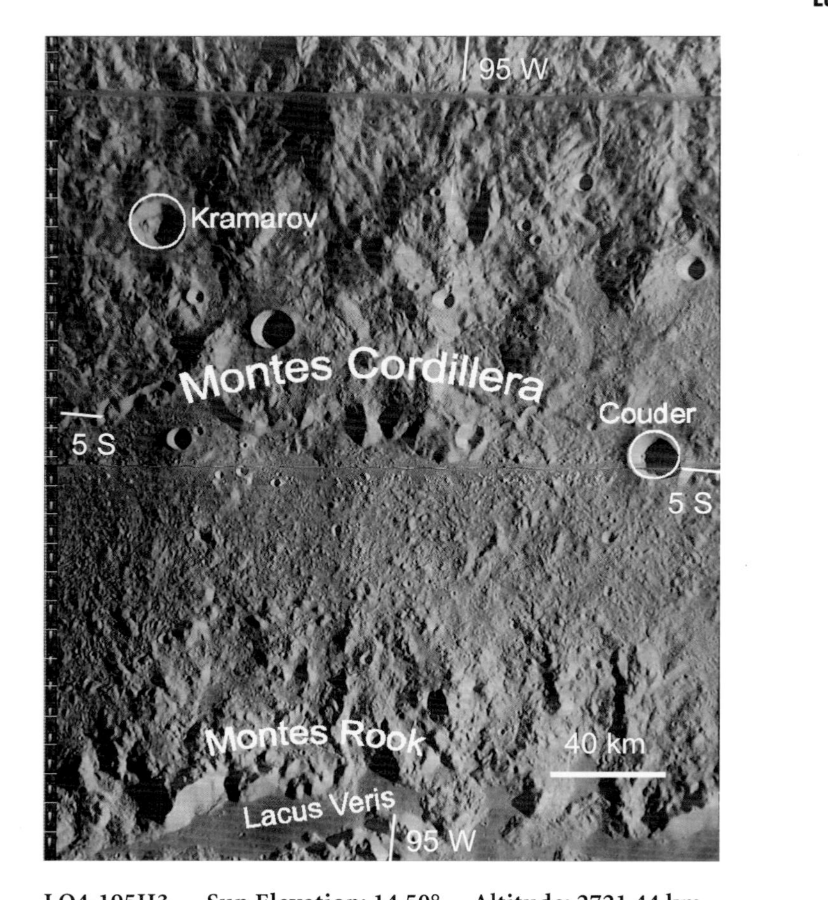

LO4-195H3 Sun Elevation: 14.50° Altitude: 2721.44 km

The area in this photo shows many differences in structure and texture. Compare the smooth mare lava of Lacus Veris with the rugged cliffs of the inner ridge of Montes Rook and the knobby, flat Montes Rook Formation between Montes Rook and Montes Cordillera. Note the steep scarp of the rim of Montes Cordillera and the thick, ropy nature of the inner Hevelius Formation of basin ejecta beyond Montes Cordillera. Craters Kramarov and Couder, about the same size and age (judging by rim sharpness), show an interesting contrast in structure. Kramarov, impacting the Hevelius Formation, has thrown more ejecta and part of its wall has collapsed. Couder, impacting the Rook Formation, has a smaller ejecta blanket and smooth wall. Its oval shape indicates a low impact angle.

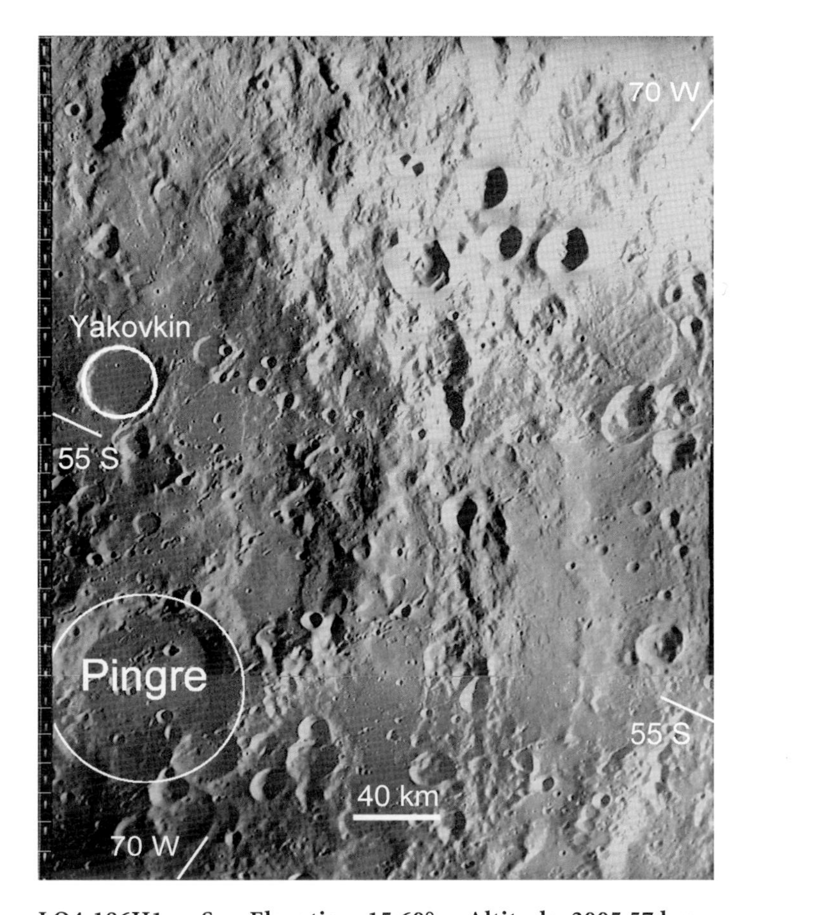

LO4-186H1 Sun Elevation: 15.60° Altitude: 3005.57 km

This area has been covered with the outer Hevelius Formation, ejecta from Orientale, as can be seen from the chain of Orientale secondary craters down through the center of the photo. On the right is a series of concentric-walled craters that may have formed as they encountered a firm layer beneath a softer layer of regolith, the ejecta from Orientale and other basins. Simulations suggest that such structures form when the apparent crater diameter is 8 to 10 times the depth of the regolith. Yakovkin appears to have received a relatively light coating of ejecta. It also may have been flooded with lava after the ejecta had been deposited; its surface seems very smooth and free of small craters.

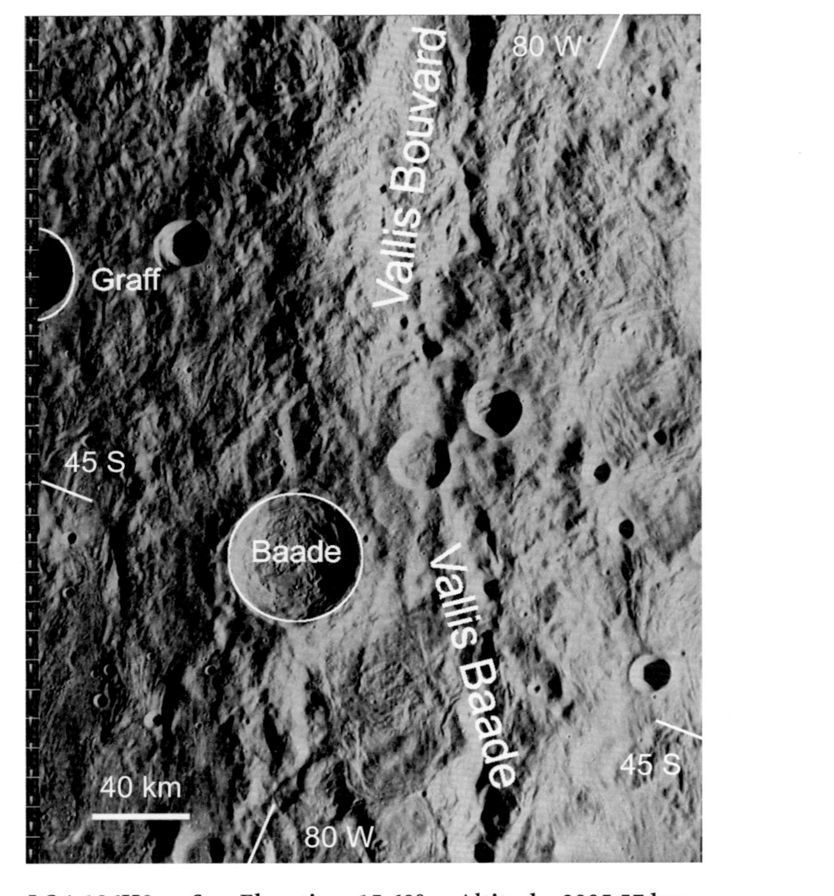

LO4-186H2 Sun Elevation: 15.60° Altitude: 3005.57 km

This area, like much of the Moon, has been repeatedly overlaid with basin ejecta. The largest scale linear features radiate from the Orientale Basin, just to the northwest. However, there are smaller-scale linear features running from southwest to northeast from the older Mendel-Rydberg Basin to the southwest. Baade, formed in a thick ejecta blanket, has a weak central peak. Vallis Bouvard and Vallis Baade are chains of secondary craters and other ejecta from Orientale that plowed these valleys in low-angle impacts. Similar valleys appear around the Imbrium and Nectaris basins.

LO4-186H3 Sun Elevation: 15.60° Altitude: 3005.57 km

Montes Cordillera is the topographic (highest) rim of the Orientale Basin. The outer range of Montes Rook is the next mountainous area toward the center of this multi-ringed basin. Between Montes Cordillera and Montes Rook is the flat knobby plain of the Montes Rook Formation. The bottom of the trough between the two rings of Montes Rook is even lower than the Montes Rook Formation. Craters Pettit, Nicholson, Wright, and Shaler show slumping of their crater walls, suggesting that they may have impacted a low-strength material, a contrast to craters impacting the melt sheet of the Maunder Formation within the Montes Rook rings.

25 S
85 W
25 S
Nicholson
Pettit
Montes Rook
40 km
Wright
85 W

LO4-187H1 Sun Elevation: 13.90° Altitude: 2722.79 km

Between the outer and inner ranges of Montes Rook is a valley of chaotic material. This hummocky zone could be a mix of rubble and melted material from the impact, together with extrusions from below. Extensive fracturing in this zone, appearing at the surface as Rimae Pettit (LO4-195H1), may indicate contraction or subsidence related to cooling and other adjustments after the impact. Pettit and Nicholson appear to have impacted the outer Montes Rook ridge while it was newly formed; their ejecta seems to have been molten and their walls either slumped to their floors or additional material landed in them. They may be secondaries excavated by ejecta thrown in a high trajectory from the Orientale event itself.

LO4-187H2 Sun Elevation: 13.90° Altitude: 2722.79 km

Lacus Veris is a small mare area formed in the trough between the two ridges of Montes Rook. Stress fractures circumferential to the basin rim along the edge of Lacus Veris. Other fractures run across and near it. These fracture lines disappear as they encounter mare material, which must have flooded after stress was released by fracturing. This fractured terrain, the Maunder Formation, is thought by most analysts to be impact melt. Crater Kopff, unlike Maunder (see LO-195H2), has a shallow, flat floor, a smooth rim without terraces, and no central peak. Its ejecta seems molten, like that of Pettit and Nicholson. Kopf may also be a secondary that impacted a still-molten melt sheet and was flooded by mare much later.

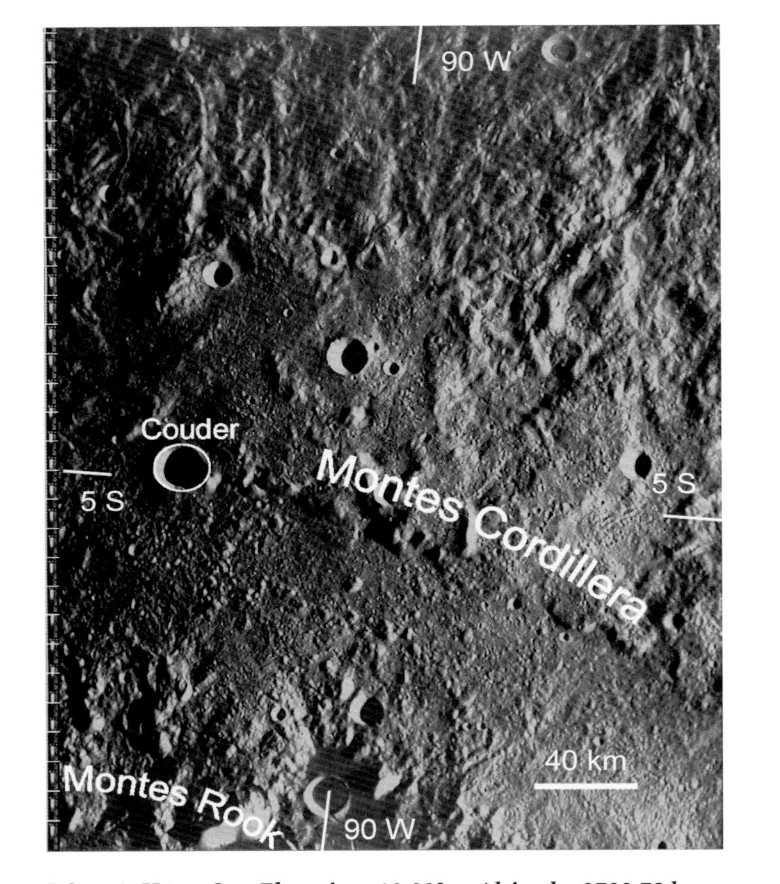

LO4-187H3   Sun Elevation: 13.90°   Altitude: 2722.79 km

This area north of the inner basin shows the knobby terrain between Montes Rook and Montes Cordillera and the striated ejecta blanket beyond Montes Cordillera. Montes Cordillera typically shows a scarp (cliff) at its inner edge. The subtle circular feature just north of the crater Couder appears to be a preexisting crater that was covered by the ejecta from Orientale.

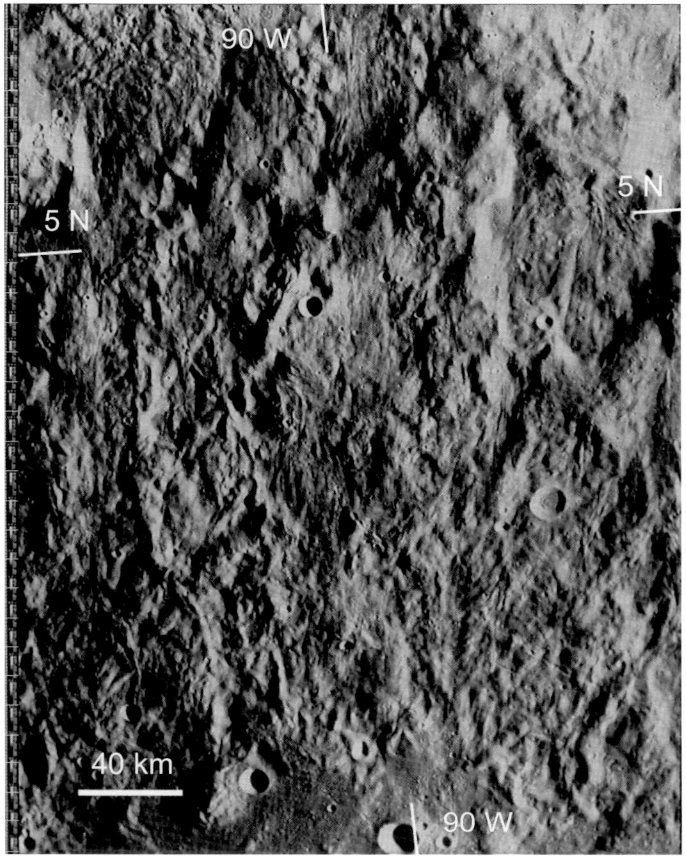

LO4-188H1 Sun Elevation: 14.50° Altitude: 2675.46 km

This area, covered by the striated inner Hevelius Formation of Orientale ejecta, has no feature with an IAU name. The radial striations are emphasized by the right-to-left sun direction. An 80-km crater on the upper right side of the photo has been submerged by the ejecta, but seems to have focused the flow lines at its far edge.

LO4-188H2 Sun Elevation: 14.50° Altitude: 2675.46 km

Einstein, Sundman, and Bohr are preexisting craters covered by Orientale deposits, including fields of secondaries.The secondary chain of Vallis Bohr and secondaries on the floor of Einstein are typical of the outer Hevelius Formation. The transition from the inner Formation is near the bottom of this picture, about one radius away from Montes Cordillera. The 80-km crater in the lower right corner is fissured, suggesting that the ejecta there may have been molten and then cracked as it cooled. Various lobate flow patterns can be seen here. The 10-km crater about 70 km southeast of Sundman is in a very dark, low-lying area, and the crater has a rare dark ray pattern in Clementine albedo data.

LO4-188H3 Sun Elevation: 14.50° Altitude: 2675.46 km

The Orientale ejecta is becoming thin here, in an area well north of the basin floor. Mare surfaces on the right are at the edge of Oceanus Procellarum. Craters Voskresenskiy and Bartels appear to have been flooded, as is common near the edge of a large mare. A possible secondary from Orientale has hit the northeast crater wall of Bartels and left debris on its floor. The scattering of craters about 10 km in diameter, with similar fresh appearances, are probably secondaries from Orientale. The floor of Moseley has a complex appearance, as if a lava flow entered it from the north.

LO4-180H2 Sun Elevation: 16.10° Altitude: 3008.99 km

Like Vallis Baade, Vallis Inghirami and its bounding ridges have been formed by the inner Hevelius Formation ejected from Orientale. Such ejecta also covered the crater Inghirami and a similar crater northwest of the head of Vallis Inghirami. Note the 21-km crater near that valley head. Its concentric crater form (a little off center in this case) suggests that it has pulverized about 2 km of deposited material from the floor of the valley before encountering more resistance. Shadowing on the valley wall indicates that the ridges rise 2 or 3 km above the valley floor.

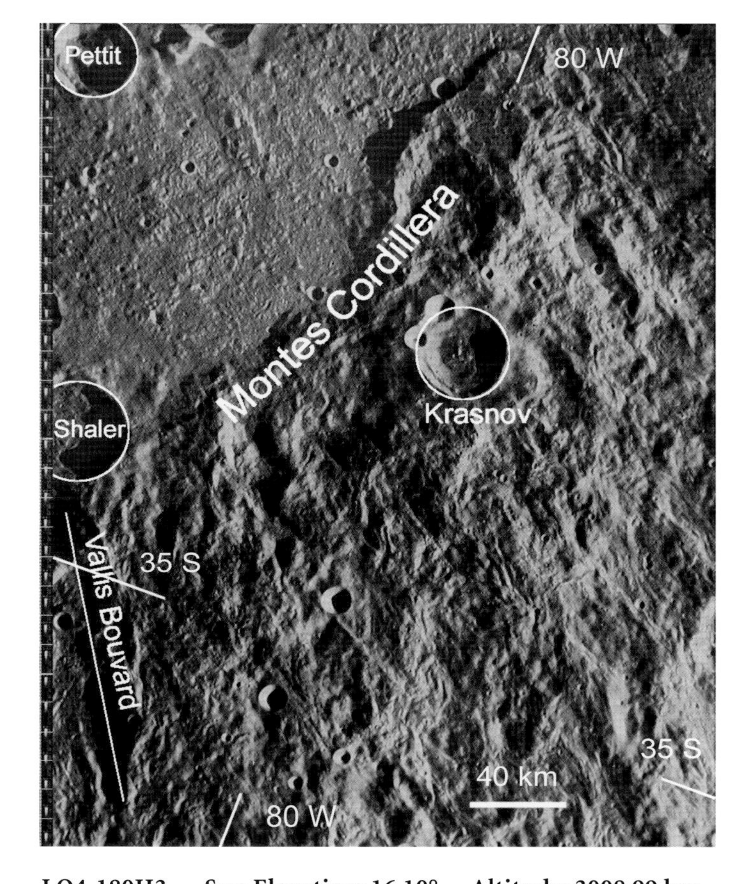

LO4-180H3 Sun Elevation: 16.10° Altitude: 3008.99 km

The crater Krasnov and the unnamed 45-km crater above it straddle the area of the scarp and provide clues to its nature. The unnamed crater is known to have formed before or in an early phase of the Orientale sequence of events because its rim has been extensively modified by ejecta. Because the rim of Krasnov is sharp, it must have formed after the Orientale event or late in its sequence. The massive deposit on the floor of the unnamed crater suggests that the rim of Montes Cordillera, whose scarp is of similar depth, is covered with deposited material. A narrow terrace on the inner wall of Krasnov may indicate either the level of the pre-Orientale surface or an interface between successive layers of ejecta.

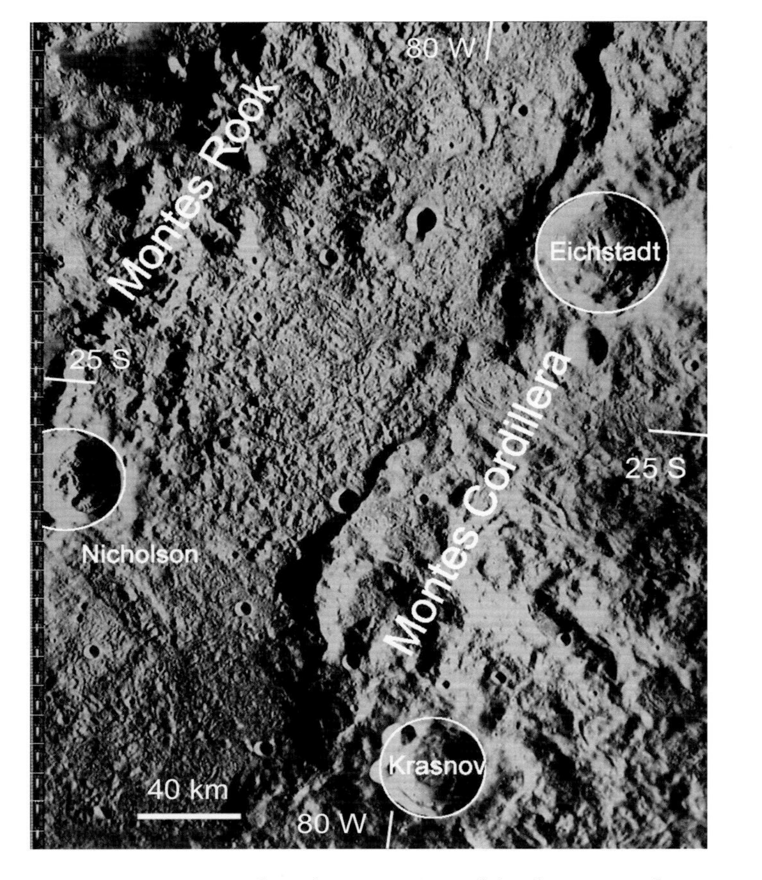

LO4-181H1 Sun Elevation: 14.80° Altitude: 2723.98 km

The inner edge of Montes Cordillera has a sharp scarp (steep slope). Such scarps may be formed by excavation of the low side or by lifting of the high side (such as by tilting a block). Strong striations appear to cross the valley between Montes Rook, reduce the scarp of Montes Cordillera south of Eichstadt, and continue across Montes Cordillera (see LO4-172H1 for the continuation to the southeast). These striations are relatively subdued in the valley between the two mountain ranges. This suggests that the valley floor was fluid (either molten or finely pulverized) and perhaps strongly shaken by a sonic or tectonic wave after the ejecta blanket was deposited.

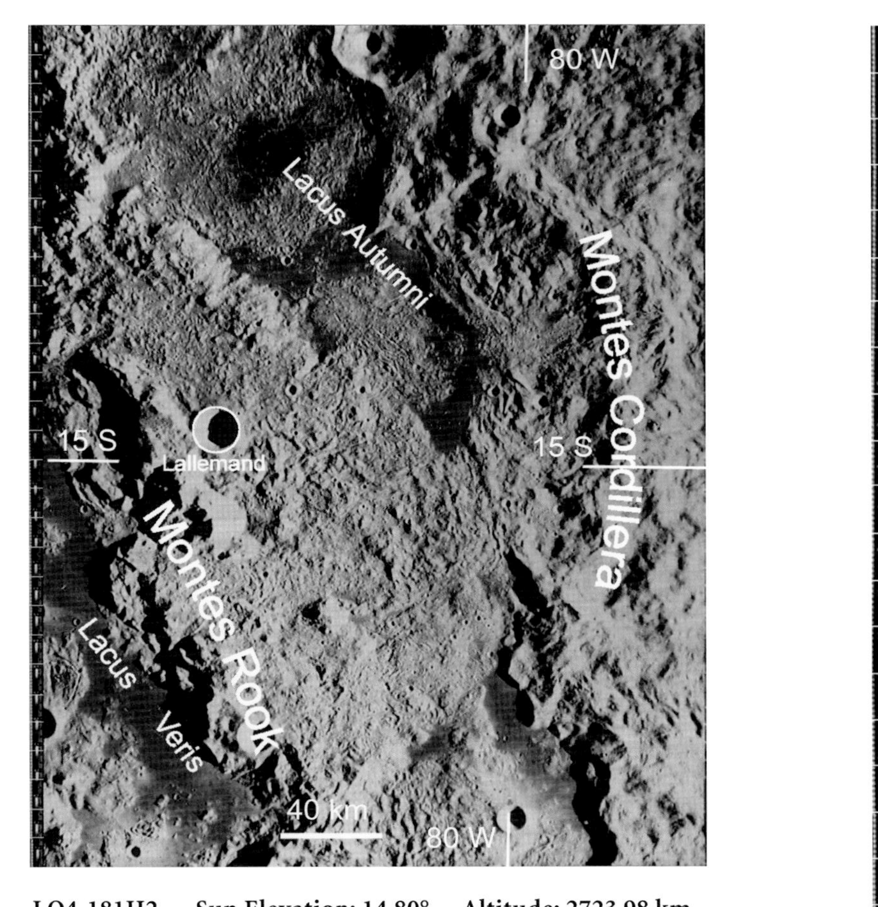

LO4-181H2 Sun Elevation: 14.80° Altitude: 2723.98 km

Mare lava flooded not only Mare Orientale in the central inner basin but also Lacus Veris in the trough between the inner and outer ridges of Montes Rook and Lacus Autumni in the valley floor between Montes Rook and Montes Cordillera. Crater counts indicate that the order of flooding is from the center of the basin (where a mascon indicates a plume of mantle material has risen) outward. Lacus Autumni has been just partially flooded. Some striations and secondary craters in Lacus Autumni may be ejecta from crater Schluter, just to the north (see LO4-181H3).

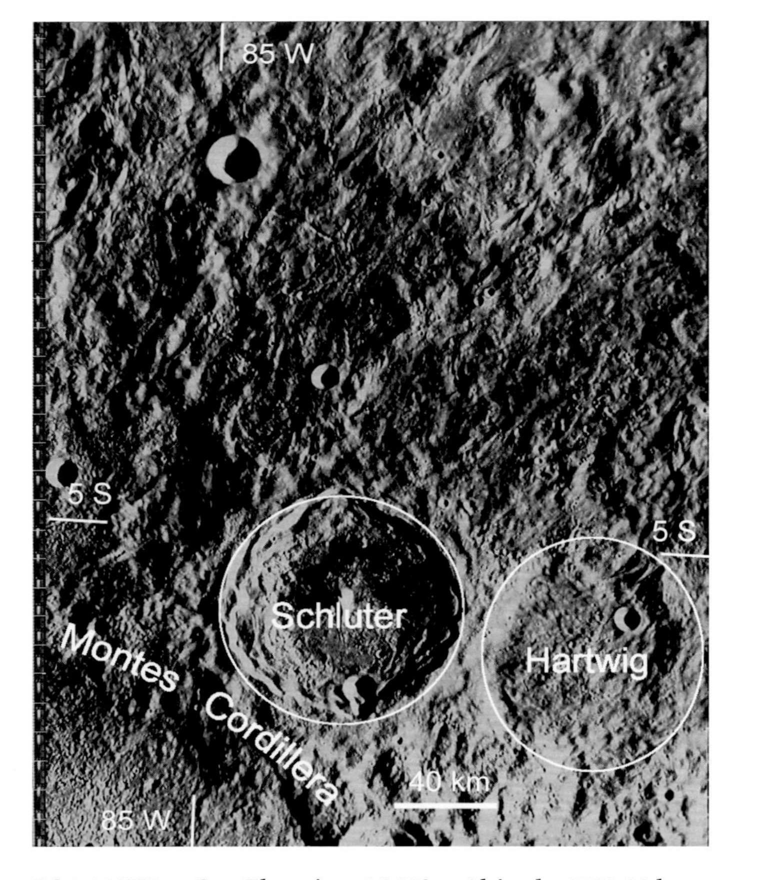

LO4-181H3 Sun Elevation: 14.80° Altitude: 2723.98 km

Craters Schluter and Hartwig, just outside the Montes Cordillera, are interesting examples of the effect of basin ejecta (Hevelius Formation). Hartwig was blanketed by this ejecta, but Schluter impacted afterward. This stratification, plus the lava flooding in part of the floor of Schluter, establishes its age as Late Imbrian.

LO4-182H1 Sun Elevation: 16.10° Altitude: 2674.44 km

This area is dominated by the inner Hevelius Formation of Orientale ejecta. The two 20-km craters in the upper half of the photo may show clues about the consistency of this ejecta. The material on their floors seems to have been molten, but with greater viscosity than the mare lava flows. The material seems to have been fractured, perhaps by cooling or by settling on the preexisting topography of the crater floors. A long thin string of secondary craters running from south-southeast to north-northwest across the lower half of the photo radiates from crater Schluter to the south (see LO4-181H3). LO4-174H1 shows much of the area on the right side of this photo without the development artifacts in this exposure.

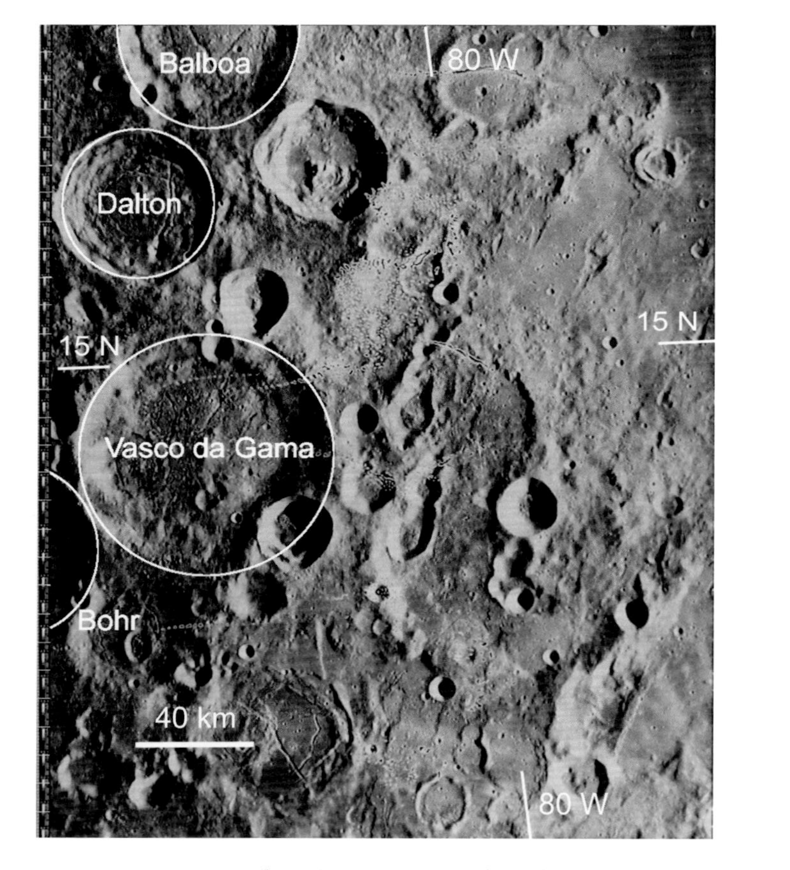

LO4-182H2   Sun Elevation: 16.10°   Altitude: 2674.44 km

This region shows clusters of Orientale secondary craters, some striation and secondary chains from the central Orientale Basin, and some deposits that appear to have been molten (with flow edges). These are typical features of the outer Hevelius Formation. Craters Balboa, Dalton, Vasco da Gama, and the unnamed 50-km crater south of Vasco da Gama all have fractures in floors. Such shallow fractured floors are common (but not universal) near the boundaries of Oceanus Procellarum, which can be seen near the upper right corner of the photo. The fracturing may be due to uplift of the floors.

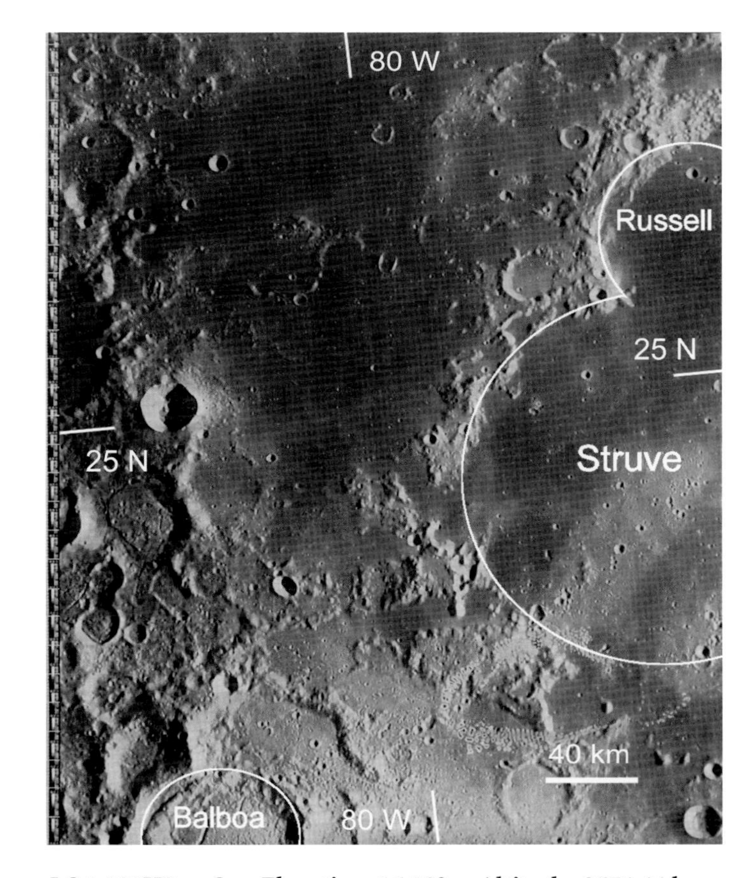

LO4-182H3 Sun Elevation: 16.10° Altitude: 2674.44 km

Several submerged craters such as Struve can be seen in this area of western Oceanus Procellarum. The depth of the mare lava can be judged by the estimated depth of flooded craters. The outer Hevelius Formation of Orientale ejecta is thinning here, as indicated by the decreasing rate of secondary craters to the northeast.

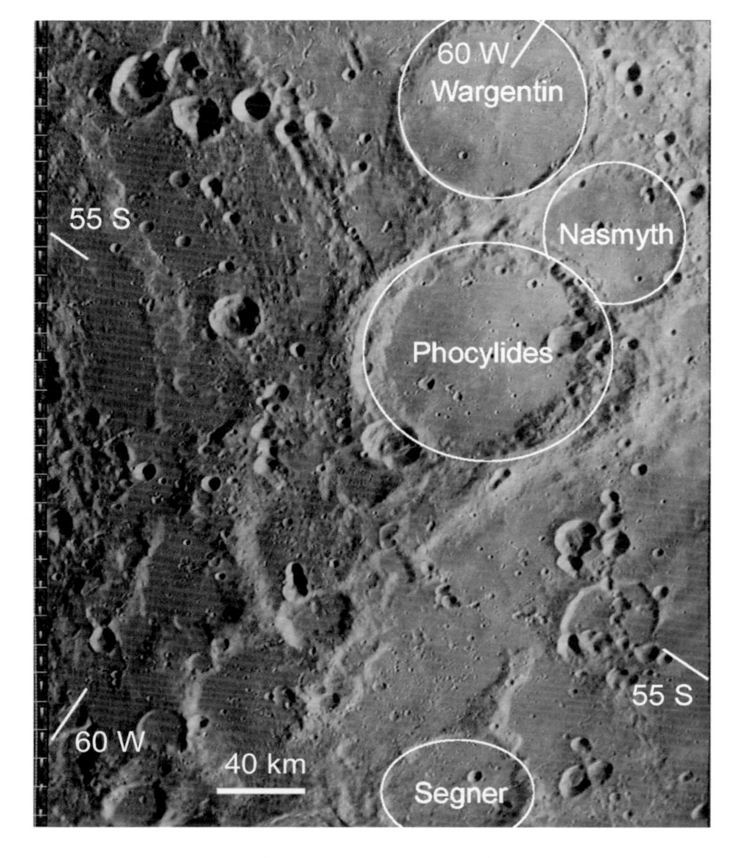

LO4-172H1 Sun Elevation: 16.30° Altitude: 3010.71 km

Molten material has flooded craters Wargentin, Nasmyth, and Phocylides to form plains, but it is not dark like typical mare lava. The material has flowed out of gaps in the southern edge of Wargentin. The melt may have been thrown out from the inner Orientale Basin. It appears to have covered some of the striated ejecta; if it were thrown out at a higher angle and a lower velocity, it would have arrived later than the material that formed the striations. The clusters of craters in the right side of this photo are secondaries from Orientale, which may have been thrown at an even higher angle than the melt.

LO4-172H2 Sun Elevation: 16.30° Altitude: 3010.71 km

To the southeast, the ejecta from Orientale meets and overlies earlier ejecta from the 160-km crater Schickard (LO4-167H2), creating a cross-hatched terrain. The floor of Inghirami is covered with ejected material that has piled up in the southeast quadrant.

LO4-172H3 Sun Elevation: 16.30° Altitude: 3010.71 km

The inner Hevelius Formation grades from the very deep material that composes Montes Cordillera to the northwest to thinner material to the southeast. In the thinner regions of the ejecta blanket, the preexisting topography can be glimpsed. The 225-km crater Lagrange has been nearly obliterated. The Orientale ejecta seems to have pushed material from the northwest rim of Lagrange onto the crater's floor. There is a gap in the Orientale ejecta on part of the floor of the old 80-km crater in the lower right hand corner of the photo.

LO4-173H1 Sun Elevation: 15.40° Altitude: 2724.03 km

There is an interesting interaction here between the old 120-km crater between Eichstadt and Lamarck and the heavy flow of Orientale ejecta that has obliterated its western rim and thrown the material onto the western half of the crater floor. When ejecta hit the small crater in the lower right hand corner of the picture, it piled up against the eastern wall, nearly up to the top of the rim.

LO4-173H2 Sun Elevation: 15.40° Altitude: 2724.03 km

The western rims of crater Rocca and two smaller craters to the south have partly shielded the western floors from ejecta, but it has accumulated on the eastern floors and against the far walls to the east.

LO4-173H3 Sun Elevation: 15.40° Altitude: 2724.03 km

Like the floor of Schluter to the west, Riccioli's floor is partly flooded with mare lava. The pervasiveness of ejecta from Orientale in this region establishes that the mare flooding must have followed the Orientale event. Riccioli could not have survived the Grimaldi Basin event (LO4-168H3), so Riccioli must have followed Grimaldi. The sequence is thus Grimaldi, Riccioli, Orientale, and mare flooding of the floor of Riccioli. Rimae Riccioli are part of a larger pattern of lineations to the north (LO4-174H1) and east (LO4-168H3). Plains-forming units north of the rim of Riccioli, the south-east floor of Riccioli, and about 80 km south of the southern rim of Riccioli are interpreted as molten or semimolten ejecta from Orientale.

LO4-174H1 Sun Elevation: 15.50° Altitude: 2673.10 km

Crater Hedin appears to be similar in age to Riccioli. It has been covered with Orientale ejecta. To the northeast, just beyond Hedin, there is a deposit with a scalloped scarp that seems to have come from the southeast, probably from Grimaldi. These deposits extend beyond Hedin to the east. In the lower half of this photo is a reference area for illustration of the inner Hevelius Formation, with its strong lineations and thick deposits of Orientale ejecta. The upper right part of the photo shows the secondary craters and thinner deposits of the outer Hevelius Formation.

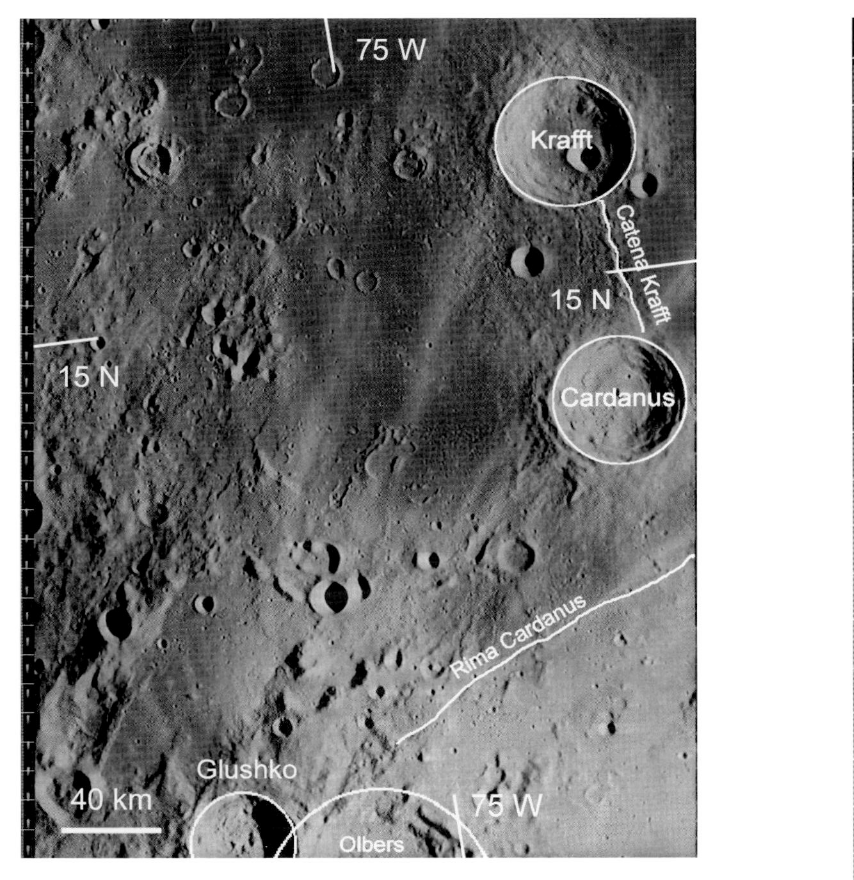

LO4-174H2 Sun Elevation: 15.50° Altitude: 2673.10 km

Clementine data show Glushko to be the source of a strong ray pattern, including the light streaks that can be seen extending to Cardanus and Krafft. Glushko has impacted a unit of plains-forming melt from Orientale. Craters Krafft and Cardanus, which are so similar in size and apparent age that they may have originated in a twin impact, have deposited their ejecta on an older mare unit within Oceanus Procellarum. Newer mare units have flooded out parts of their fields of secondary craters. The crater chain Catena Krafft is composed of secondaries from crater Krafft.

LO4-174H3 Sun Elevation: 15.50° Altitude: 2673.10 km

Craters Struve, Eddington, and Russell are ringed plains, old craters flooded by mare lava within Oceanus Procellarum. The outer Hevelius Formation from Orientale has left deposits on the crater rims. Note the furrow on the western rim of Russell and the ridge on the rim wall between Struve and Eddington. The light areas may have been deposited on the floor of Struve by Orientale and not flooded over with mare lava. Or, the Orientale ejecta may have fallen on preceding lava flows.

LO4-167H2   Sun Elevation: 17.40°   Altitude: 3009.04 km

The striated inner Hevelius Formation grades into the smoother outer deposits in this photo. These deposits sometimes create smooth plains in areas of mild preexisting topography, such as crater floors. The ejecta from the earlier (Pre-Nectarian) crater Schickard is so rugged that its form can be seen below the newer ejecta blanket. Flooding of the floor of Schickard with lava seems to have preceded deposit of Orientale ejecta.

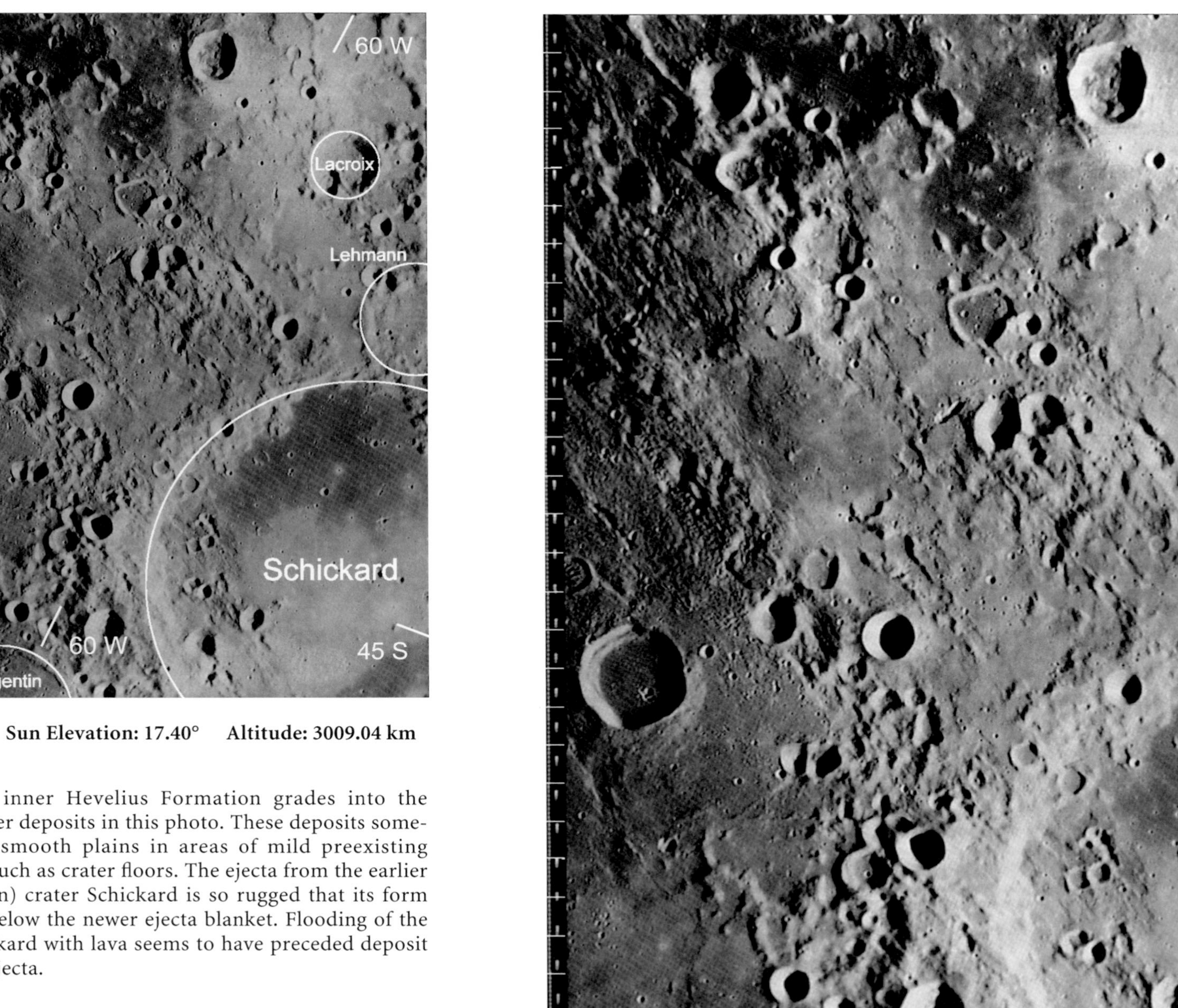

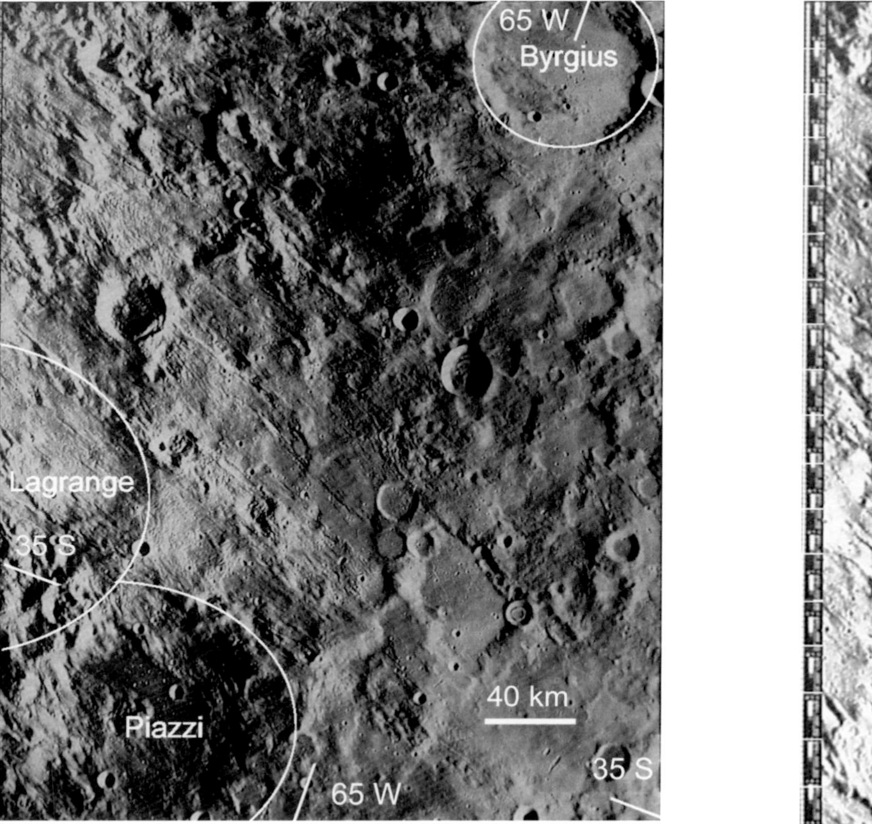

LO4-167H3 Sun Elevation: 17.40° Altitude: 3009.04 km

The dark, smooth plains areas south of Byrgius seem to be isolated lava lakes, associated with no particular topographic features. These units escaped heavy ejecta from Orientale, but narrowly: note the long furrow from Orientale just to the south of the dark plains and the heavily striated terrain further to the south as well. The fresh crater on the east rim of Byrgius is very young; Clementine albedo data show a strong ray pattern from it, which can be seen on the floor of Byrgius and beyond. Some of these rays can be seen crossing the dark plains unit near the right edge of the photo. The floor of Piazzi also shows a small area of smooth dark plains. The remainder of that floor is massively filled with Orientale Hevelius Formation material and material from the north-western rim of Piazzi.

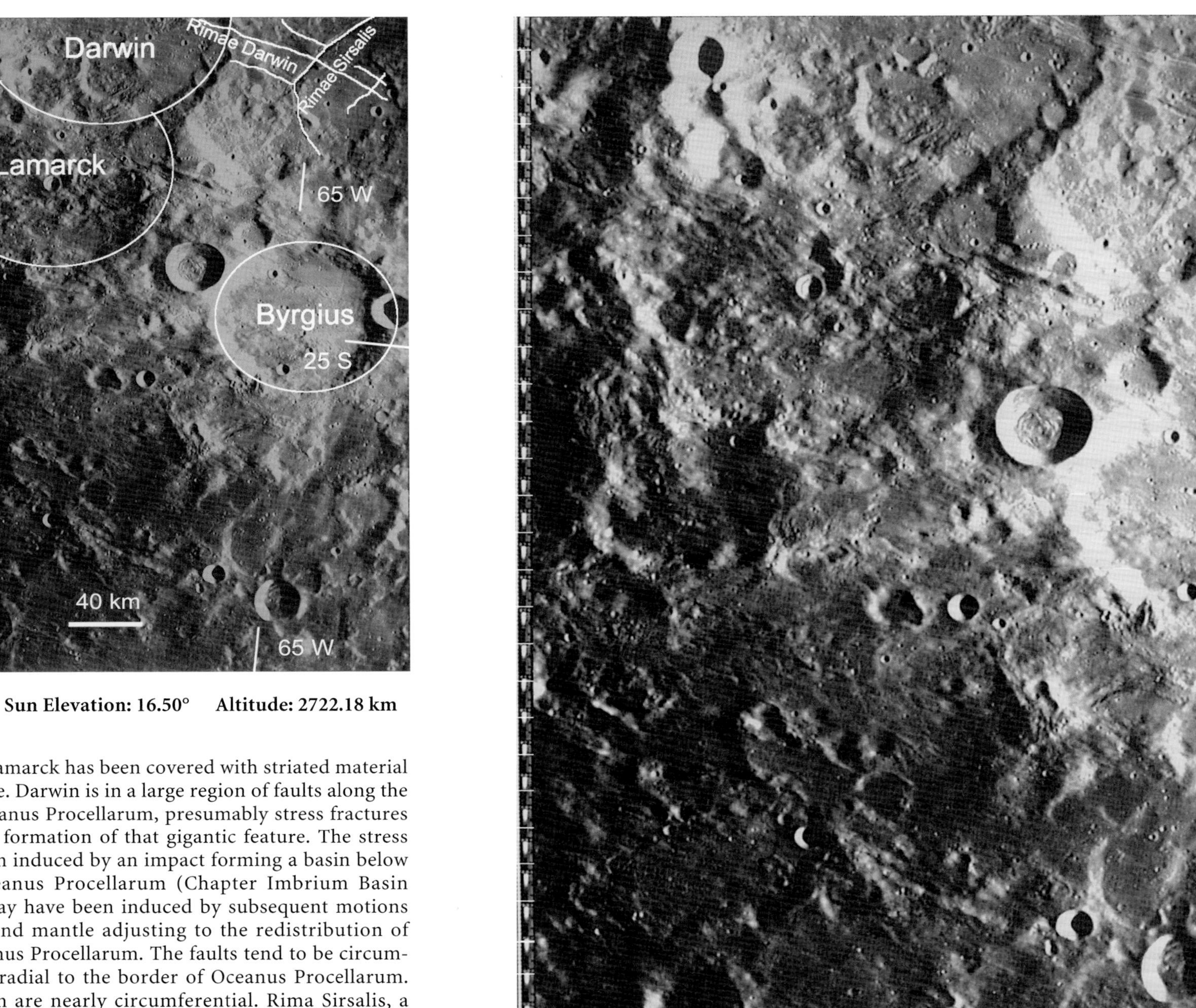

LO4-168H1 Sun Elevation: 16.50° Altitude: 2722.18 km

The floor of Lamarck has been covered with striated material from Orientale. Darwin is in a large region of faults along the border of Oceanus Procellarum, presumably stress fractures related to the formation of that gigantic feature. The stress may have been induced by an impact forming a basin below southern Oceanus Procellarum (Chapter Imbrium Basin Region) or may have been induced by subsequent motions of the crust and mantle adjusting to the redistribution of mass of Oceanus Procellarum. The faults tend to be circumferential and radial to the border of Oceanus Procellarum. Rimae Darwin are nearly circumferential. Rima Sirsalis, a radial fault more than 300 km long, has its southwestern terminus in a network of faults (Rimae Sirsalis) interspersed with Rimae Darwin.

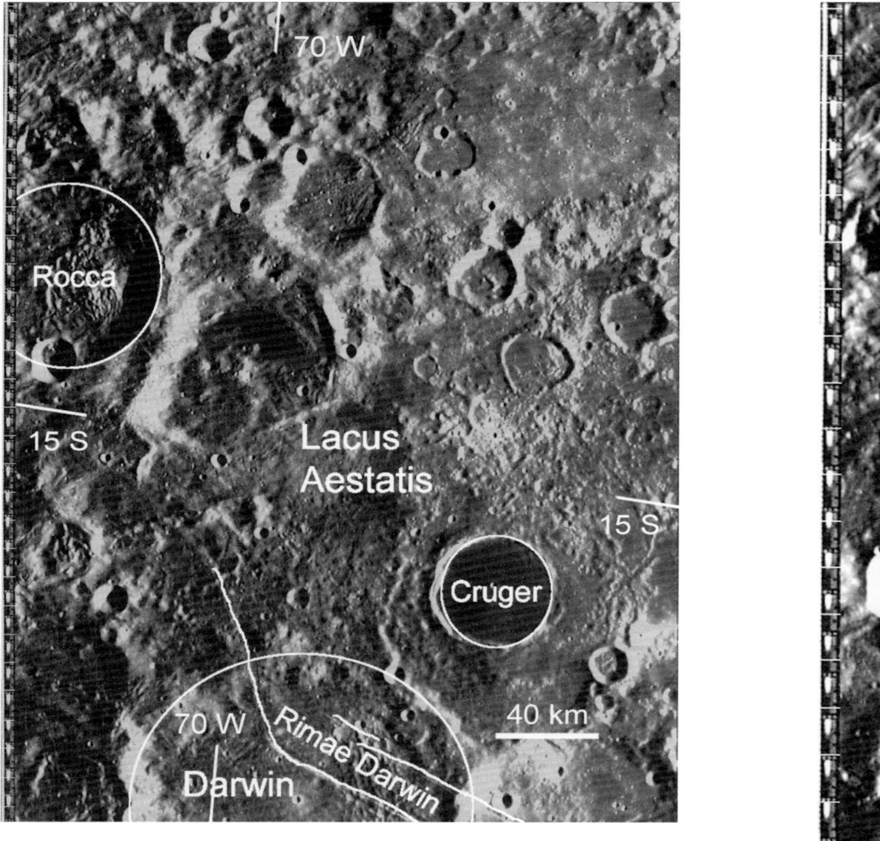

LO4-168H2 Sun Elevation: 16.50° Altitude: 2722.18 km

Diverse surfaces of the outer Hevelius Formation are shown here. Underneath this ejecta blanket is topography from older impacts such as the two craters whose floors have been flooded after the Orientale event to form Lacus Aestatis. The Rimae Darwin can be seen to modify the striations from Orientale, so the rimae must have been formed later. A characteristic texture of the outer Hevelius Formation is the pitted terrain formed by intense arrays of secondary craters and secondary crater chains in the area north of Cruger. Cruger's ejecta blanket covers and smooths the sprinkling of Orientale secondaries. Cruger is therefore Late Imbrian. The floor of Cruger was later resurfaced by mare lava.

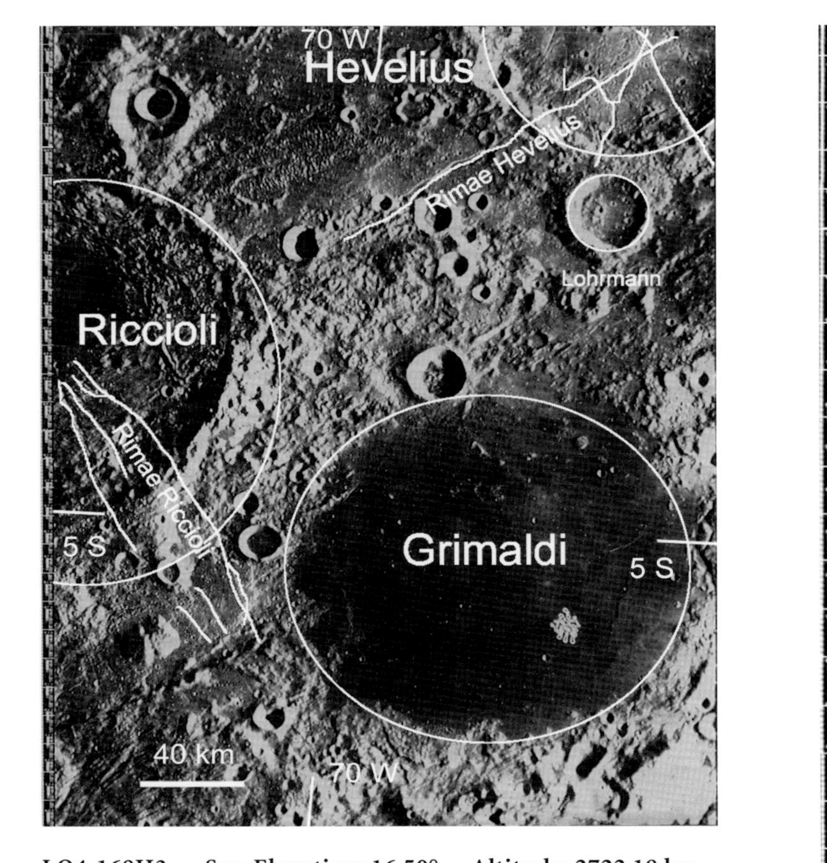

LO4-168H3 Sun Elevation: 16.50° Altitude: 2722.18 km

The Grimaldi Basin is older than Orientale, as can be seen by the Orientale striations in its rings. However, its mare floor was formed after Orientale. The smooth area to the north of Grimaldi at the radius of Lohrmann lies between two rings of the Grimaldi Basin (see LO4-168H2 for the southern part of these rings and a similar smooth area between them). This area may have had a texture similar to the plains between the Montes Rook and Montes Cordillera before being covered with Orientale ejecta. A heavy lobe of plains deposits (probably from Orientale) covers the western rim of Grimaldi, crossed by the southeast ends of the Rimae Riccioli. Rimae Hevelius and Rimae Riccioli are part of the fracture pattern around the rim of Oceanus Procellarum. The flowery pattern in the southeast quadrant of Grimaldi is a development artifact.

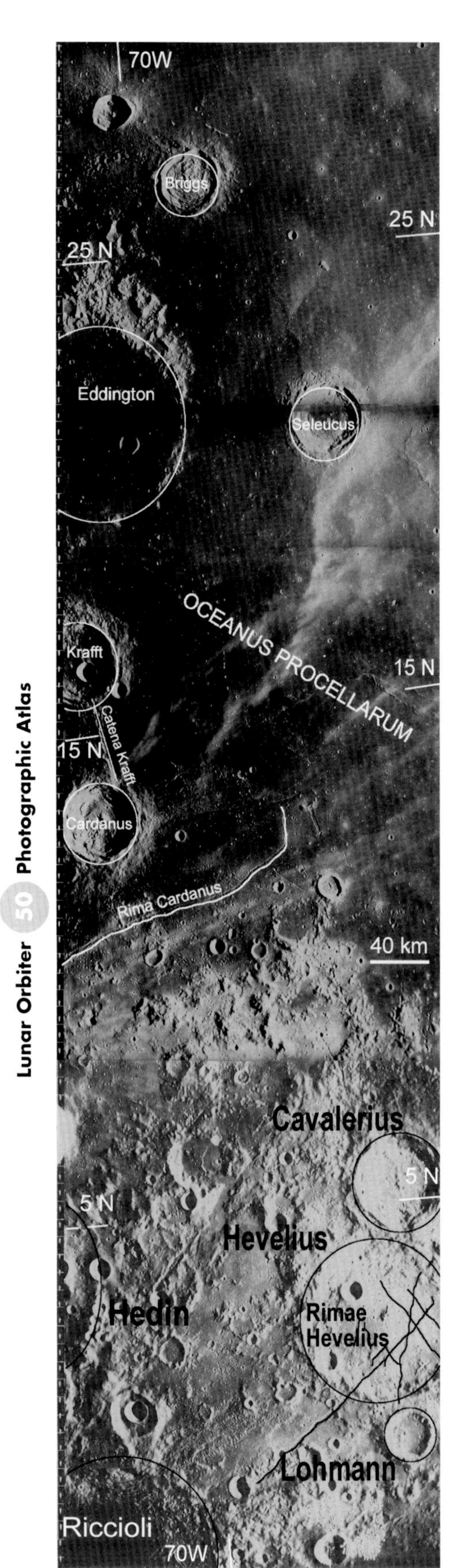

LO4-169H
Sun Elevation: 17.10°
Altitude: 2672.11 km

Where Oceanus Procellarum meets highlands, the younger mare invades low-lying parts of the older highlands just as water meets a shoreline on Earth. Ejecta and secondaries from Krafft and Cardanus overlie the mare, so they are younger yet.

The lava flooding Eddington is of very low viscosity; it finds its way into every crevice. The flow may be not only horizontal but may also seep up from below, through the fractured rock of the crater floor.

The bright rays come from Copernican crater Glushko (LO4-174H2).

The linear trough just east of Hedin (radial to Orientale) terminates in a plains deposit, as if the trough were plowed by a body of material that was (or became) molten. The long fault that crosses the floor of Hevelius and extends nearly to the rim of Riccioli has a graben form (flat floor) in the highlands outside of Hevelius, but is narrower (a rille) as it crosses the crater floor at a lower elevation; this is what one would expect of a fracture zone that has a V shape in the vertical plane.

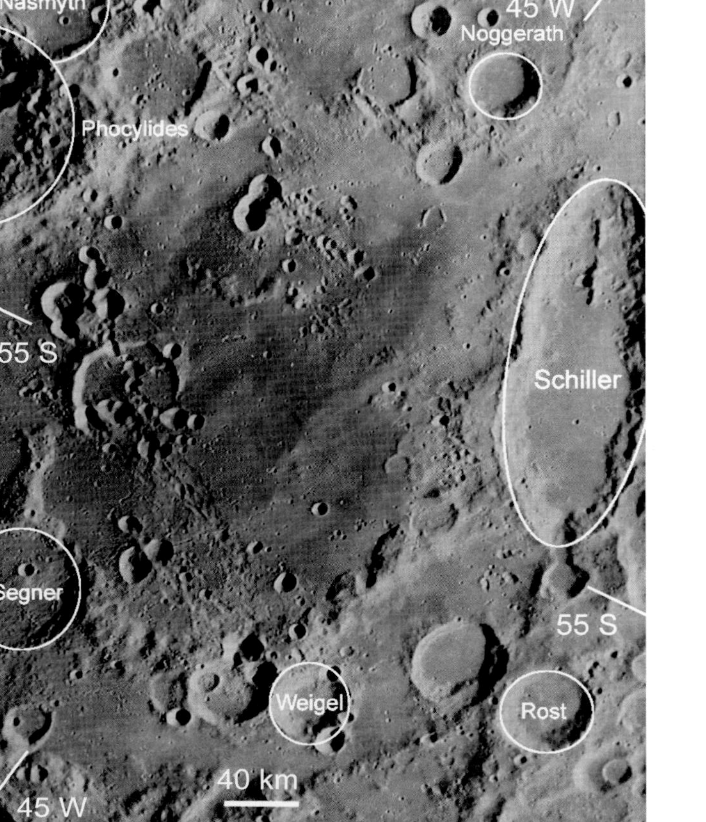

LO4-160H1 Sun Elevation: 17.30° Altitude: 3012.04 km

Schiller is a rare elongate crater. If it was formed by impact, it must have been by a small number of primary impactors arriving at a shallow impact angle, perhaps as low as 3°. Schiller is mostly in the trough between two raised rings of the Pre-Nectarian Schiller-Zucchius Basin. Orientale ejecta has impacted mare that has flooded this basin before the Orientale impact. Craters Weigel, Rost, and Noggerath are examples of old craters that have received impacts on their rims. Note the manner in which the impacts cause the crater walls to collapse onto the crater floor.

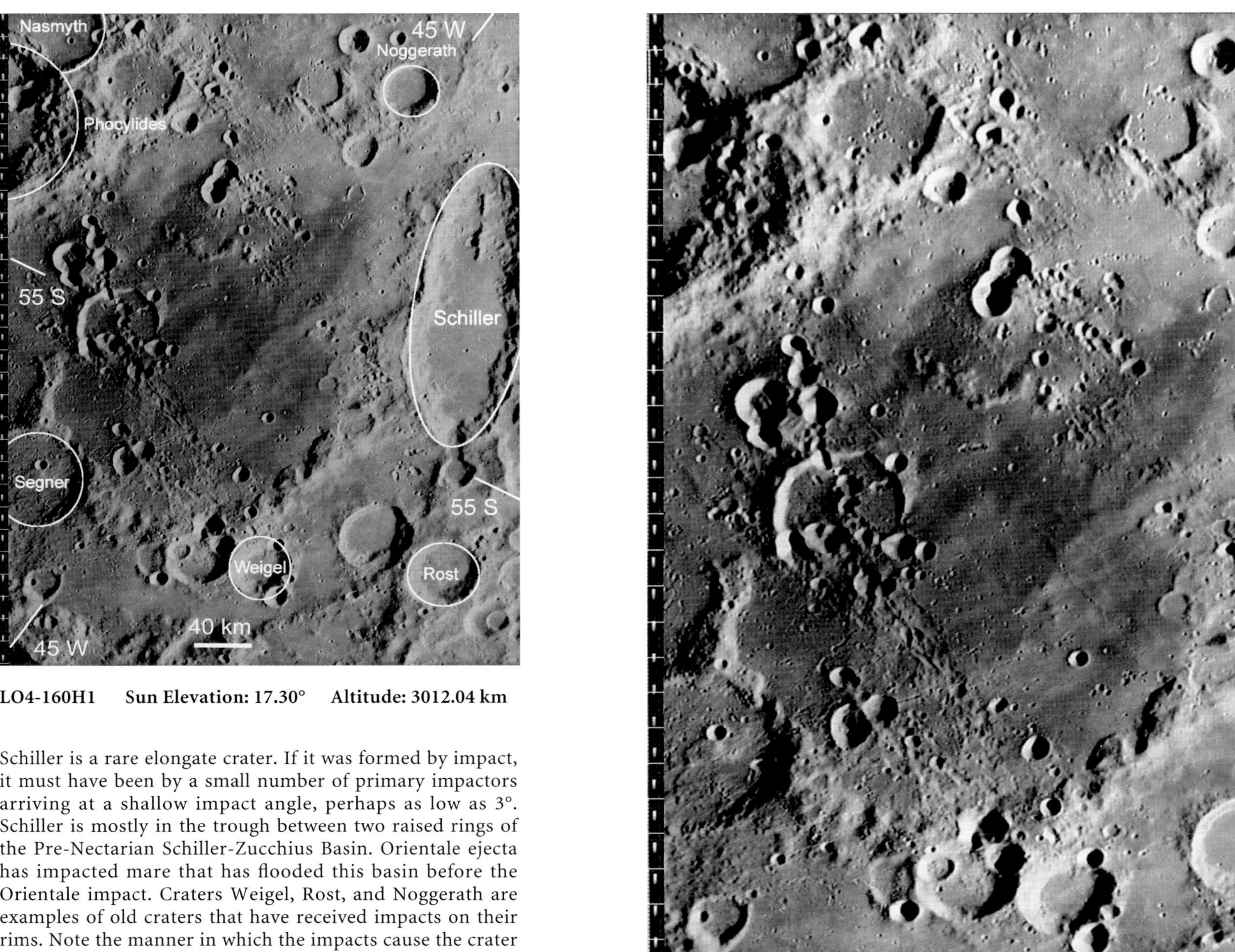

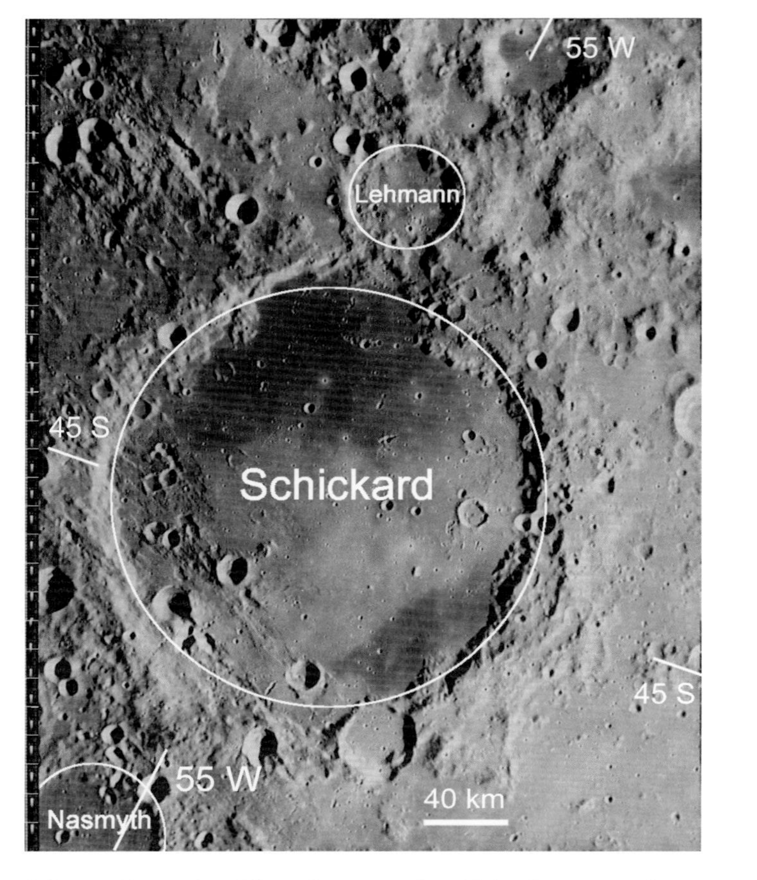

LO4-160H2   Sun Elevation: 17.30°   Altitude: 3012.04 km

The lighter material in the interior of crater Schickard is mapped as Orientale plains-forming material like that on the floor of Wargentin and Phocylides (LO4-172H1). The darker material inside Schickard is mapped as mare basalt, which appears to predate the plains-forming material. Lehmann was formed in the same period as Schickard and was overrun by the plains-forming material.

LO4-160H3 Sun Elevation: 17.30° Altitude: 3012.04 km

Southwest of Vieta is an irregular lake of dark mare material, the westernmost member of a circular pattern of such features (see Chapter 6). These lakes are similar to those in the rings of the Orientale Basin, such as Lacus Veritas and Lacus Autumnis. However, there is no central depression or mare unit within this mysterious ring of lakes. The area around them appears to be covered with plains-forming material of the outer Hevelius Formation.

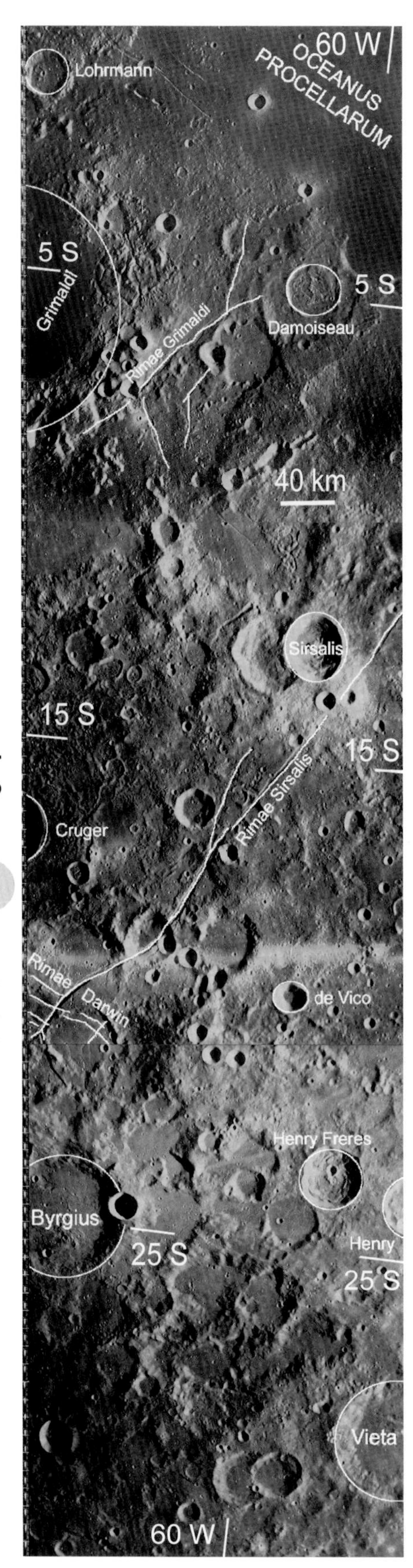

LO4-161H
Sun Elevation: 16.50°
Altitude: 2722.66 km

This chaotic region has been overlaid by ejecta from the Grimaldi Basin and then by the Orientale Basin. Damoiseau appears to have been mostly flooded with a heavy, viscous material, probably molten or semimolten material from Orientale. Some of this material has backflowed over and down the northeast crater wall. The lowest part of the floor may have been flooded with mare like the flat-floored craters nearby. The floor material has fractured because of the cooling of the molten material.

Rimae Grimaldi and Rimae Sirsalis are part of the fracture pattern around the border of Oceanus Procellarum. They can be seen to cross and modify the secondary craters of Orientale, establishing their younger age. The main rima of the Sirsalis feature is the longest such feature on the Moon, leaving its mark on crater floors, rims, and ejecta until it disappears into Oceanus Procellarum (LO4-156H2). It is overlaid only by the youngest crater in its path, the 13-km crater just to the south of crater Sirsalis.

The 22-km crater on the east rim of Byrgius is Copernican, with an extensive ray field. Some of these rays can be seen extending eastward. Henry Freres has a slumped wall, possibly caused by the impact that caused the 8-km young crater in its floor.

Chapter 6

Humorum Basin Region

6.1. Overview

Basins, Maria, and Highlands

Figure 6.1 shows the Humorum Basin Region. The Humorum Basin is of the Nectarian Period, younger than the Nubium Basin and older than the Orientale Basin. It is sufficiently distant from later basins that it has not been extensively modified, except by mare flooding of its floor and outer trough. Its neighboring basins have been more modified, partly by Humorum itself. The highland regions to the southeast are relatively isolated from basins and preserve the heavily cratered surface of the crust. One would like to say that they are "pristine" highlands, but in consideration of the fact that the highland surface is formed from intense cratering bombardment, that term is misleading. The point is that this highland area has not been covered by such a heavy layer of basin ejecta that earlier cratering events have been obscured.

Figure 6.1. LO4-137M. The Humorum Basin Region, the middle and lower area of this photo, includes the multi-ringed Humorum Basin itself, the Nubium Basin to the east, southern Oceanus Procellarum to the north, and the heavily cratered highland area to the south. The bright-rayed crater Tycho is also within this highland region. Northeast of the Humorum Basin Region is the Imbrium Basin and the bright craters (left to right) Aristarchus, Kepler, and Copernicus.

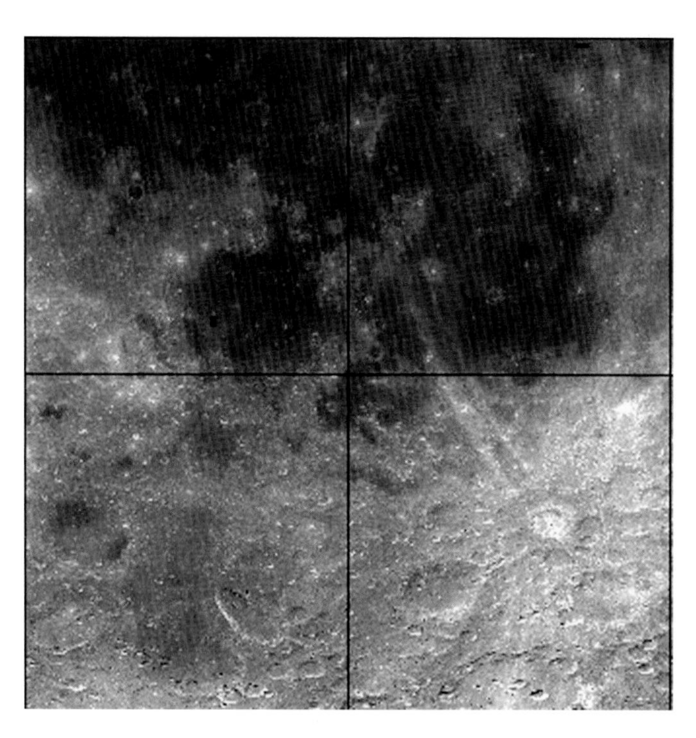

Figure 6.2. Clementine albedo display of the Humorum region. The floors of the basins in this region are flooded with dark mare material. Mare Humorum is left and above the center of the image and Mare Nubium is to the upper right. Oceanus Procellarum is near the top of the image. Tycho's bright ray pattern can be seen across both highland and mare surfaces. This is a section of the PIGWAD[1] Mercator projection; the center of this figure is at 30° west longitude and 30° south latitude and the squares are 30° on a side. *Sources:* NRL and USGS, with permission.

The juxtaposition of so many basins in the vicinity of Oceanus Procellarum has allowed rising mare lava to form a nearly continuous sea, interrupted only by the rims of basins and craters. The extent of this mare material is clear in the Clementine albedo map of Figure 6.2.

The albedo map distinguishes between the bright crust material, the dark mare material from the upper mantle, and the rays, which are newly exposed crust material thrown out from young craters. The rays will darken with exposure to the solar wind, which frees metallic iron from the surface mineral grains in a process known as maturing. Rays from impacts into highlands that fall on mare are different in composition than mare, so the brightness contrast persists long after maturing (Hawke, 2004). In additional time, gardening by subsequent crater bombardment will cause the ray material to blend with the mare surface as well.

Apollo Landings

Apollo 12, the second mission of humans on the Moon, explored and returned samples from a shallow mare surface of southeast Oceanus Procellarum (LO4-125H3). The mission confirmed the character of maria established for Mare Tranquillitatis by Apollo 11. However, this mare proved to be 500 million years younger and much richer in KREEP. Some samples appear to be from rays of Copernicus and establish the time of that impact as 810 million years ago.

Apollo 12 landed within walking distance of Surveyor 3, establishing a new level of landing accuracy that permitted future missions to be targeted for even more interesting landing sites. Parts of Surveyor 3 were returned to Earth and examined for the character and rate of micrometeoroid impacts.

Apollo 14 explored and sampled a geologic structure called the Fra Mauro Formation (LO4-120H3). The Fra Mauro Formation is material ejected from the Imbrium Basin. The samples, together with samples from Apollo 15, establish the time of the Imbrian impact as 3.85 billion years ago. Other samples returned by Apollo 14 are older, in the range of 3.87 to 3.96 billion years, and may come from impacts in the subsurface below the Imbrian ejecta.

6.2. High-Resolution Images

Table 6.1 shows the high-resolution images of the Humorum Basin Region in schematic form.

The following pages show the high-resolution subframes from south to north and west to east; that is, they are in the order LO4-155H2, LO4-155H3, LO4-156H, LO4-148H1... LO4-108H3.

Subframes LO4-155H1, LO4-142H1, LO4-131H1, LO4-119H1, and LO4-107H1 are not included in print because they are redundant with adjacent exposures to the east and west (they are included in the enclosed CD). The subframes of exposures 156H, 132H, and 113H have each been combined into full rectangular images.

Latitude Range	Photo Number										
0–27 N	162	157	150	144	138	133	126	121	114	109	102
0–27 S	161	156	149	143	137	132	125	120	113	108	101
27 S–56 S	160	155	148	142	136	131	124	119	112	107	100
56 S–90 S	154			130		118		106		094	

Longitude at Equator	62 W	58 W	49 W	41 W	35 W	30 W	23 W	16 W	10 W	3 W	4 E

Table 6.1. The cells shown in white represent the high-resolution photos of the Humorum Basin Region (LO4-XXX H1, -H2, and -H3, where XXX is the Photo Number). The Orientale Basin Region is to the west, the Imbrium Basin Region is to the north, the Eastern Basins Region is to the northeast, the Nectaris Basin Region is to the east, and the South Polar Region is to the south.

[1] Planetary Interactive G.I.S.-on-the-Web Analyzable Database (USGS)

LO4-155H2 **Sun Elevation: 17.1°** **Altitude: 3011.10 km**

The ridge in the lower right corner of this photo radiates from the Schiller-Zucchius Basin. Subtle striations in the upper left radiate from Orientale.

Rimae de Gasparis
30 S
Vieta
de Gasparis
50 W
Fourier
30 S
40 km
50 W

LO4-155H3 Sun Elevation: 17.1° Altitude: 3011.10 km

The striations and crater chains through Vieta and Fourier are radial to Orientale. The brightness in the vicinity of Fourier is due to rays from small primary craters of the Copernican Period. There are at least four; one is on the northeast rim of Vieta and the others are about 50 km north of Fourier, about 35 km northeast of Fourier, and about 35 km southeast of Fourier. Their ray patterns are shown clearly in albedo data from the Clementine mission.

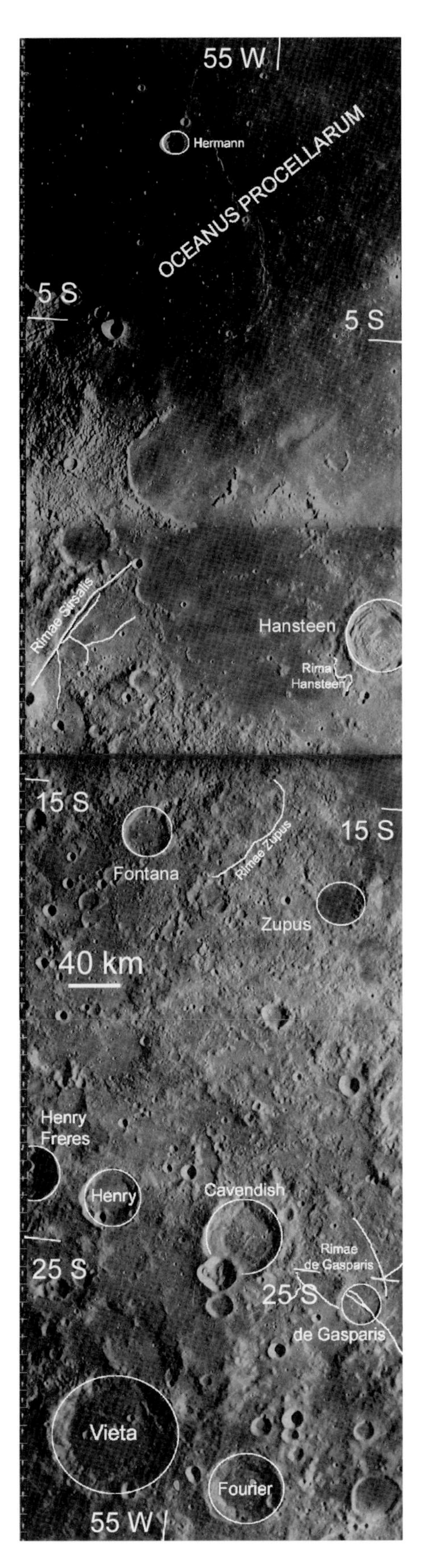

LO4-156H
Sun Elevation: 16.6°
Altitude: 2721.57 km

This frame shows the transition between the highlands and southern Oceanus Procellarum. The linear feature that runs southeast from Hermann is an example of a wrinkle ridge, also known as a mare ridge; such features often appear in a large mare surface. Their origin is unclear, but it may be related to cooling and solidification of the lava.

As is common along the "shore" of Oceanus Procellarum, radial fractures are revealed by rimae (rilles) at the surface. Flooding by lava, cooling, and possible isostatic adjustment are likely sources of the fracturing.

Note the 80-km buried crater at the edge of Oceanus Procellarum near Rimae Sirsalis; only part of its rim remains visible, like the smile of the Cheshire Cat. Striations between Fontana and the mare support the hypothetical basin under southern Oceanus Procellarum, as discussed in the introduction to the chapter on the Imbrium Basin Region (Chapter 7).

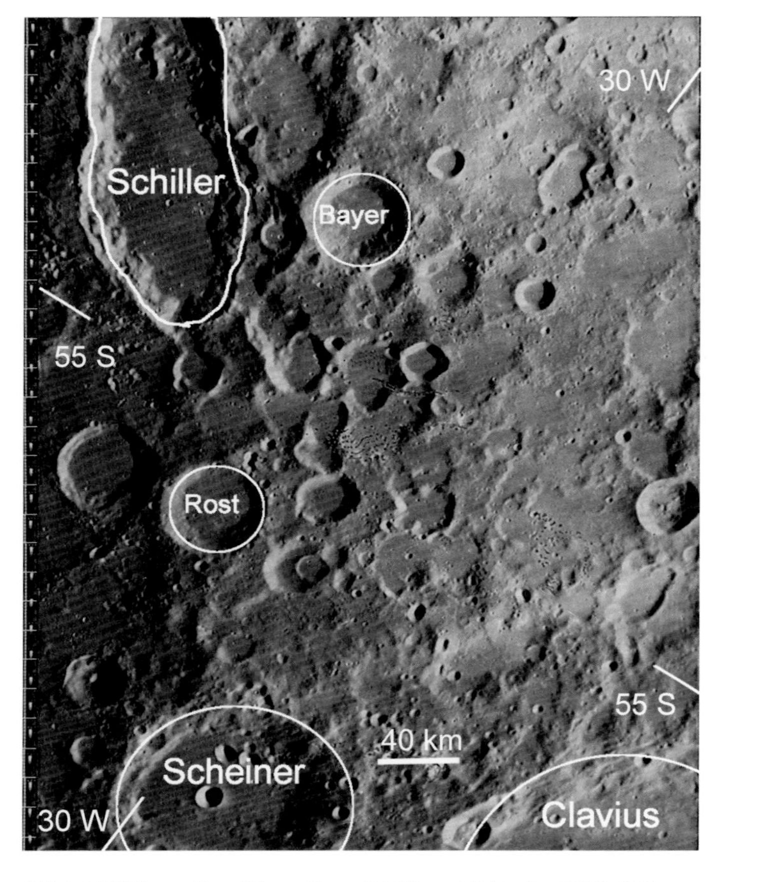

LO4-148H1 **Sun Elevation: 17.9°** **Altitude: 3009.38 km**

The furrow and ridge running northeast of Scheiner is probably ejecta from the Schiller-Zucchius Basin. Similar, less pronounced linear features run parallel to the furrow in the middle part of the picture. Ejecta from Clavius partly overlies the furrow. The floor of the furrow is very flat and smooth; it may be the result of molten material in the basin ejecta. Bayer is a concentric ring crater that may have penetrated through the Schiller-Zucchius ejecta layer into more primitive terrain below.

LO4-148H2 Sun Elevation: 17.9° Altitude: 3009.38 km

The degraded state of Mee indicates that it is a relatively old, Pre-Nectarian crater. The ejecta blanket in the lower right half of the photo has been deposited by the Humorum Basin to the north. Chains of small secondary craters radiate from the Orientale Basin far to the northwest.

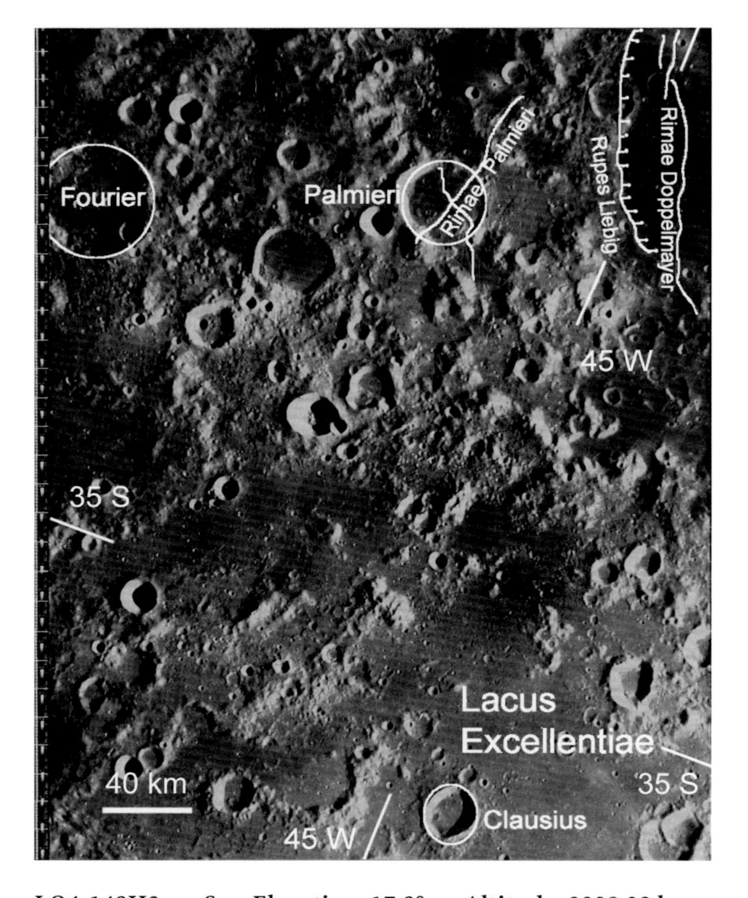

LO4-148H3 Sun Elevation: 17.9° Altitude: 3009.38 km

This picture features an unusual range of feature types. Lacus Excellentiae is part of a ring of mare lakes in a trough of the Humorum Basin. An outer ring of Humorum can be seen passing between Fourier and Clausius, interrupted by valleys blasted by Humorum ejecta. Rimae Palmieri and Rimae Doppelmayer are examples of the radial and circumferential stress fractures near the edge of Mare Humorum, whose edge can be seen in the upper right.

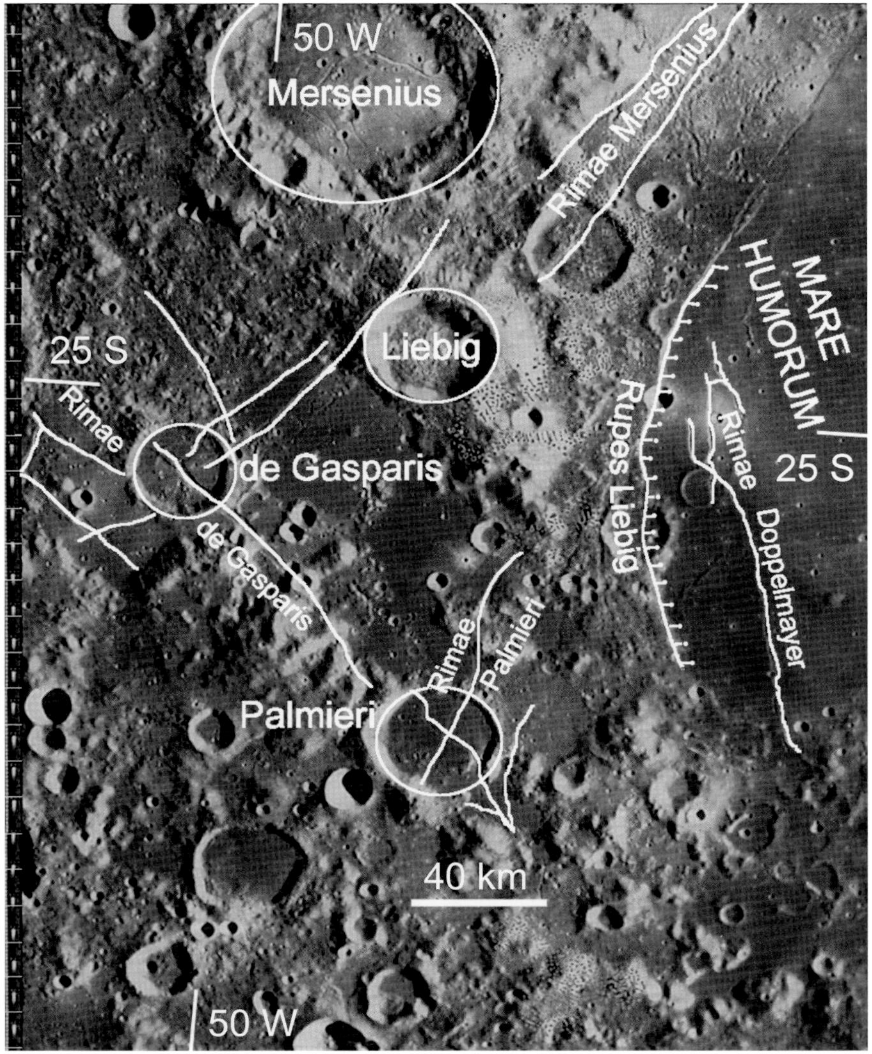

LO4-149H1 **Sun Elevation: 17.5°** **Altitude: 2720.28 km**

This picture illustrates the structure of the Humorum Basin. The inner basin has been flooded with lava. Rupes Liebig (rupes means a scarp, a cliff) may have been formed as lava rose partway across the shelf (which may have been similar to the Maunder Formation of Orientale) and then subsided as the source plume relaxed. As often is the case where a mare has a thin edge, the shelf is covered with dark plains-forming material. The topographic ring surrounds Mare Humorum west of this flat plain. A trough outside this first ridge has been partly flooded with lava. A second circumferential ridge, lower and less continuous than the first, lies outside the trough. Most of the rima features reveal stress faults from the mare flooding and cooling processes.

LO4-149H2 | Sun Elevation: 17.5° | Altitude: 2720.28 km

It has been suggested that Mons Hansteen could be an Earth-style volcano (it looks much like Mt. Fuji, for example). However, that would require a source of high-viscosity lava only a few kilometers away from Billy, which has been flooded with typical low-viscosity lava. Another explanation would be that it is a remnant of an interior basin ring (like Montes Rook, which is rich in such pyramidal blocks).

LO4-149H3 Sun Elevation: 17.5° Altitude: 2720.28 km

This is the southeast section of Oceanus Procellarum. Many further pictures of this gigantic mare are covered in the chapter on the Imbrium Basin (Chapter 7). Interesting features of this picture are the segments of crater walls that have survived inundation by the mare lava. Some of the walls are so high that it seems that the missing segments must have melted into the mare floor. The right side of this picture contains large swirls of development bubbles.

LO4-142H2 **Sun Elevation: 19.5°** **Altitude: 3006.60 km**

Thick ejecta and secondary craters of the Humorum Basin cover this area. This terrain is called the Vitello Formation (crater Vitello is in LO4-142H3); it is the equivalent of the Hevelius Formation of the Orientale Basin or the Fra Mauro Formation of the Imbrium Basin. The two Eratosthenian craters that obscure Hainzel may have landed essentially simultaneously, as indicated by the straight segment between them. Hainzel had already been formed when they landed, although the interval may have been short.

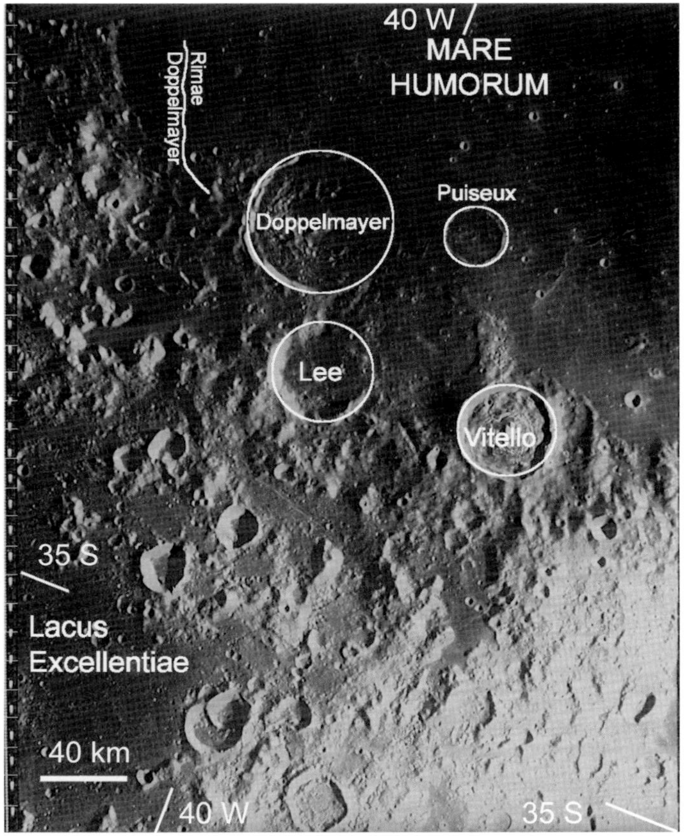

LO4-142H3 Sun Elevation: 19.5° Altitude: 3006.60 km

One ring of Humorum passes through or near Vitello. Another, probably the highest ring, passes just north of Lacus Excellentiae, which floods a trough external to that ring. The material on the floor of Vitello appears to have been molten before hardening and then fracturing, very much like Kopff in Orientale (LO4-187H2). Doppelmayer is slanted; it may have formed partly on the basin floor and partly on a terrace or scallop of the rim. As a result, it is only partly flooded. If Puiseux was flooded through a very small break in its northern rim (and not from below), the viscosity of the lava must have been very low to spread evenly over the crater floor.

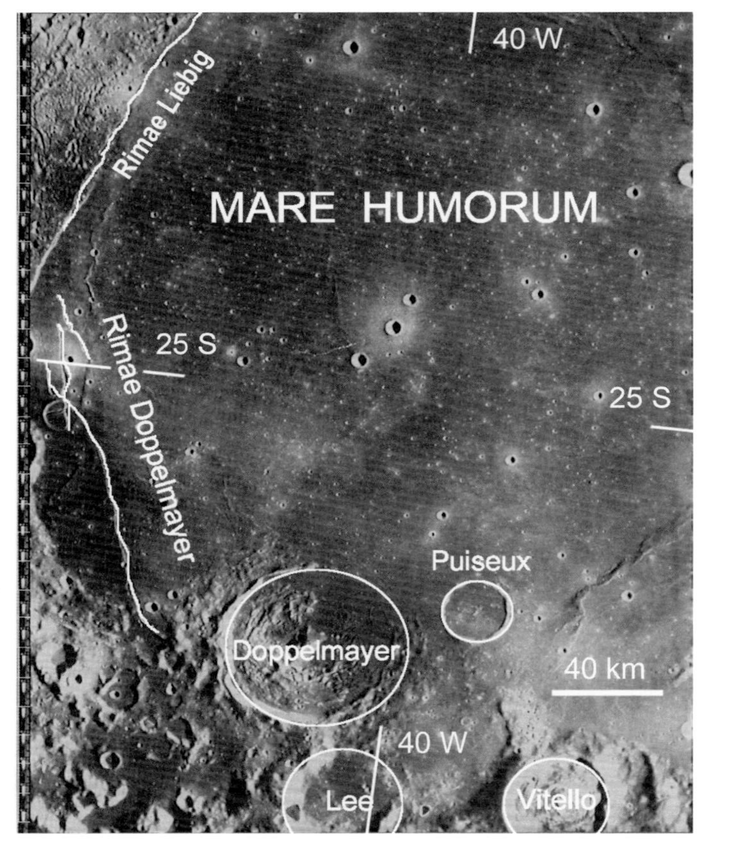

LO4-143H1 Sun Elevation: 18.8° Altitude: 2719.10 km

Although the Humorum Basin does not appear to be as large as the Orientale Basin, the central mare is about the same size, probably because the crust is thinner in the Humorum region. As a result, the inner detail and perhaps inner rings are lost because of the flooding of lava. Instead there is a very smooth, but not featureless, surface. The scarp at the western edge of the mare (Rupes Liebig) suggests that the mare surface rose to a level higher than its present elevation and then subsided a bit, partly as a result of contraction on cooling. Stresses induced by the subsidence find surface expression in Rimae Doppelmayer and wrinkle ridges such as the one near the left edge of the picture. The bright streaks across the mare are rays from Tycho.

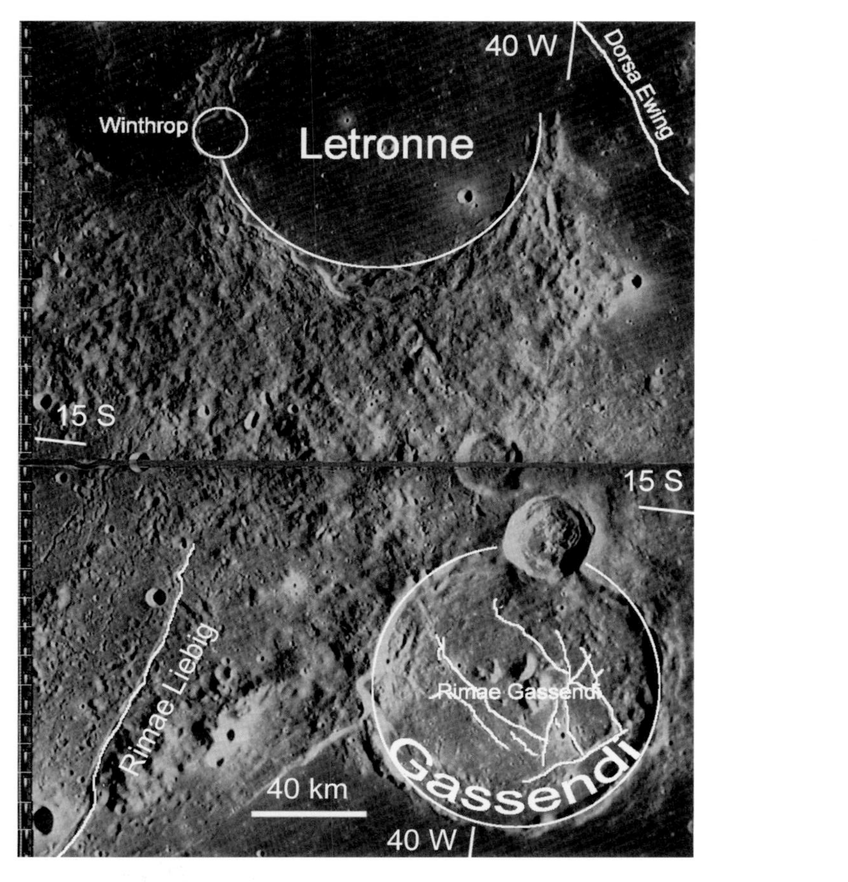

LO4-143H2 Sun Elevation: 18.8° Altitude: 2719.10 km

Gassendi was formed by an impact precisely on the scarp bounding Mare Humorum. Its floor, slightly lower than the nearby lava surface, appears to have been a melt sheet, heavily fractured upon cooling or isostatic adjustment. A nearly continuous rim has protected it from flooding with lava except for a small breach in the southern rim, which has allowed some lava to enter a sector of the floor. The floor of Gassendi has been heavily fractured, possibly by upward pressure of lava on a melt sheet. Letronne has been fully breached and flooded from the north by Oceanus Procellarum.

40 W
OCEANUS
Surveyor 1
Flamsteed
5 S
PROCELLARUM
5 S
Dorsa Rubey
40 km
40 W

LO4-143H3 Sun Elevation: 18.8° Altitude: 2719.10 km

The cross marks the June 1966 landing of Surveyor 1, whose soft landing confirmed the safe bearing strength of the lunar soil, previously affirmed by Luna 9. The Surveyor 1 camera showed the local surface to be flat, with boulders near a crater in the distance. Surveyor 1 landed in a remnant of a crater rim left behind by an inundated "ghost crater," informally called the Flamsteed ring. Dorsa Rubey is an example of a wrinkle ridge, a common feature of mare surfaces. A smaller ridge forms an interior ring within the vestigial rim of the Flamsteed ring.

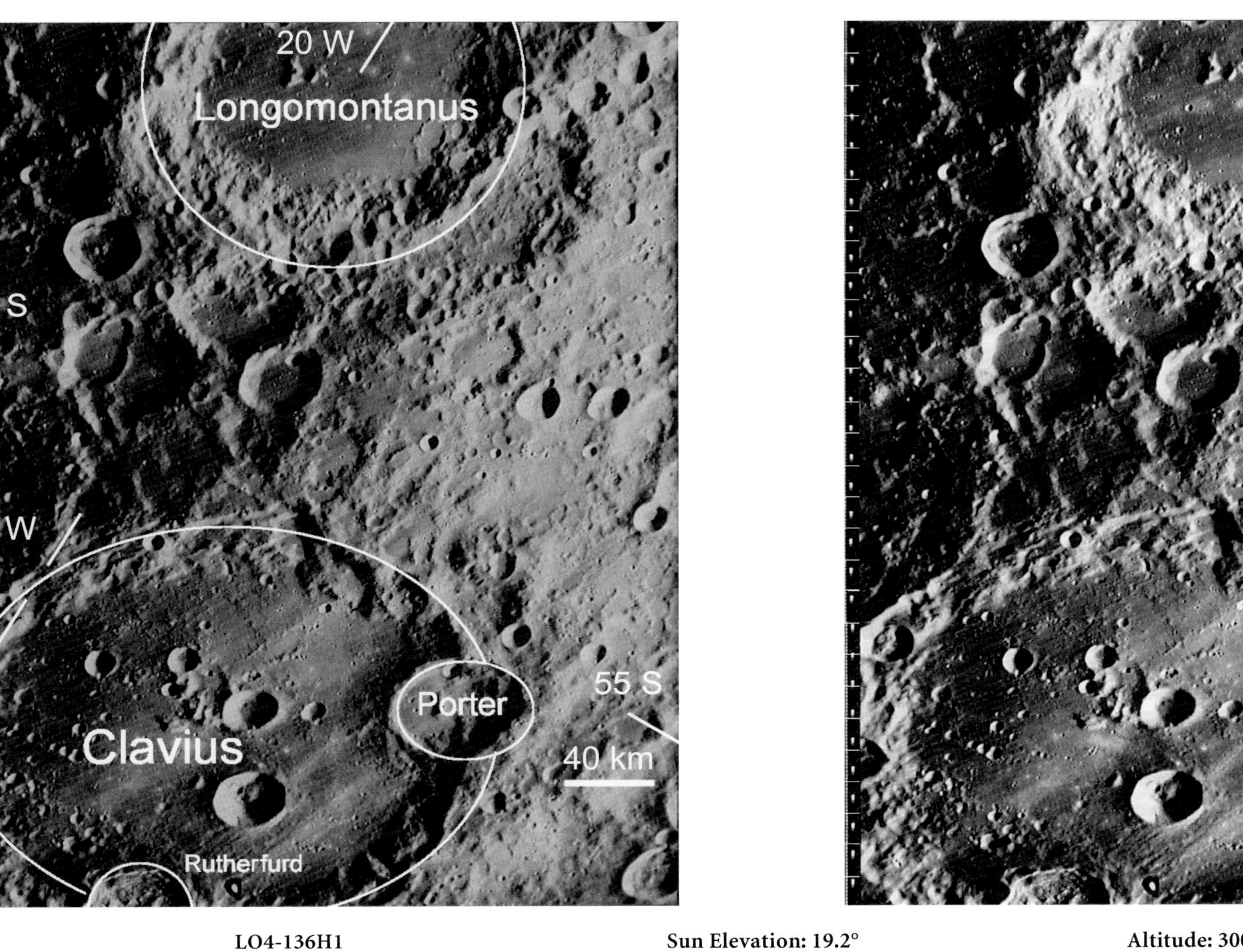

LO4-136H1 Sun Elevation: 19.2° Altitude: 3002.96 km

Clavius, at 225 km in diameter, is in a transition region between craters and basins. The rim shows multiple collapse features, presenting a scalloped appearance. The peak structure is complex, an intermediate step to an inner ring. The streaks on the floor of Clavius and Longomontanus are rays from Tycho. Porter has hit the northeastern corner of Clavius, piling the material of the rim of Clavius neatly to that side.

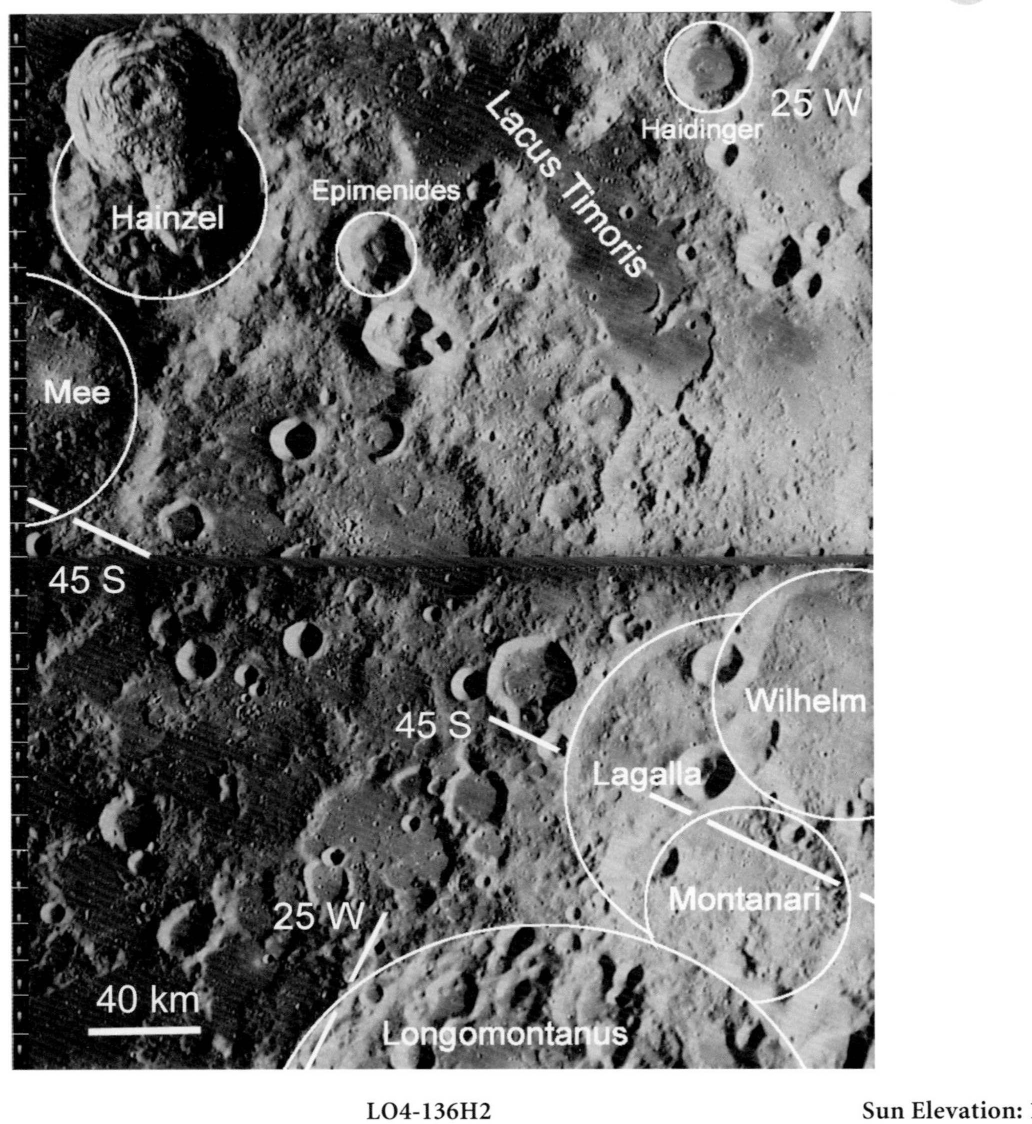

LO4-136H2 Sun Elevation: 19.2° Altitude: 3002.96 km

The gouges in the northern rim of Longomontanus are from the direction of the Nubium and Imbrium Basins and are likely to be secondaries from one of those basins. The formation of the gouges suggests that the incidence angle of the secondary impactors was about the same as the angle of the rim before impact. The valley of Lacus Timoris may have been gouged out by ejecta from the Humorum Basin.

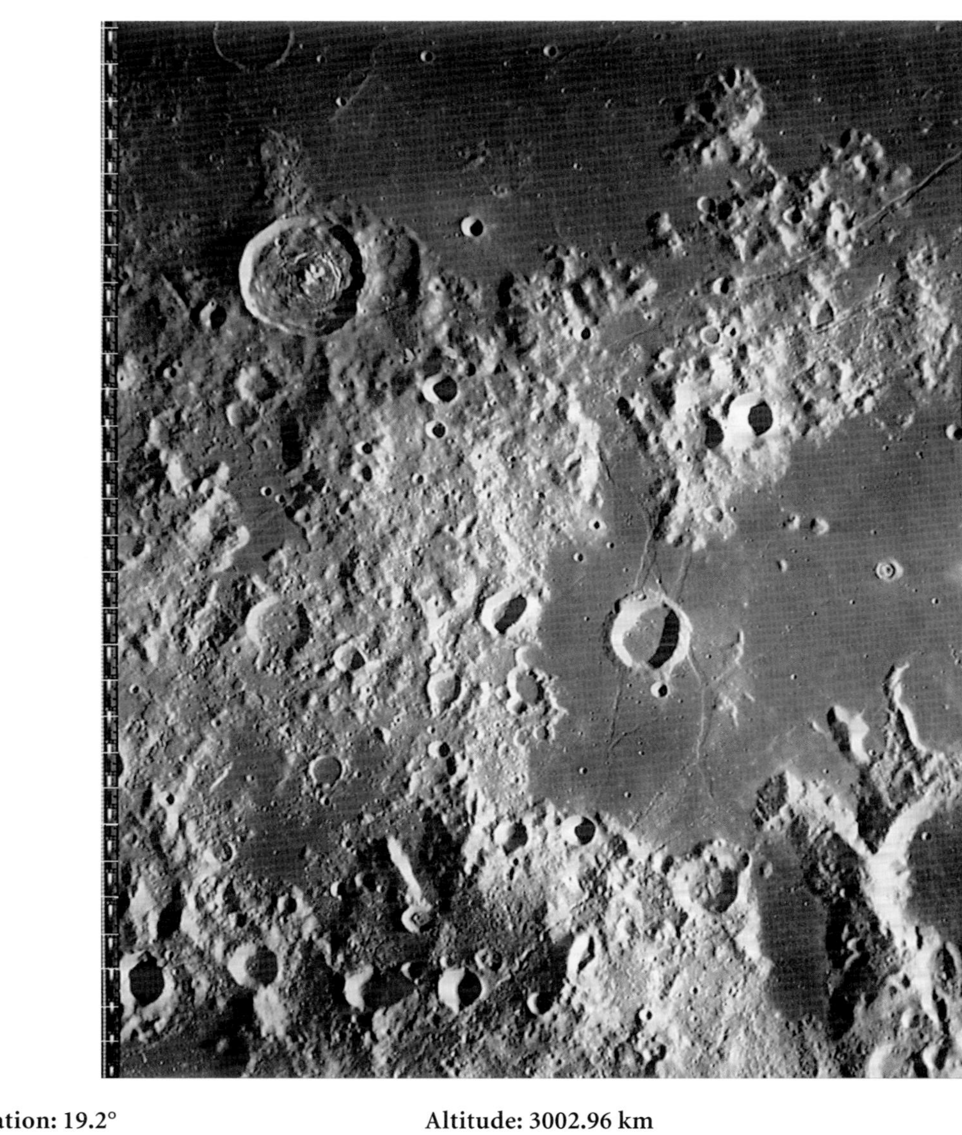

LO4-136H3 Sun Elevation: 19.2° Altitude: 3002.96 km

Palus Epidemiarum may be similar to the lakes in the trough ring around basins, simply an unusually large area of mare in the trough. Lava from Palus Epidemiarum seems to have flowed into Mare Humorum through a sinuous channel that passes through a valley in the ring that bounds Mare Humorum.

The radial and circumferential Rimae Ramsden are typical of the surface features overlying stress fractures near the edge of maria. Rupes Kelvin is the boundary of the ring that bounds Mare Humorum; this ring is similar to the outer range of Montes Rook. Another higher ring passes through Capuanus.

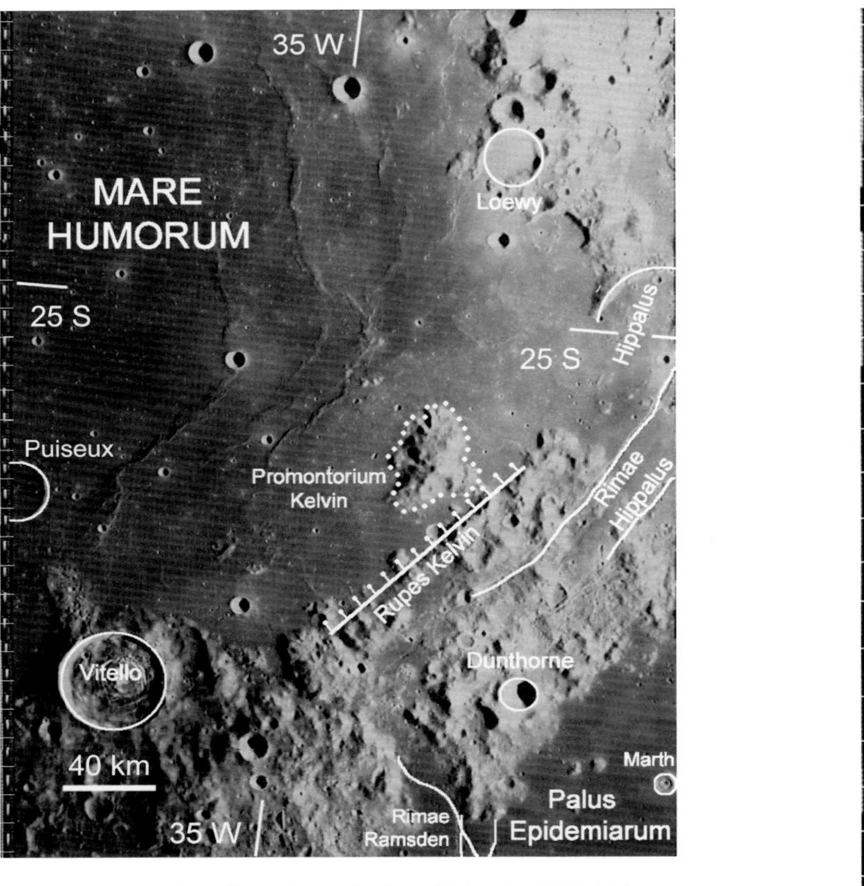

LO4-137H1 Sun Elevation: 18.8° Altitude: 2718.01 km

Wrinkle ridges (dorsa) such as those in the middle of this picture often appear on mare surfaces. Here they form a series of circumferential ridges. They could have been formed by extrusion of the mare lava along cracks in the cooled and hardened surface. Scarps in the mare surface (between the ridges and the rim) indicate that the direction of low-viscosity lava flowed from the center toward the edge of the mare.

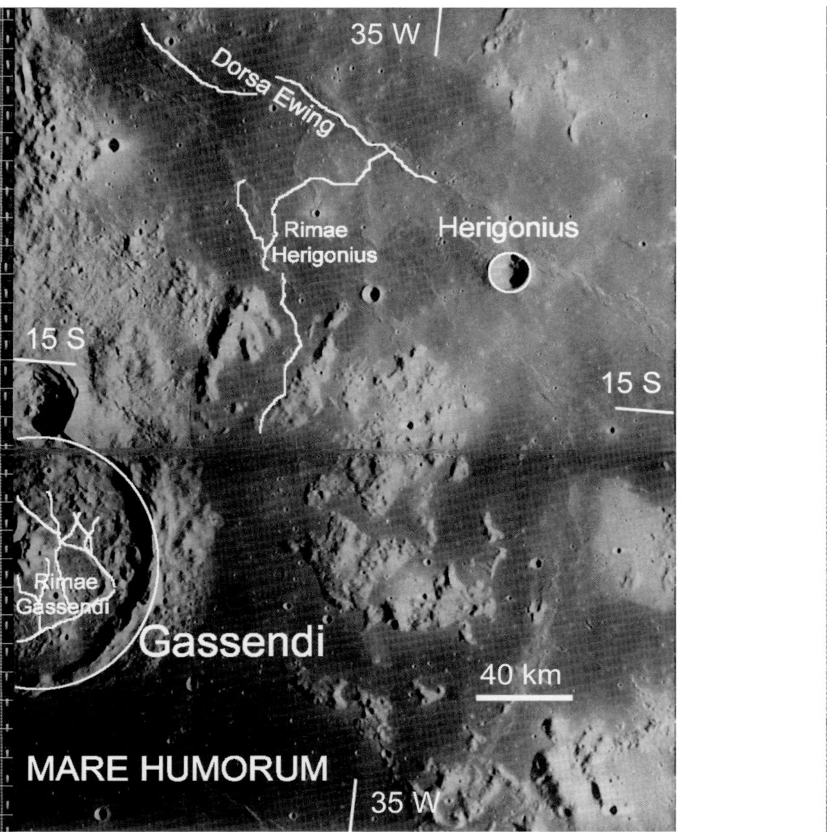

LO4-137H2 Sun Elevation: 18.8° Altitude: 2719.10 km

This photo shows a breach in the rim of Mare Humorum that allowed the mare surface to connect with that of Oceanus Procellarum in the northeast corner of this photo. Perhaps the lava rose within Mare Humorum to a point where it overflowed the wall and opened the breach. The darker (younger?) surface of Mare Humorum seems to have extended through the breach toward Oceanus Procellarum. Scarps indicating flow boundaries and the sinuous channels of Rimae Herigonius may be related to the movement of lava from Mare Humorum into (or onto) Oceanus Procellarum. Dorsa Ewing and other nearby wrinkle ridges may be related to underlying topography; the mare is shallow here (see the ghost craters in LO4-137H3).

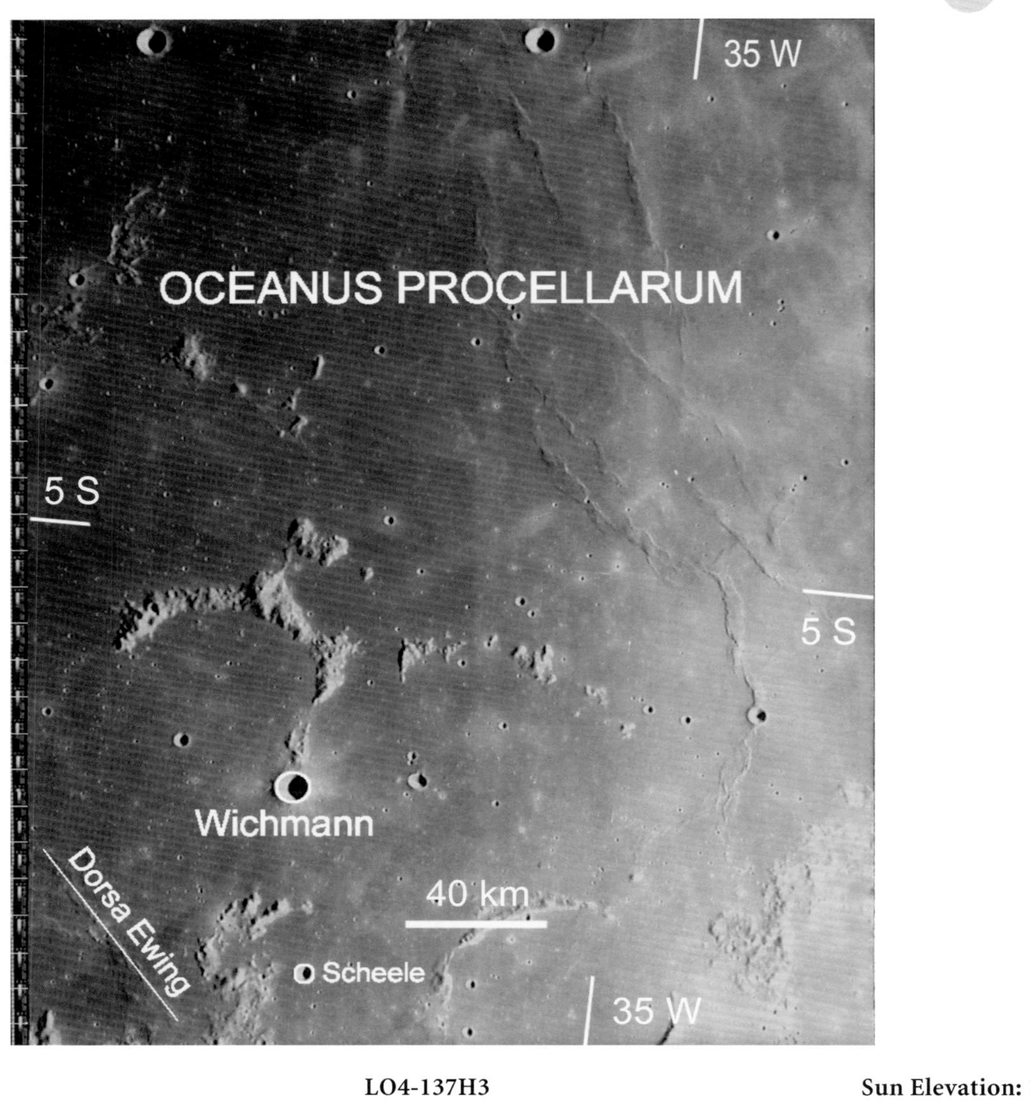

LO4-137H3 Sun Elevation: 18.8° Altitude: 2719.10 km

There was probably an ejecta blanket from the Humorum Basin here, but it was covered by the mare surface of Oceanus Procellarum, just as it nearly erased the large "ghost" craters.

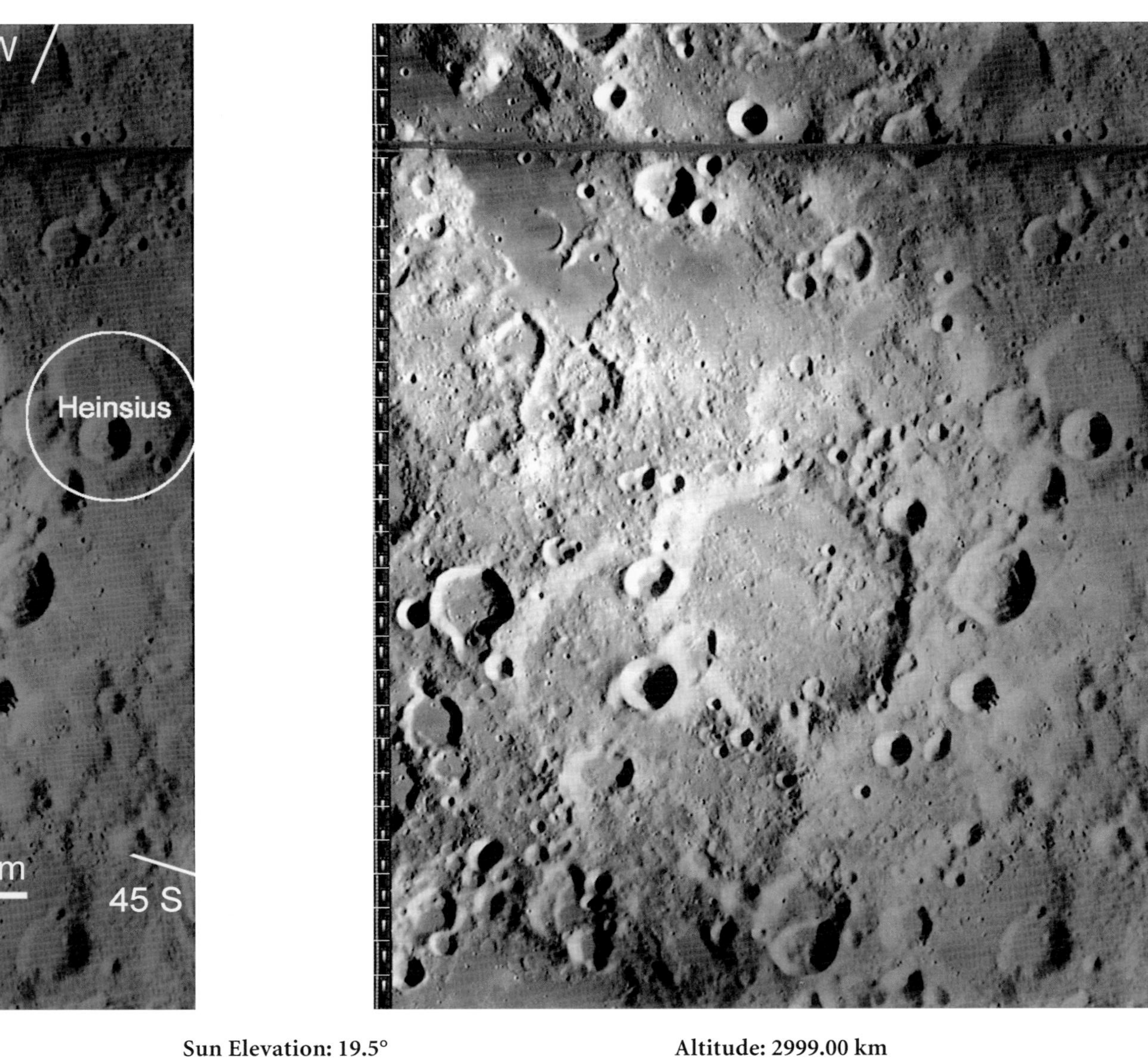

LO4-131H2

Sun Elevation: 19.5°

Altitude: 2999.00 km

The striations between Haidinger and Heinsius are ejecta from the Humorum Basin, similar in nature to the inner Hevelius Formation of the Orientale Basin. Crater chains in Wilhelm are radial to both the Nubium Basin and the more remote Imbrium Basin. This area is also overlain with rays from Tycho.

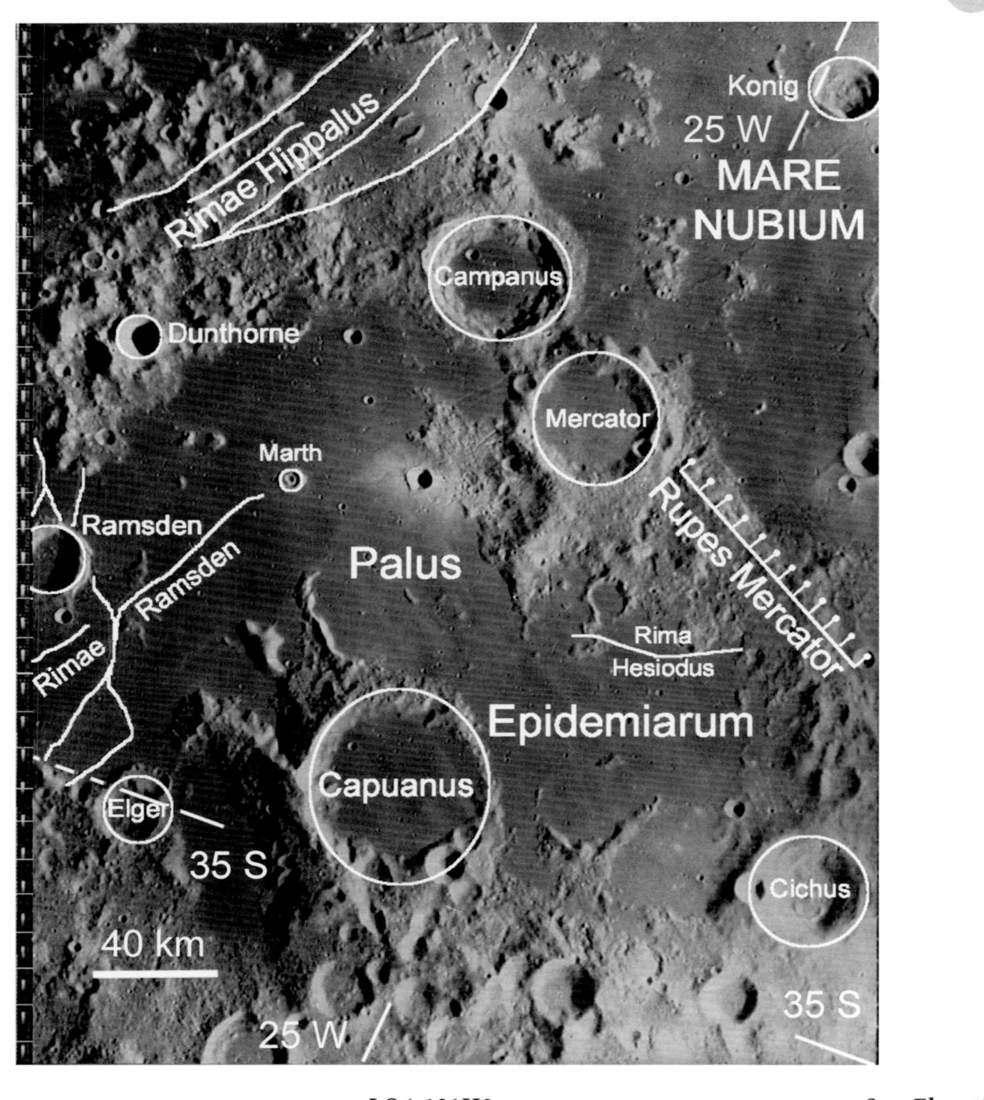

LO4-131H3 Sun Elevation: 19.5° Altitude: 2999.00 km

The deeper (smoother, relatively featureless) part of Palus Epidemiarum is the largest of a set of small tracts of mare formed in a trough ring of the Humorum Basin. Its size may be influenced by an intersection of the Humorum trough with an earlier trough of the Nubium Basin. Although these small mare units could all be termed “lacus” (lake), the extension of Palus Epidemiarum to the southeast appears to be relatively shallow, so this feature has been named “palus” (swamp). Rupes Mercator is part of the degraded rim of the Nubium Basin. Mercator has impacted this rim, and then somewhat later Campanus as well. The relative ages of Campanus and Mercator are indicated by the “worn-away” appearance of Mercator and its ejecta.

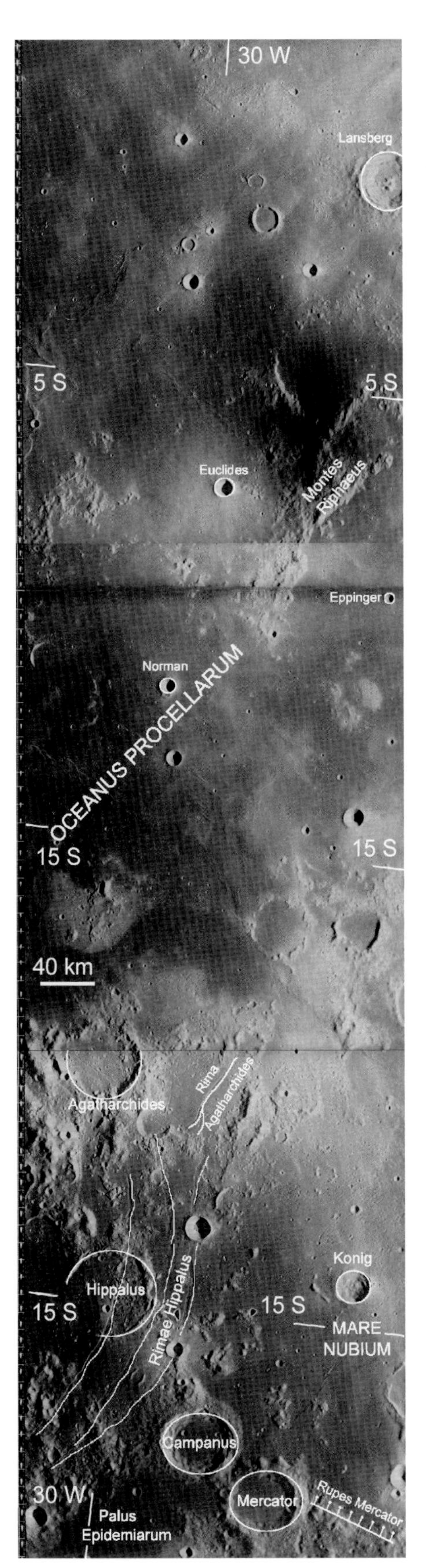

LO4-132H
Sun Elevation: 18.9°
Altitude: 2717.39 km

Montes Riphaeus is a remnant of very old crust. It may be an arc of the rim of a large crater containing Mare Cognitum. The intersecting ray patterns come from Copernicus to the northeast and Kepler to the northwest. The small, sharp, rayed craters such as Euclides could be secondaries from Copernicus.

In the lower left of this picture, the dark mare is probably in a trough of the Humorum Basin. The remains of an outer ring appear as an island above it.

Rimae Hippalus are nearly concentric with the Humorum Basin, but they trend outward toward the north instead of curving around to the northwest. Rima Agatharchides is part of the total system of rimae, which is influenced by Mare Humorum and Mare Nubium and the underlying basin structures. These rimae traverse older features like the rim of Hippalus, but are interrupted by newer craters such as the one about 30 km southeast of Hippalus. The ejecta from the newer craters fills in the rimae. The bounding ring of the Humorum Basin has been suppressed here, possibly because of interactions with the preexisting ring and melt sheet of the Nubium Basin.

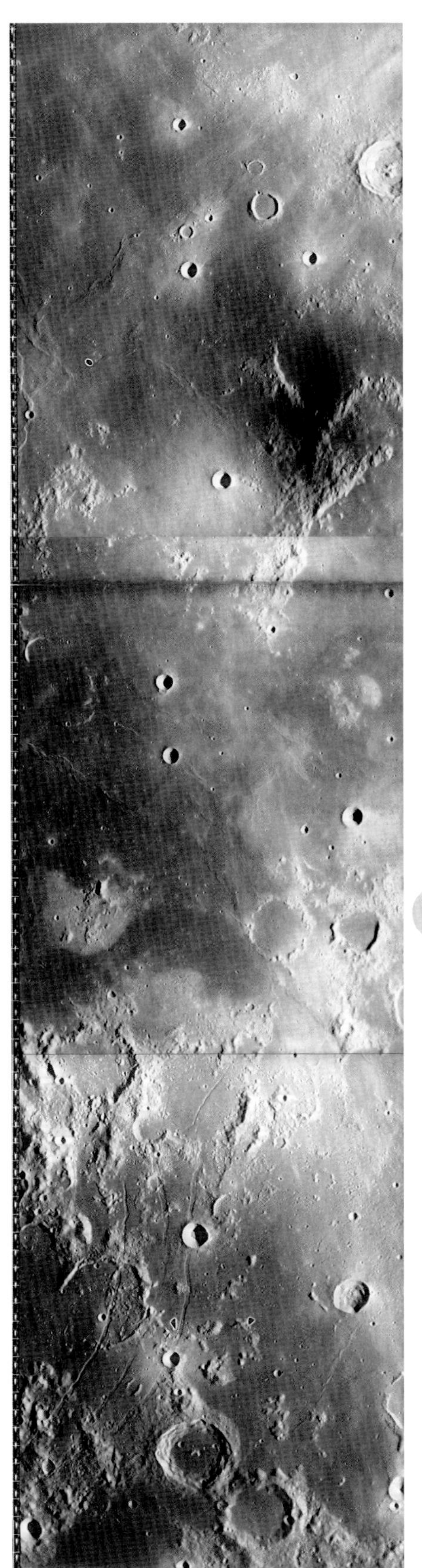

LO4-124H1 Sun Elevation: 19.9° Altitude: 2994.47 km

Which came first, Clavius or Maginus? The relevant brightness of the floor of Maginus is due to its proximity to Tycho, so that is not a clue. It is difficult to see strong evidence whether the ejecta of Clavius overlies that of Maginus or the reverse, possibly because of the overburden of ejecta from numerous other craters and basins. Maginus shows more cratering around its rim than Clavius. This establishes a significantly older age for Maginus, which is mapped as Pre-Nectarian whereas Clavius is mapped as Nectarian. Striations on the rim of Maginus in the direction of Clavius are mapped as ejecta from Clavius.

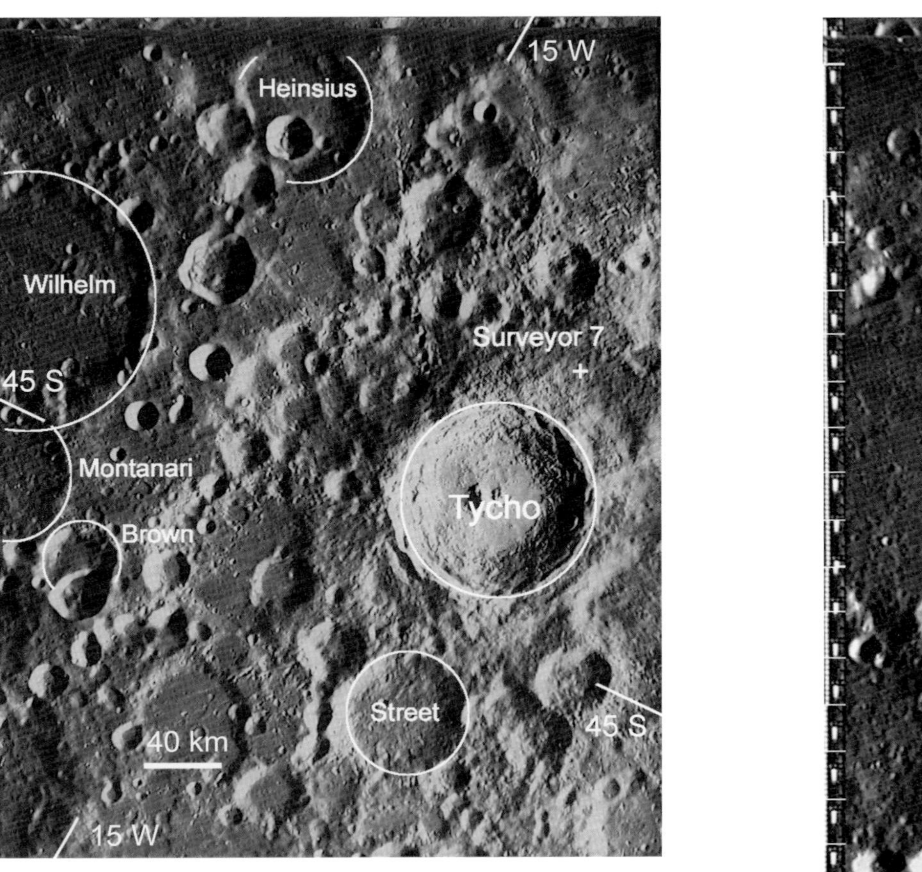

LO4-124H2 **Sun Elevation: 19.9°** **Altitude: 2994.47 km**

The floor, rim, and rays of Tycho are very bright, indicating its young age (Copernican Period). Consequently, there has not been time for significant degradation of its topography. The central peak is typical of craters of this size, formed of material thrown directly upward by the compressional shock wave rebounding from original crust below the boundary of fractured crater material. Surveyor 7 was boldly targeted for the fresh ejecta from Tycho, landing in January 1968. The Surveyor 7 camera showed a surface of blocks up to a meter in size. The surface was roughly textured, but with fewer craters than were seen in the mare landing sites. The lower rate of cratering results from the young age of the terrain.

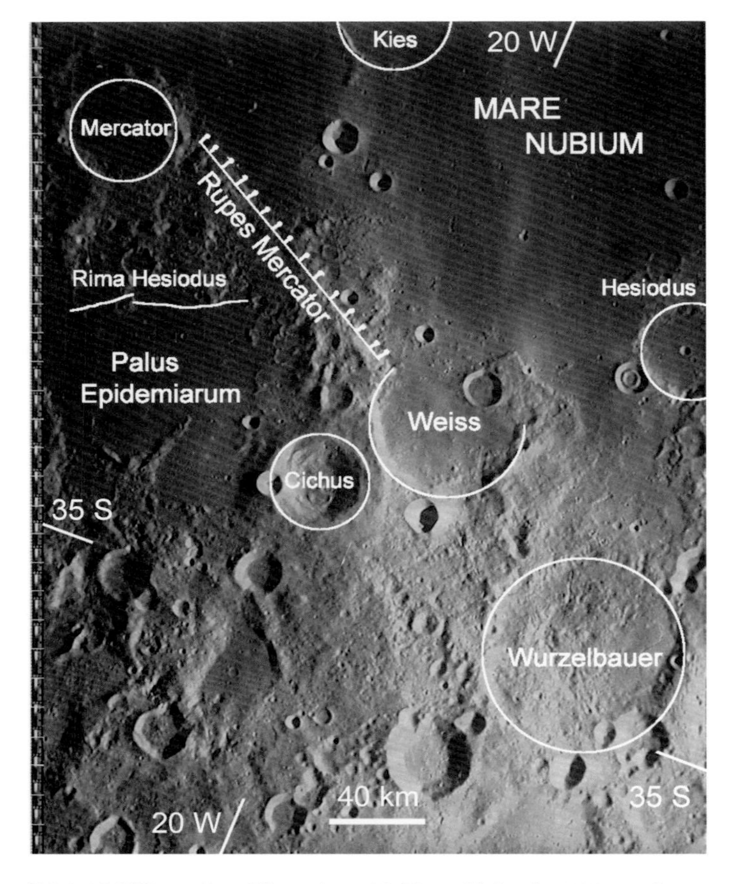

LO4-124H3   Sun Elevation: 19.9°   Altitude: 2994.47 km

The Nubium Basin is much older than the Humorum Basin. Extensive flooding, influenced by its proximity to Oceanus Procellarum, has hidden much of its ridge and trough ring structures. Rupes Mercator is a part of a ridge, probably the topographic ring, of the Nubium Basin. This ring is elliptical in shape, an indication of an oblique impact.

LO4-125H1 Sun Elevation: 18.9° Altitude: 2716.97 km

The concentric structure of Konig and the slightly smaller crater to the east of it suggest that they have bottomed out in a layer below the surface of Mare Nubium. The smooth top layer of the crater wall of Bullialdus and ghost crater rims such as Kies suggest a similar shallow depth to the mare, in the range of 1 to 2 km. The ray from Tycho that extends across the mare also covers the ejecta of Bullialdus. This ejecta is thick for about one crater radius outside of Bullialdus and then breaks up into chains of small secondary craters. Rima Hesiodus is a typical stress fracture near the edge of Mare Nubium.

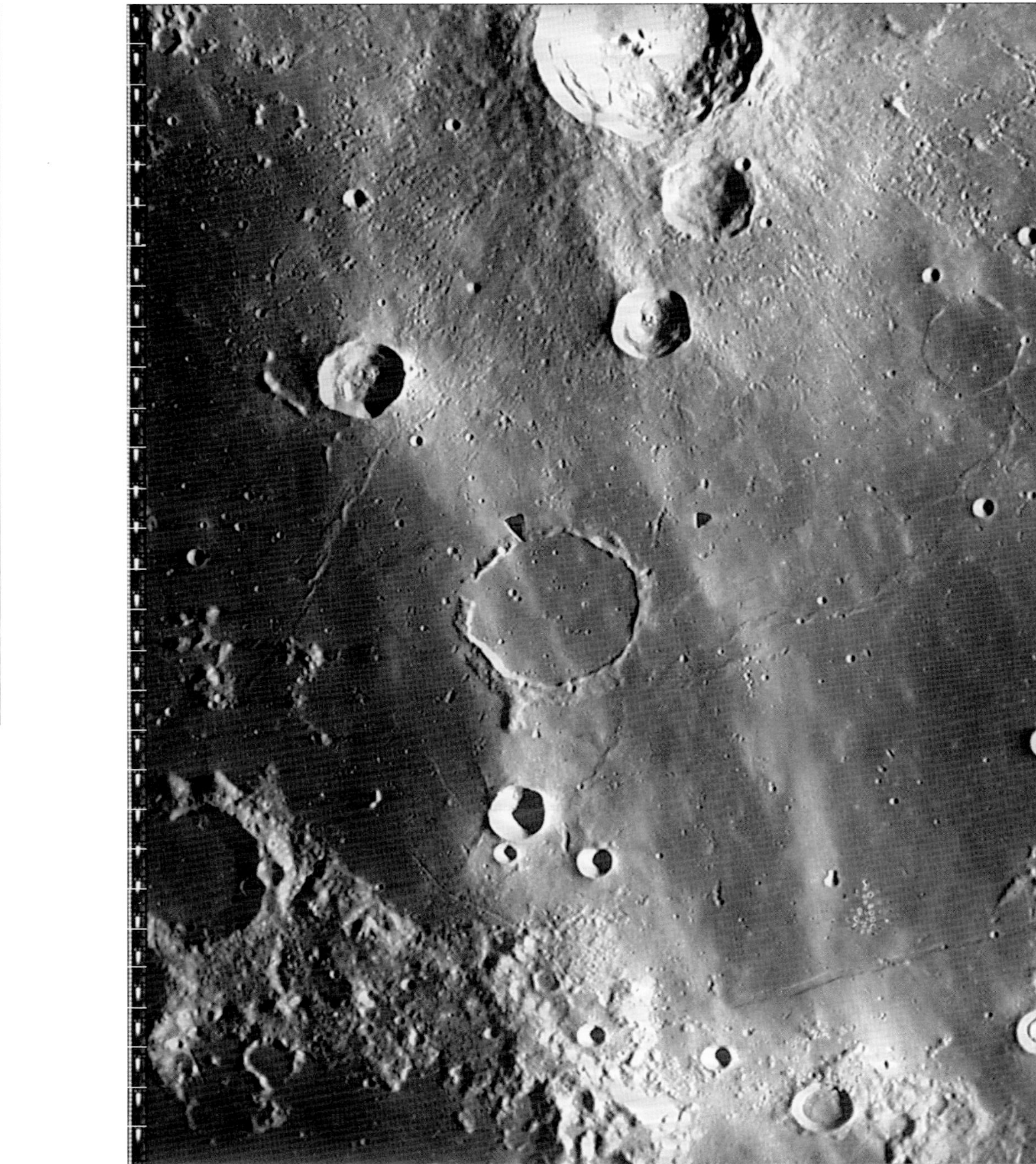

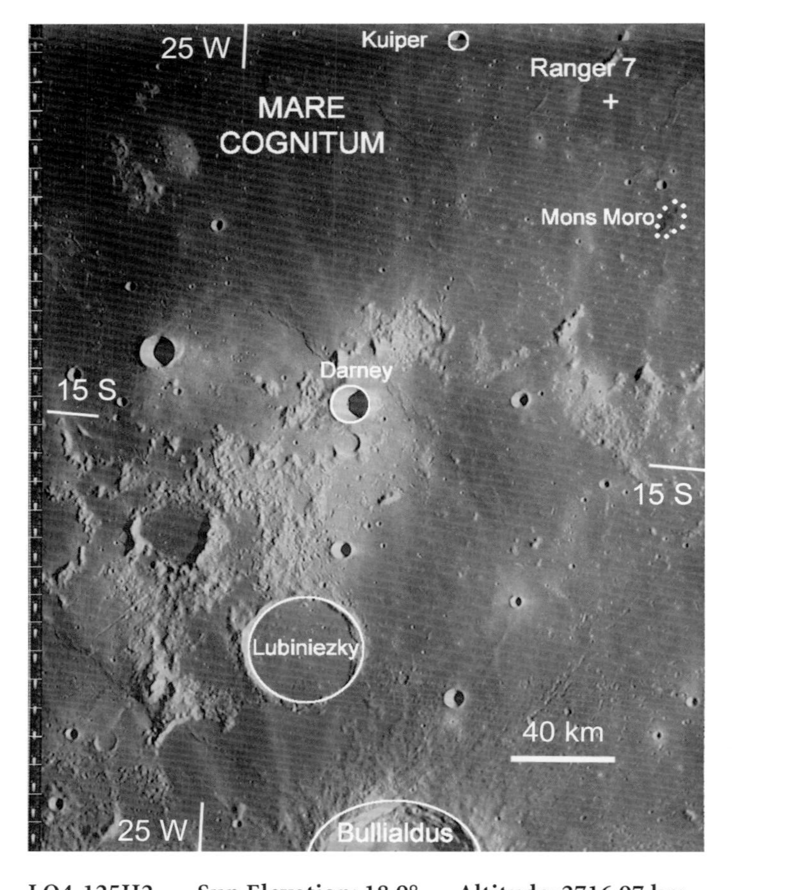

LO4-125H2 Sun Elevation: 18.9° Altitude: 2716.97 km

Mare Cognitum (known sea) was so named because Ranger 7, the first fully successful Ranger, transmitted a nested set of pictures. The resolution of the last picture was sufficient to show craters as small as 1 m in diameter, more than two orders of magnitude better than the resolution of Earth-based photography. The pictures established that the surface smoothness of mare surfaces similar to that photographed would be suitable for an Apollo landing and astronaut excursions. The presence of blocks in some of the Ranger 7 pictures indicated significant bearing strength, but doubt still remained in some minds on this subject. Darney is on the rim of a 240-km crater that underlies Mare Cognitum.

LO4-125H3 **Sun Elevation: 18.9°** **Altitude: 2716.97 km**

The southeast range of Montes Riphaeus is a sector of the rim of the crater that underlies Mare Cognitum. The rays in the upper right corner of this picture come from Copernicus. Mare Insularum (sea of islands, named for the old topography that protrudes) may be an extension of Oceanus Procellarum or may be part of an outer trough of the Imbrium Basin. Luna 5 crashed (May 1965) because its retro-rocket failed to fire. Surveyor 3, the second Surveyor to successfully land (April 1967), confirmed the suitability of this area for an Apollo landing, which was accomplished by Apollo 12 (November 1969). The touchdown of Apollo 12, the second manned landing, was about 160 m from Surveyor 3; the astronauts visited the Surveyor and removed parts for examination on earth.

LO4-119H2 Sun Elevation: 20.3° Altitude: 2990.55 km

Although the floor, rim, and rays of Tycho are very bright, the inner ejecta (out to about one diameter from the rim) is only as bright as the typical highlands. This dimmer ejecta is probably relatively coarse and from a deeper layer of the target. Furrows in the ejecta blanket are limited to the darker area, but chains of small secondary craters extend through both darker and lighter areas. The light ejecta material is probably very fine grained. Such material darkens with exposure to the solar wind (which frees iron particles within the grains) and from meteoroid gardening (which mixes the underlying soil with the grains). The Tycho impact occurred too recently (only about 100 million years ago) for the rays to have had time to darken.

LO4-119H3 **Sun Elevation: 20.3°** **Altitude: 2990.55 km**

Pitatus is about the same size as Tycho and shows a similar central peak. However, it is much older (Nectarian). The chain of secondaries from the Imbrium Basin is superposed on the rim of Pitatus, and its floor was flooded along with that of the Nubium Basin. Rays from Tycho appear on both the floor of Pitatus and Mare Nubium. Rimae Pitatus, around the edge of the floor of Pitatus, have been attributed to fractures caused by uplift of the floor, perhaps in the course of its flooding. It appears that lava has flowed from the floor of Pitatus into Hesiodus through the wide channel in the rims.

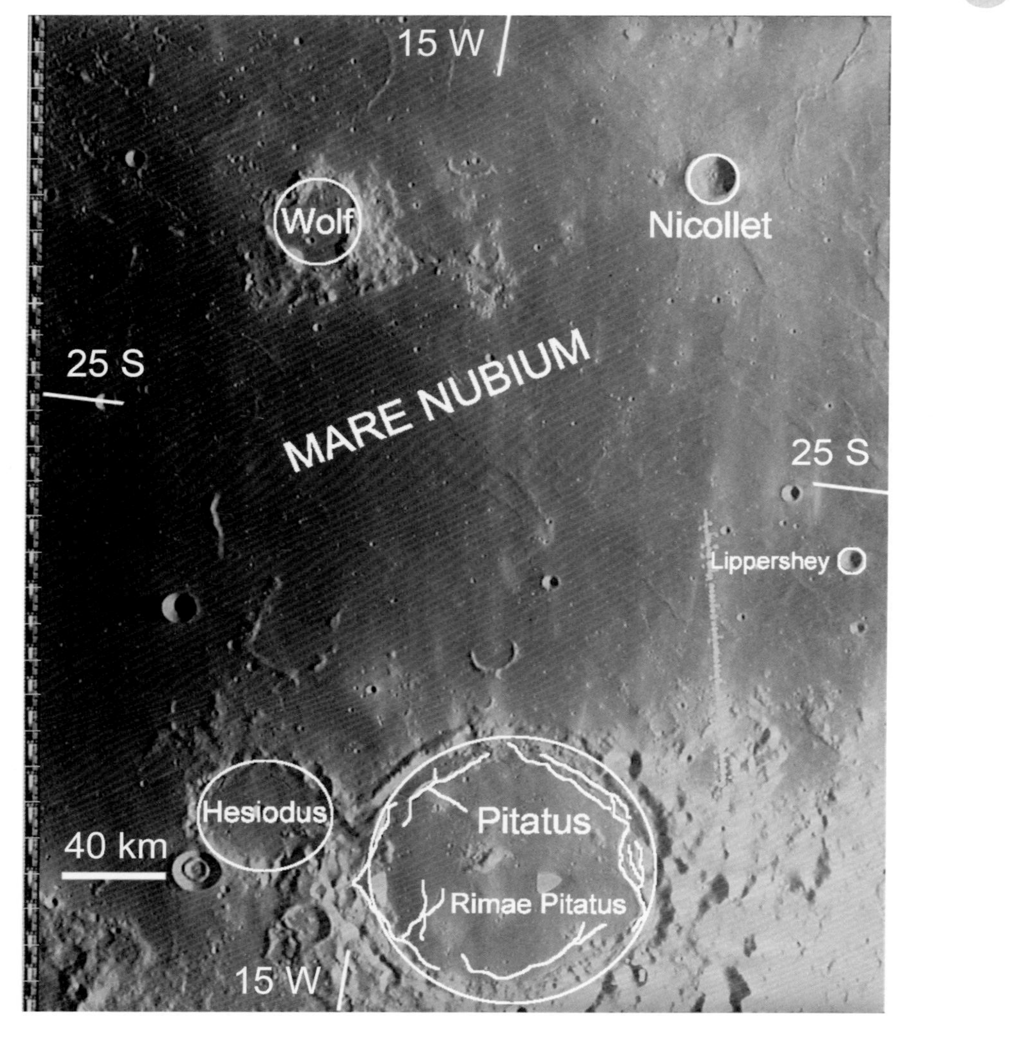

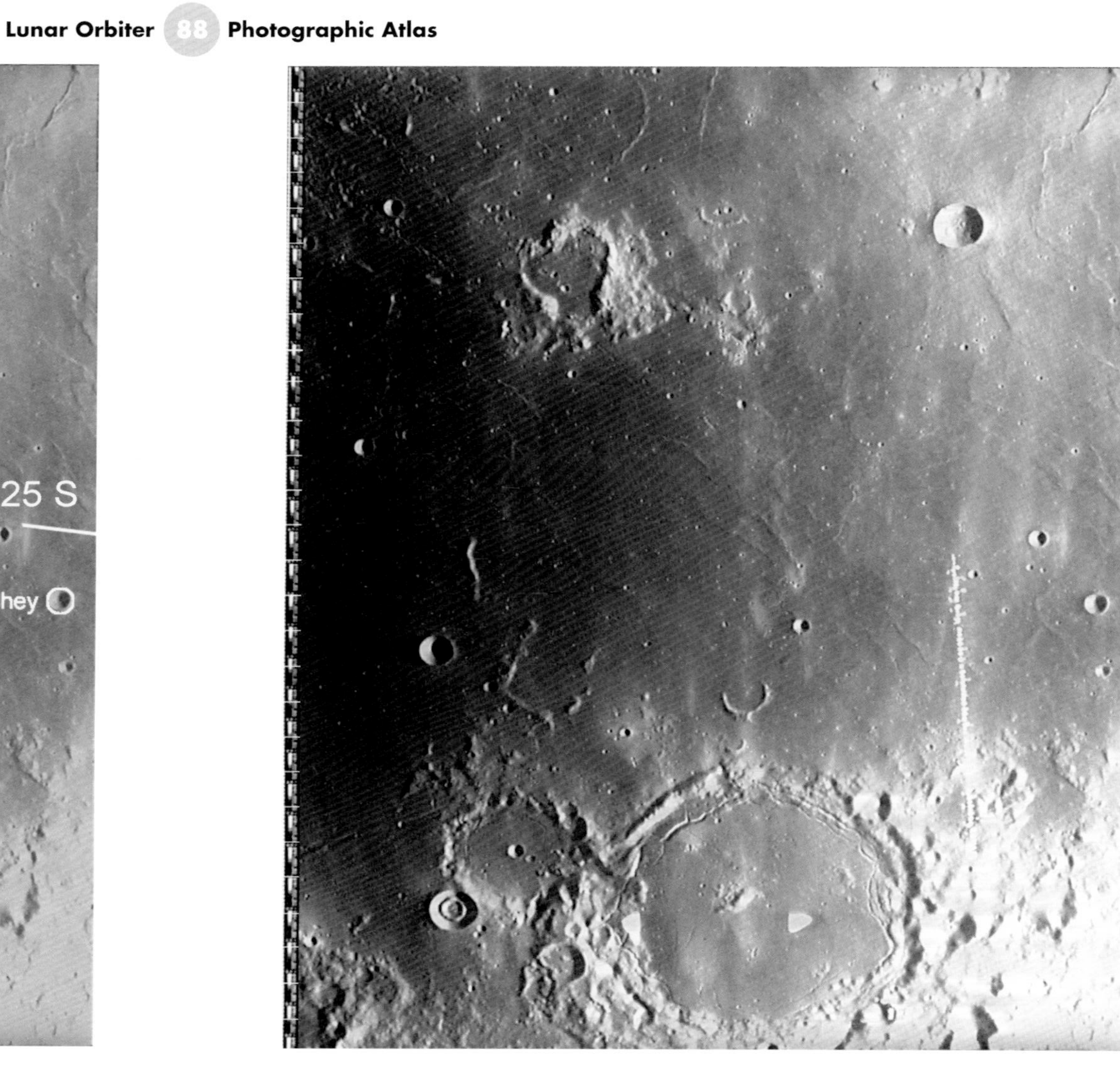

LO4-120H1 Sun Elevation: 20.1° Altitude: 2717.50 km

Wolf, shaped like a Valentine heart, would require a salvo of four craters if it were shaped by impact. An alternate explanation is that it is a caldera formed by internal processes during the formation of the mare. The concentric inner ring of the small crater to the southwest of Hesiodus may indicate that the crater “bottomed out” in the relatively shallow layer of lava near the edge of Mare Nubium. Note that similar-sized and larger craters such as Nicollet do not show similar morphology in the more central, presumably deeper, parts of the mare.

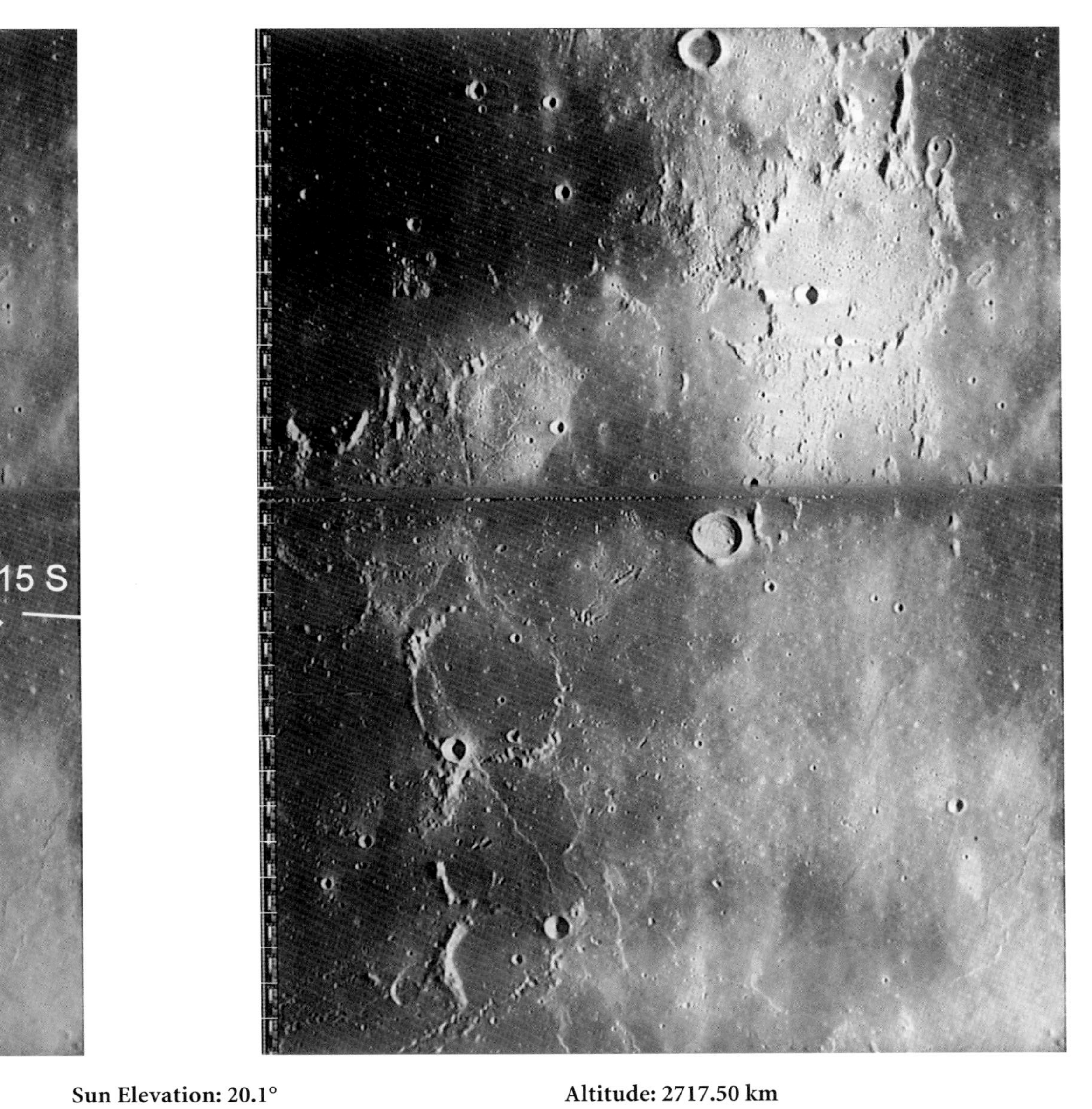

LO4-120H2 **Sun Elevation: 20.1°** **Altitude: 2717.50 km**

The northern rim of the Nubium Basin, as inferred from the nearly circular shape of the ring around Mare Nubium, runs through the ridges near Rimae Opelt and south of Guericke. The ejecta blanket from the Nubium Basin probably survives under the ridges and troughs in the vicinity of Guericke, but the striations are from Imbrium. Copernicus is the source of the bright rays running north to south through this picture. Mare Nubium appears to be connected to Mare Cognitum (the dark area north of Rimae Opelt) through a broad channel east of Rimae Opelt.

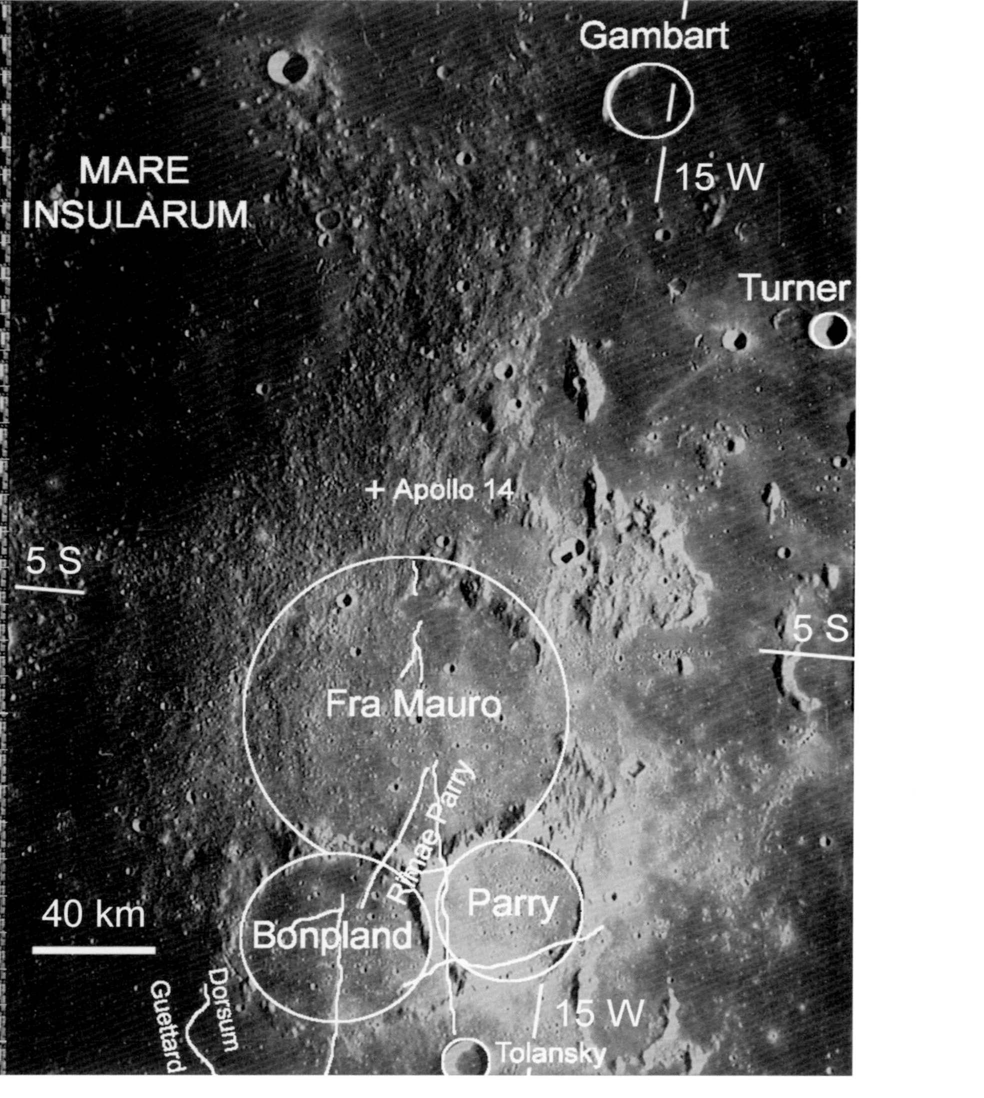

LO4-120H3 Sun Elevation: 20.1° Altitude: 2717.50 km

The heavily furrowed lumpy deposits north and south of Fra Mauro are called the Fra Mauro Formation, now known to be ejecta from the Imbrium Basin, after analysis of rocks returned from the Apollo 14 mission. The Fra Mauro Formation corresponds (as a terrane type) to the inner Hevelius Formation of the Orientale Basin. Among the 43 kg of rock returned by Apollo 14 are samples from near the rim of the 370-m Cone crater. This crater brought up rocks from the Fra Mauro Formation that had not been ground up in the "topsoil" of the regolith. Analysis of these rocks has established the time of the Imbrium impact as about 3.8 billion years ago.

LO4-112H1 Sun Elevation: 20.4° Altitude: 2986.04 km

The area around Jacobi, beyond the Tycho rays near the top, is about as far from mare, basins, and large craters as one can get on the near side of the Moon. It is called "intercrater terrane" (a terrane is an area with specific formation characteristics) and is believed to be Pre-Nectarian in age. Clementine altitude data show a depression equivalent to a 200-km crater centered on the northeast rim of Jacobi (Cuvier would be beyond the rim of such a crater). Such an ancient, degraded crater may explain the relative flatness of this area. Lilius is large enough (61 km) to have a central peak like that of Tycho. Central peaks occur in craters with diameters in the range of 20 to 200 km.

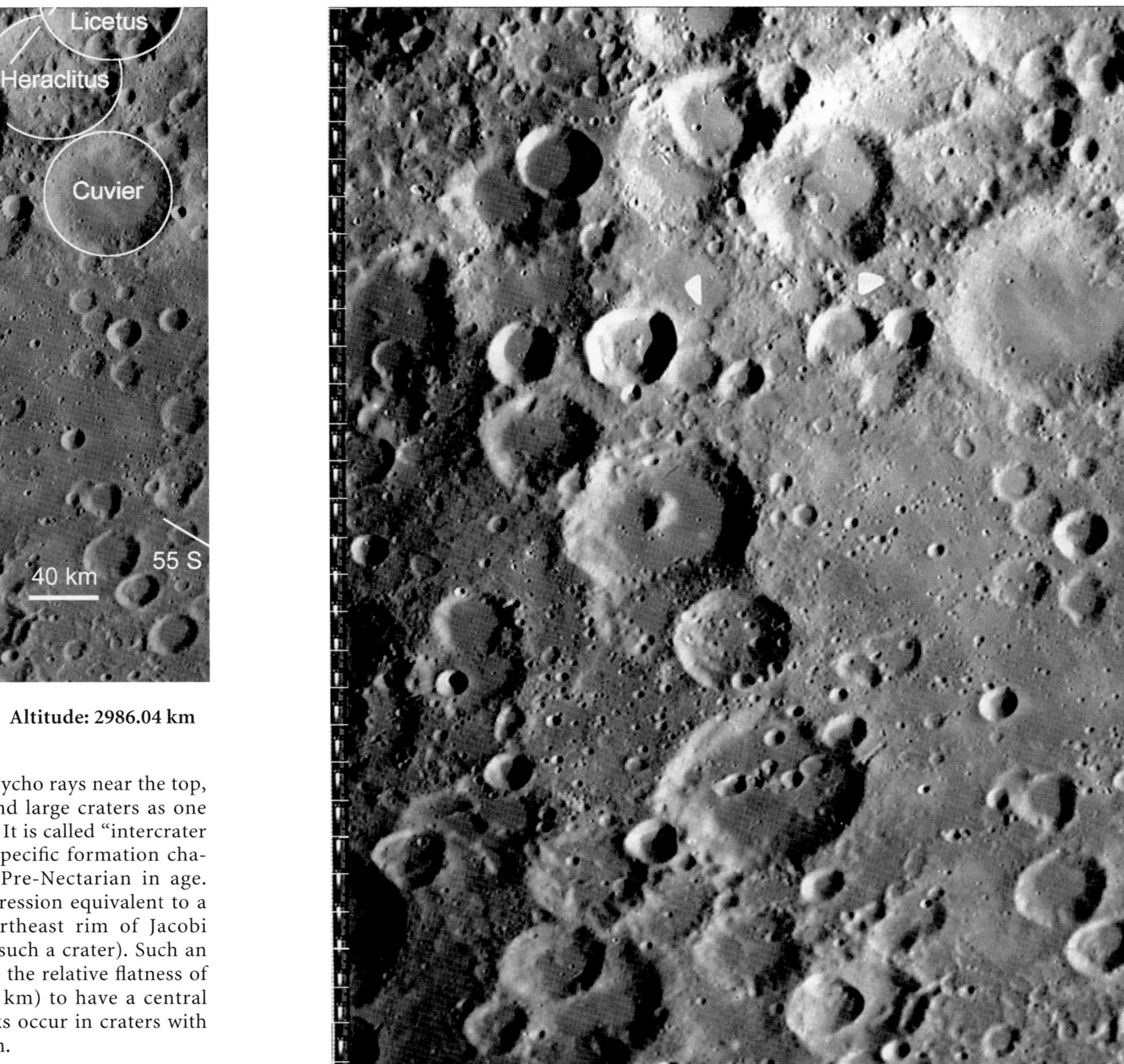

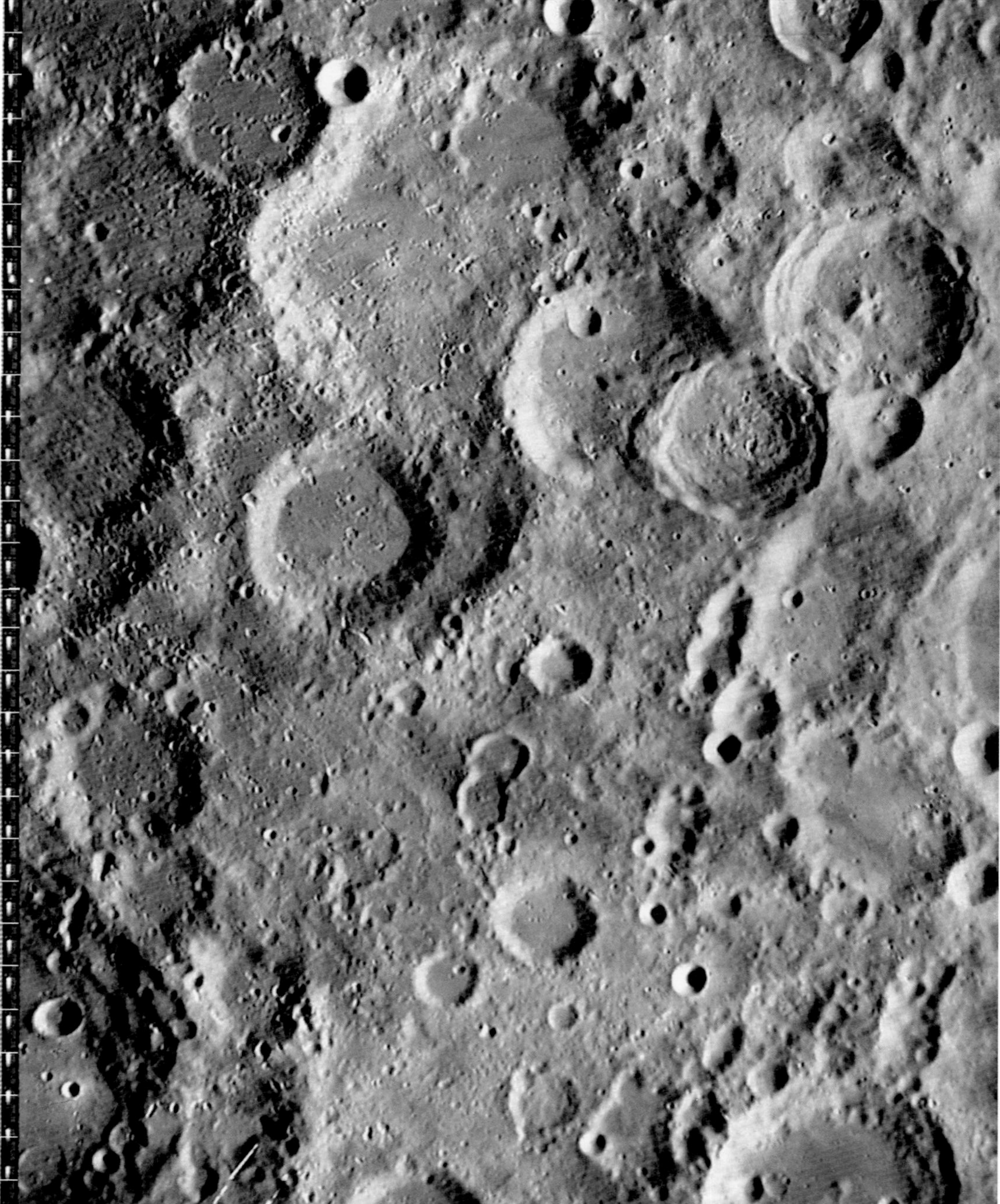

LO4-112H2 Sun Elevation: 20.4° Altitude: 2986.04 km

Soft ray patterns and chains of very small secondary craters radiate from Tycho just to the west of this picture. Chains of elongate low-angle-of-incidence craters between Nasireddin and Licetus could have come over the South Pole from the Schrodinger Basin. They could also have come from the southwestern sector of Imbrium. It is interesting to study the sequence of the cluster of large craters in the upper half of this picture. Nasireddin has clearly impacted and collapsed the walls of Huggins and Miller; Huggins has collapsed the wall of Orontius. Saussure seems younger than Orontius. A crater even older than Orontius is represented only by part of its rim, just to the east of Saussure.

LO4-112H3 Sun Elevation: 20.4° Altitude: 2986.04 km

The large chain of secondaries from the Imbrium Basin between Pitatus and Deslandres lies on the inner ejecta blanket of the Nubium Basin. Ejecta from the Nubium Basin appears to have eroded the walls of Deslandres and deposited fluid (molten or fragmented) material on the floor of Hell and the smaller crater south of Hell. However, the pattern of ejecta seems to disappear on the floor of Deslandres. The crater about 80 km to the northeast of Hell looks like craters that have been flooded with mare lava, but the material is much lighter than mare. The walls of the Deslandres depression can be seen as square, rather than round. Is the Deslandres depression an impact crater or the result of tectonic forces producing uplift and sinking? Could tectonic shaking have resurfaced the smooth part of the floor of Deslandres?

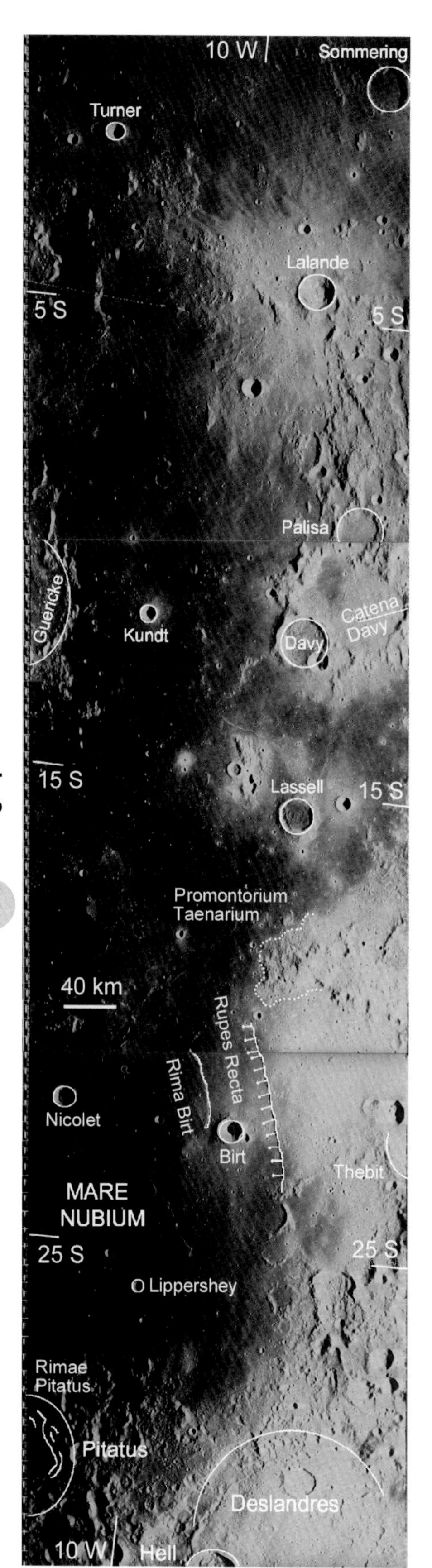

LO4-113H
Sun Elevation: 20.1°
Altitude: 2718.02 km

Lalande has a spectacular ejecta pattern because it impacted an area with inherently light-colored material and deposited it on the dark surface of Mare Insularum. The deep troughs to the southeast of Lalande and radial to it are too large to be caused by Lalande and are from the Imbrium Basin.

An ancient 160-km flooded crater may underlie the mare surface in the center of this picture. Promontorium Taenarium could be a remainder of the rim of this crater, associated with the intersection between that rim and a ring of the Nubium Basin. The mountain northwest of Lassell would be the top of a complex central peak of the 160-km flooded crater.

Rupes Recta (straight wall) is hard to explain. It is not curved like the scarps on the inside of basin rings, and it does not appear to be directly related to the shore of Mare Nubium. It may be a continuation of the degraded scarp near Deslandres. Albedo patterns on the eastern border of Mare Nubium are also unusual. Perhaps this is dark mantling material, pyroclastic fountain material that occurs near mare borders and spreads ballistically beyond the mare.

LO4-107H2 **Sun Elevation: 21.6°** **Altitude: 2981.85 km**

Stofler is one of many large craters whose floor is remarkably flat and smooth, a terrain type that has been called "smooth terra plains." The floor appears to be younger than the crater rim, which shows degradation. One explanation of smooth terra plains is that they are flat surfaces that have been covered by the outer (smoother) type of basin ejecta. This type of ejecta starts at least one radius away from the basin's topographic ring and usually ends between two and three radii away. To have this type of terrain, a flat-floored crater must be within three radii of some subsequent basin and must not be within one radius of any subsequent basin. Further, the crater must not have been flooded with mare lava or suffered some other event that seriously disrupted its floor. Tectonic shaking may further smooth and flatten surfaces with such deposits.

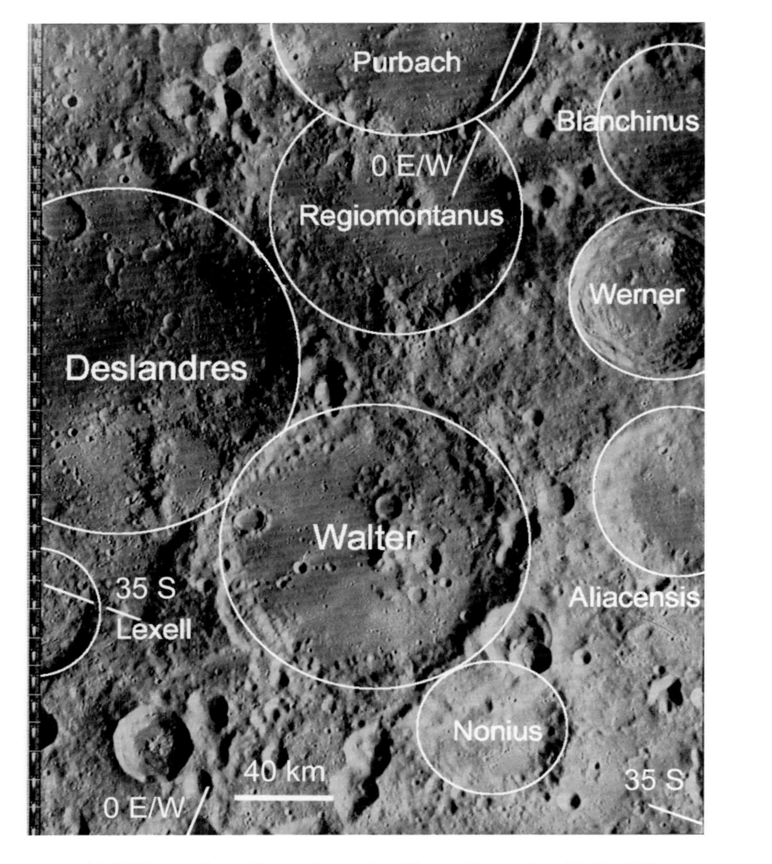

LO4-107H3 Sun Elevation: 21.6° Altitude: 2981.85 km

Here is a cluster of craters, all of similar size (70 to 140 km), that are clustered together. Their ages vary, but all probably followed the Nubium Basin event. Their centers are within 120 km of a central point, in uniform target material. The overlapping ejecta from the these craters resulted in the chaotic intercrater areas, with rims impinging on each other and material piling up between and sometimes within the craters. Subsequently, the Imbrium Basin event has deposited a uniform layer of material that has smoothed previously undisturbed parts of the crater floors. Imbrium also dropped a few secondaries here. Werner arrived later; its floor has not been smoothed. Finally, Tycho has dropped its rays over the cluster.

LO4-108H1 Sun Elevation: 21.1° Altitude: 2719.01 km

Thebit (Late Imbrium) and Werner (Eratosthenian) are free from evidence of smooth terra plains and Imbrium secondaries and exhibit little rim erosion. On the other hand, Purbach, Regiomontanus, Deslandres, and La Caille (Pre-Nectarian) show smooth terra plains units and some secondaries, both from the Imbrium event.

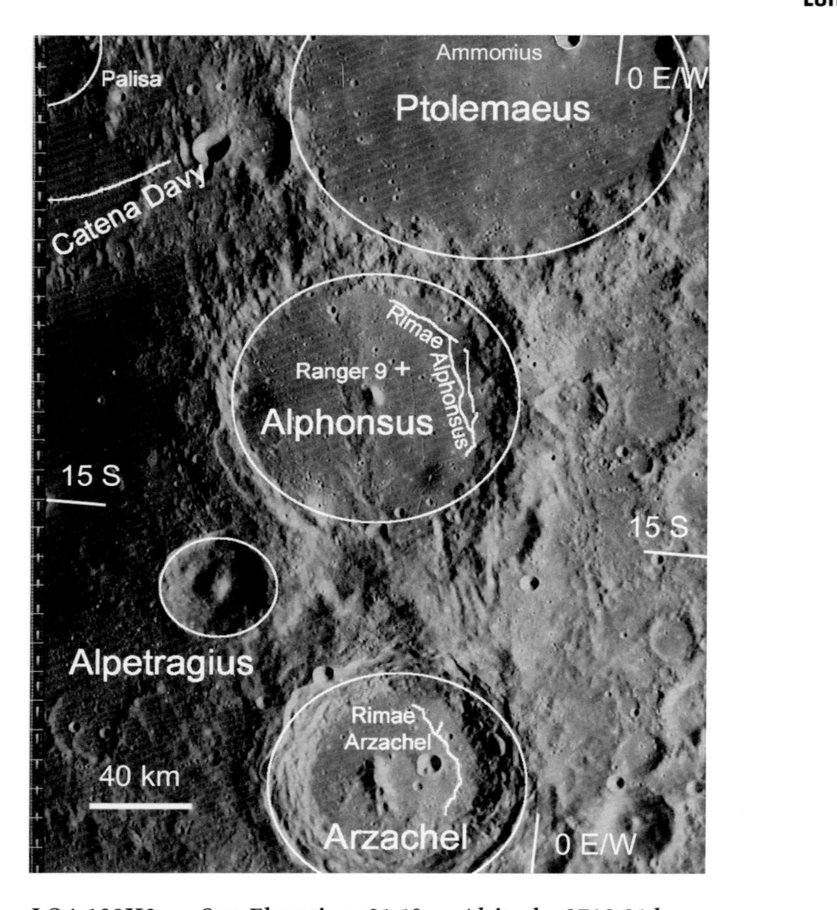

LO4-108H2   Sun Elevation: 21.1°   Altitude: 2719.01 km

The rims of Ptolemaeus and Alphonsus show striations from the Imbrium Basin, but the effects on their floors have been all or partly erased by lava flows that have reduced the depths of these craters. These craters show an unusual high level of thorium in Lunar Prospector data. Clementine albedo data show Ptolemaeus to be dark, but not as dark as nearby Oceanus Procellarum and Sinus Medii. The craters with dark halos in Alphonsus may be sources of fountains of dark glass beads. Ranger 9 photos showed that parts of the floor of Alphonsus were smooth like mare surfaces at scales significant for landing and surface operations.

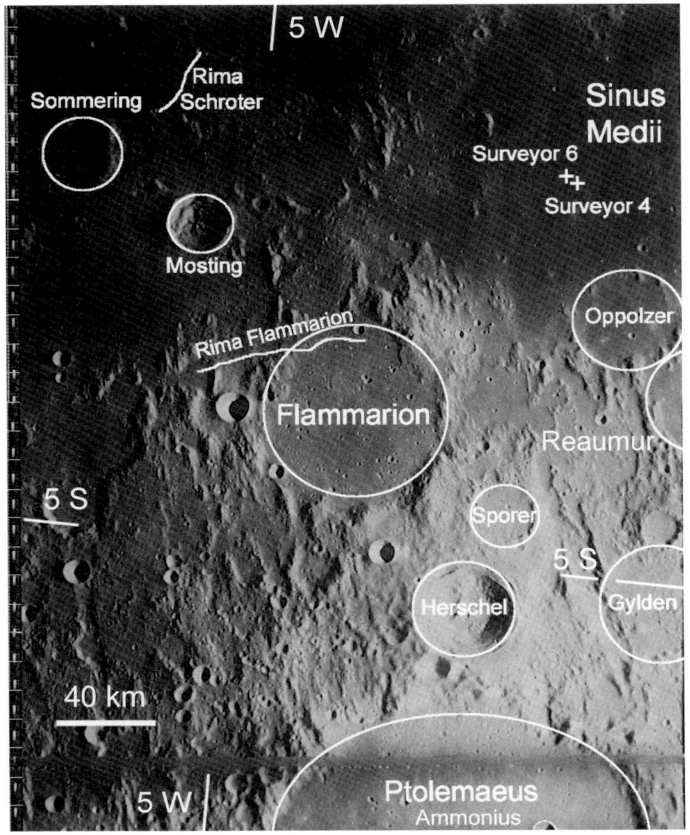

LO4-108H3 Sun Elevation: 21.1° Altitude: 2719.01 km

The Apollo program was very interested in Sinus Medii ("Central Bay") as a landing site because its central position, relative to mare sites to the east and west, provided additional launch window opportunities. Surveyor 4 failed before touchdown, but Surveyor 6 landed successfully (November 1967). Its alpha-scattering instrument confirmed results from that of Surveyor 5, showing that the element composition of mare surfaces corresponded to the broad range of rock types called basalt. This implied that at least a major portion of the Moon had undergone a melt phase that supported differentiation of minerals. Sinus Medii is not large enough to be a basin; it is a crater that has been flooded with lava. It may be influenced by the second external trough of the Imbrium Basin, the same trough that underlies Mare Frigoris.

Chapter 7

Imbrium Basin Region

7.1. Overview

The Imbrium Basin Region includes a number of interesting features such as Oceanus Procellarum, Copernicus, Kepler, the Aristarchus plateau, and Vallis Schroteri (Schroter's Valley) as well as Mare Imbrium, the Fra Mauro Formation of Imbrium ejecta, and surrounding maria. Because these features are so large, yet closely related, the region stretches from the western edge of the visible Moon to the central meridian and from the equator to 60° north.

The Imbrium Basin

The Imbrium Basin (Figure 7.1) has had a larger influence on the current appearance of the near side surface of the Moon than any other single feature. Although mare flooding hides much of the structure of Imbrium, enough remains to associate its major features with those of Orientale and other basins. Specifically, the basin has an inner circular depressed floor, largely flooded with mare. A scarped rim bounds this depression, and beyond the rim is a trough and then additional raised rings. Flooding by surrounding mare has obscured much of the ejecta blanket except for the southeast sector. Here the Fra Mauro Formation can be seen to be similar to the Hevelius Formation of the Orientale Basin. Chains of secondary craters that radiate from the central part of the Imbrium Basin extend beyond this region nearly to the eastern limb.

When the mountain ranges that form the topographic ring of the Imbrium Basin were named, they were not perceived to be parts of a continuous circle; the sectors of the ring were given the names of familiar mountain ranges on Earth. Clockwise from the north they are Montes Alpes (Alps, France, Switzerland, and Italy), Montes Caucasus (Caucasus, Georgia), Montes Apenninus (Apennines, Italy), Montes Carpatus (Carpathians, Romania), and Montes Jura (Jura Mountains, Switzerland, France).

Oceanus Procellarum

Oceanus Procellarum (Figure 7.2) contains the largest distinctly named area of mare on the Moon. It has often been considered as evidence of a possible enormous ancient basin, the Procellarum Basin, perhaps three times the size of the Imbrium Basin and underlying that basin as well as Oceanus Procellarum. This hypothesis is supported by the determination that the crust under this area is thinner than in the rest

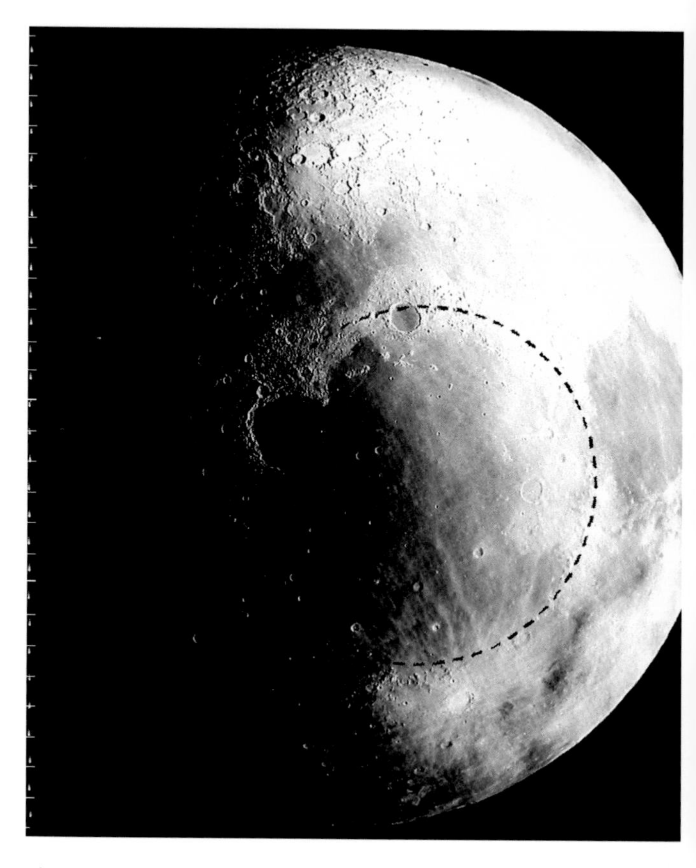

Figure 7.1. LO4-134M. The inner edge of the topographic ring of the Imbrium Basin is shown as a dashed circle, actually an ellipse here because the center of the photo does not coincide with the center of the basin. Within this ring, most of the area is covered with mare material (Mare Imbrium). Beyond this ring is a trough partly filled with mare (Mare Frigoris to the north and Mare Insularum and Mare Vaporum to the south). The diameter of the topographic rim is 1160 km.

of the Moon, perhaps caused by the removal of the top layer of crust by the massive impact. Additional evidence of the uniqueness of this region is provided by remote sensing, and by sample collection, which shows the region to be rich in KREEP minerals and thorium.

However, there are problems with that hypothesis. In particular, there is little sign of massive radial ridges and troughs of an ejecta blanket or external rings that would be expected to accompany such a large basin. Also, there are positive features (mountains) that protrude from the mare that are hard to reconcile with a large, flat central floor, even allowing for

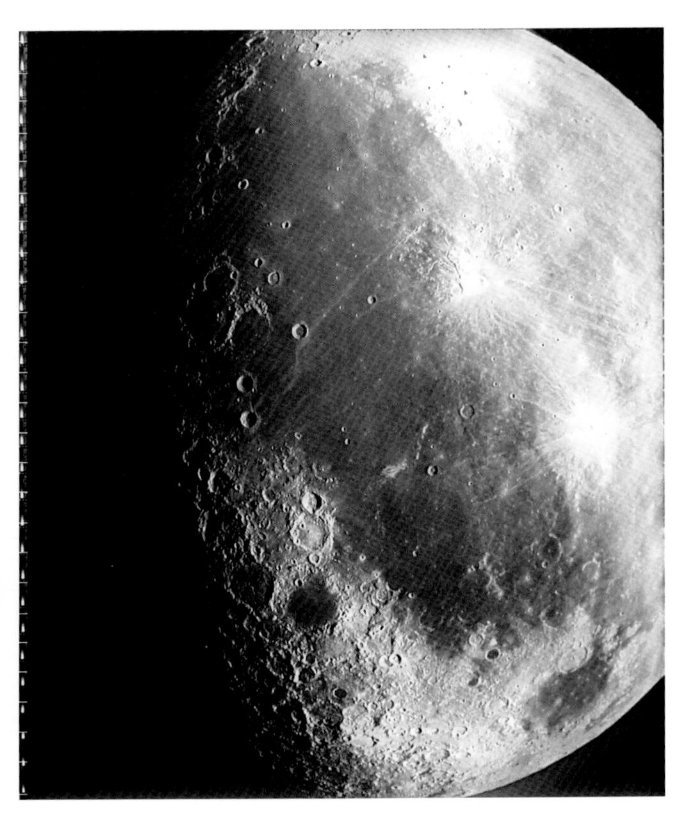

Figure 7.2. LO4-162M. The large dark area in this picture is Oceanus Procellarum. Mare Imbrium is in the upper right part of the picture. The bright-rayed craters are Aristarchus and Kepler (below and to the right of Aristarchus). The Grimaldi Basin is to the lower left of Oceanus Procellarum and the Humorum Basin is to the lower right.

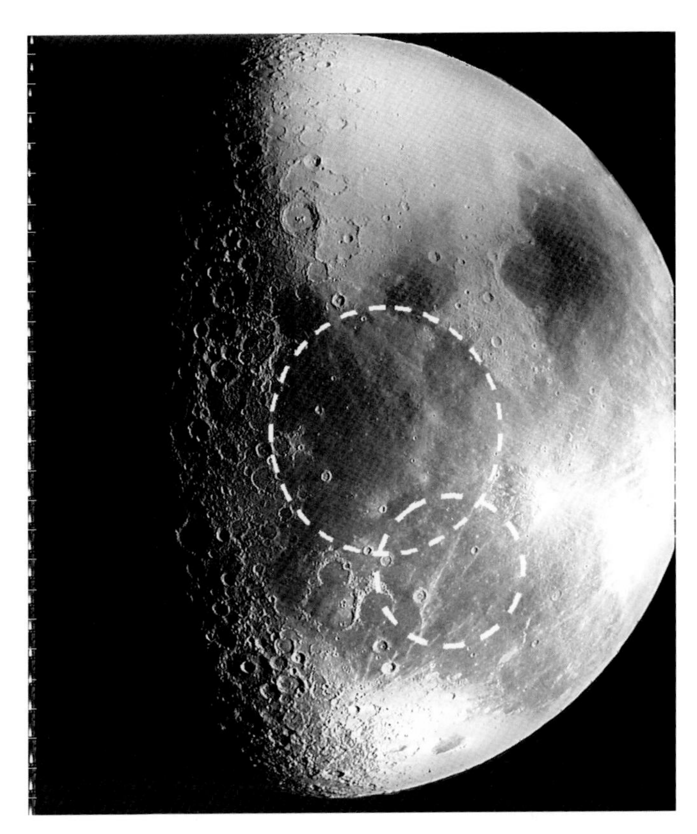

Figure 7.3. LO4-175M. These two possible basins, called here the Lavoissier-Mairan Basin (to the north) and the Cardanus-Herodotus Basin (each named after craters near their proposed main rings) are inferred from circular depressions revealed in Clementine elevation data. These basins may have attracted the lava that flooded the northern part of Oceanus Procellarum.

internal rings. It has also been suggested that Oceanus Procellarum is an outer trough of the Imbrium Basin. If so, it is an extraordinarily large area of mare to fill a trough.

Examination of elevation data from Clementine and earlier spacecraft missions suggests a third explanation of Oceanus Procellarum. Mare lava shrinks as it cools, leaving depressions where it is deepest. Roughly circular depressions suggest multiple basin-sized depressions beneath Oceanus Procellarum (McEwen, 1994). Supporting evidence of the depth of the flooded depressions has come from analysis of partially flooded crater rims (De Hon, 1979). Further, arcs of two circular rims can be seen in the northern part of this mare (Figure 7.3).

The proposed Lavoissier-Mairan and Cardanus-Herodotus Basins have influenced the formation of two plateaus, which help in determining the basin sizes and locations. The plateau west of the Montes Jura (Jura Mountains) was formed by ejecta from the proposed Lavoissier-Mairan Basin, the Imbrium Basin, and a large crater under Sinus Roris. The Aristarchus plateau may have been built up by the eastern intersection of the main rings of these two proposed basins (Byrne, 2004). Analysis of data from Lunar Prospector has revealed patches of high-thorium material surrounding these proposed basins, especially the northern one (Gasnault, 2002).

A similar examination of the topography of the southern part of Oceanus Procellarum (Figure 7.4) shows a depression that may reveal a basin. In this case, the lava has left exposed remnants of crater rims and perhaps part of a complex central peak or inner ring. There are indications that a basin in this area could be south of the position shown in Figure 7.4 (the possible Flamsteed-Billy Basin of Wilhelms, 1987), but striations beyond the mare to the southwest are a better fit to the location shown. Clementine data show a distinctly higher elevation of the mare floor between the Cardanus-Herodotus Basin shown in Figure 7.3 and the Reiner-Letronne Basin in Figure 7.4.

The possible existence of these basins does not refute the contribution of an outer trough of the Imbrium Basin to the depression beneath the northeastern part of Oceanus Procellarum. Nor do these basins in themselves explain the thinness of the crust and the unusual distribution of radioactive, rare earth, and related elements (KREEP) in this part of the Moon. This leaves room for the Procellarum Basin hypothesis, so long as the basin formed when the crust and early mantle were still soft enough for isostatic adjustment.

Fra Mauro Formation

At one time the Fra Mauro Formation was thought to contain numerous volcanic features, but other geologists believed it to be ejecta from the Imbrium Basin. The formation is so widespread that dating the Imbrium impact helps establish ages of many other features. Analysis of samples from Apollo 14 (Humorum Basin Region) and Apollo 16 (Nectaris Basin Region) established that this formation was deposited from the Imbrium Basin impact. See the chapters for these regions for further discussion of these Apollo landings and

Figure 7.4. Part of LO4-162M. The dashed circle outlines a possible basin, the Reiner-Letronne Basin that is inferred from a depression in Clementine elevation data. This basin may have attracted the lava that flooded the southern part of Oceanus Procellarum. The Grimaldi Basin is to the west and the Humorum Basin is to the southeast. The southeastern part of this possible basin is in the Orientale Basin Region and the southeast part is in the Humorum Basin Region.

their contributions to our understanding of the Imbrium Basin.

7.2. High-Resolution Images

Table 7.1 shows the high-resolution images of the Imbrium Basin Region in schematic form.

The following pages show the high-resolution subframes from south to north and west to east. That is, they are in the order LO4-189H1, LO4-189H2, LO4-189H3, LO4-183H1, LO4-183H2 ... LO4-162H1, LO4-162H2... LO4-110H3.

High-resolution photos of Mission 4 did not cover the extreme northwestern part of this region. Therefore, the first photo in this chapter is LO4-189M, which completes the near side coverage in this area.

Subframes LO4-183H3, LO4-170H3, LO4-158H3, LO4-145H3, LO4-134H3, LO4-122H3, and LO4-110H3 have not been printed because of redundancy but are in the enclosed CD. The subframes of LO4-152, LO4-144, LO4-138, LO4-121, and LO4-114 have been merged into full frames.

Lattitude Range	Photo Number															
56 N–90 N			190	176		164		152		140	128		116		104	
27 N–56 N		189	183	175	170	163	158	151	145	139	134	127	122	115	110	103
0–27 N		188	182	174	169	162	157	150	144	138	133	126	121	114	109	102
0–27 S	195	187	181	173	168	161	156	149	143	137	132	125	120	113	108	101

Longitude at Equator		89 W	82 W	76 W	68 W	62 W	56 W	49 W	41 W	35 W	30 W	23 W	16 W	10 W	3 W	4 E

Table 7.1. The cells shown in white represent the high-resolution photos of the Imbrium Region (LO4-XXX H1, -H2, and -H3, where XXX is the Photo Number). The Orientale Basin Region is to the southwest, the Humorum Basin Region is to the south, the Eastern Basins Region is to the east, and the North Polar Region is to the north.

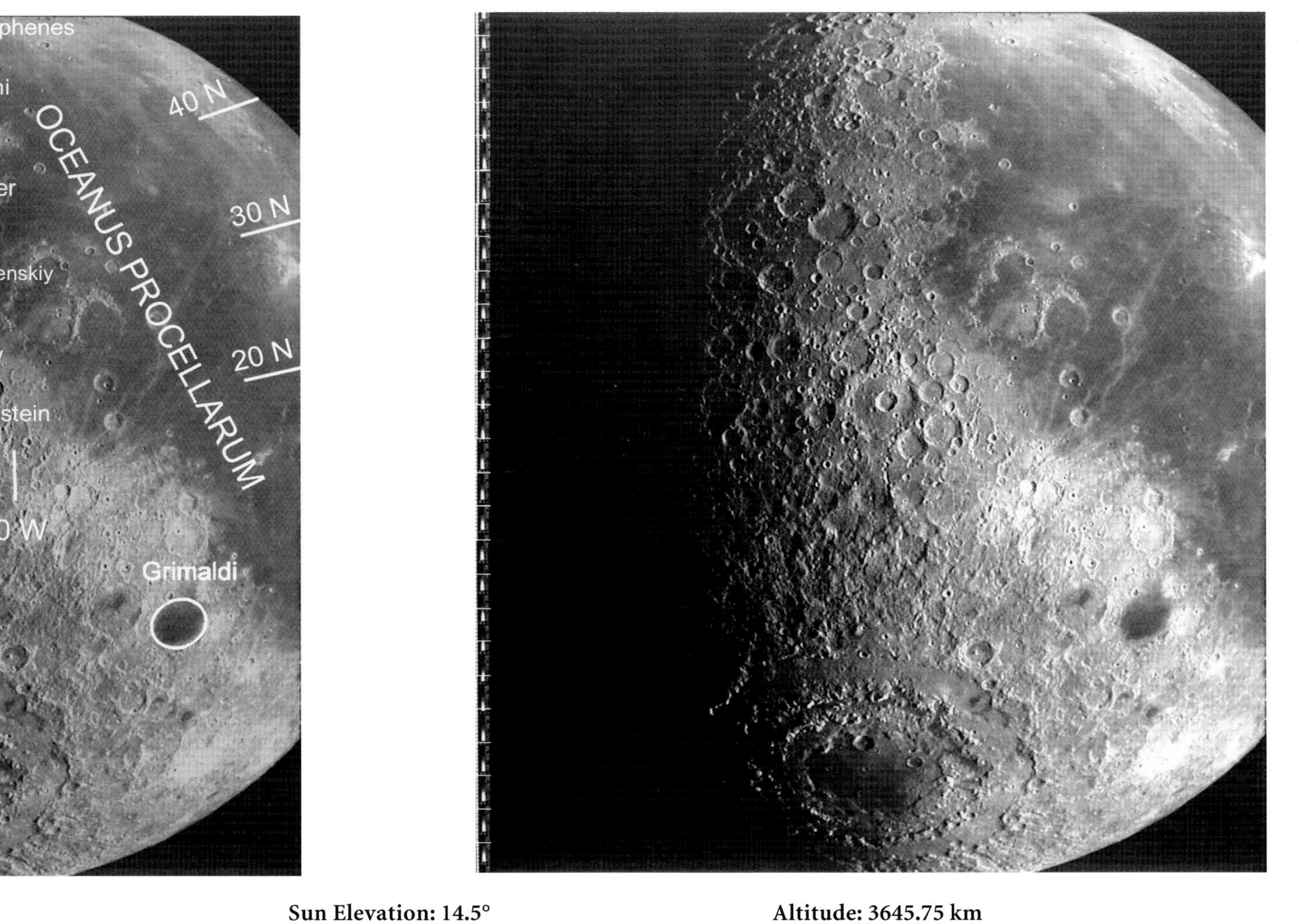

LO4-188M **Sun Elevation: 14.5°** **Altitude: 3645.75 km**

This medium-resolution photo is provided to fill in features in the western limb area of the Imbrium Basin Region (and some of the adjoining far side) that were not photographed in high resolution. The limb area from Volta north is better seen in LO4-190M, shown in the North Polar Region. The Orientale Basin extends its ejecta blanket and secondaries beyond Einstein. The rays in northern Oceanus Procellarum come from Harpalus, south of Pythagoras in the North Polar Region.

LO4-189H
Sun Elevation: 18.4°
Altitude: 2877.84 km

Striations on and within the rims of Repsold, Gerard, von Braun, and Lavoisier may have come from the possible Lavoissier-Mairan Basin. The east-west ridges and valleys northeast of Repsold may be ejecta from a crater under the nearby bay of Oceanus Procellarum. In the 50-km crater west of von Braun, there is a moat of lava around the inside of the rim.

The bulging floors of these craters are heavily fractured, probably caused by tensile failure of the floor material, a melt sheet, as it is forced to expand. The fractures on the floor of Repsold cross crater rims without deflection; they are responses to deep stress.

The eastern edge of Oceanus Procellarum shows a thinning layer of mare, leaving many crater rims. Some craters such as Aston and Voskesenkiy are internally flooded with mare, which must have come from below. The degraded 55-km crater southeast of Voskesenkiy shows striations that may have come from the possible basin in northern Oceanus Procellarum.

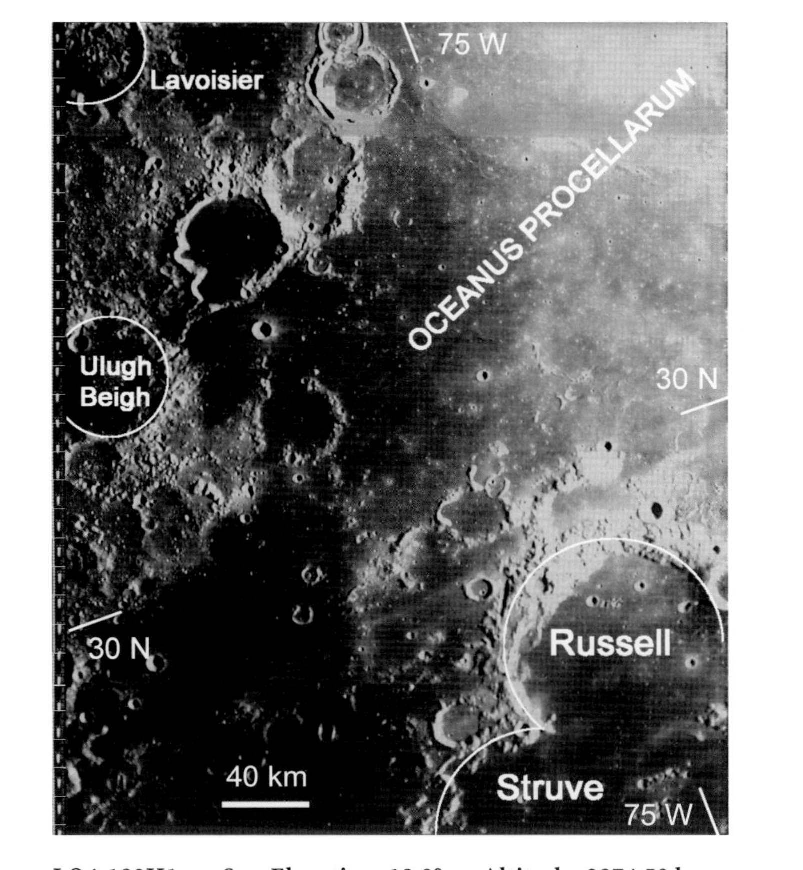

LO4-183H1 **Sun Elevation: 19.8°** **Altitude: 2874.59 km**

This photo shows the transition between the highland region to the west, through the shallow mare partly flooding craters, to very deep mare. The deep mare completely covers large craters and perhaps even an entire basin.

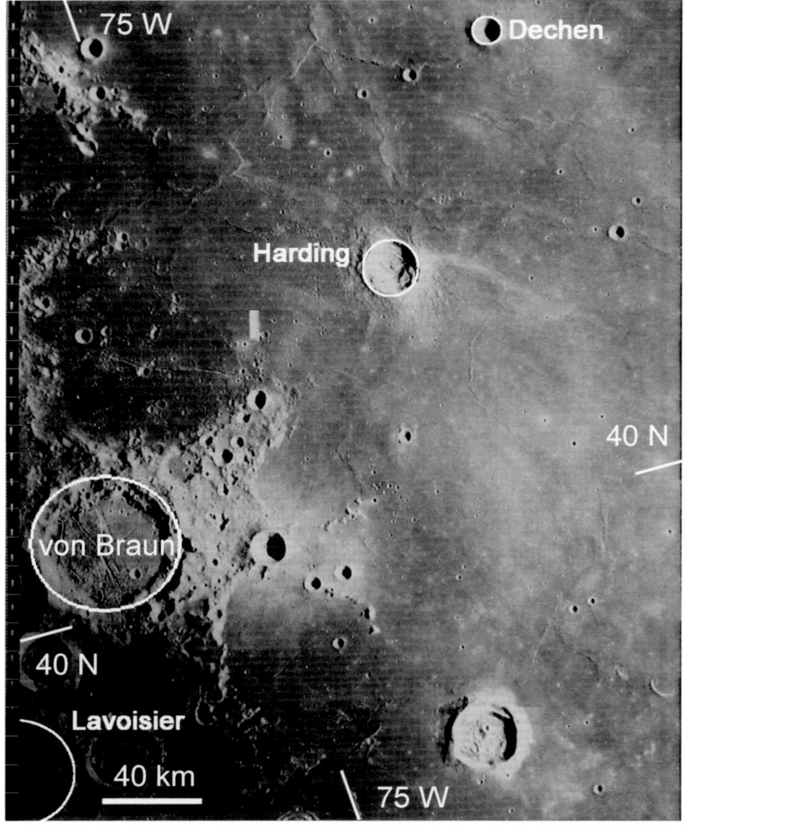

LO4-183H2 Sun Elevation: 19.8° Altitude: 2874.59 km

The highland area east of von Braun illustrates how complex topography can be sculpted by a series of circular craters, partly flooded with mare. The material in the western half of the floor of von Braun could be ejecta from the possible basin beneath northern Oceanus Procellarum. Striations in the rim of the 90-km flooded crater north of von Braun and in the residual highland material in the upper left corner are essentially radial to that possible basin.

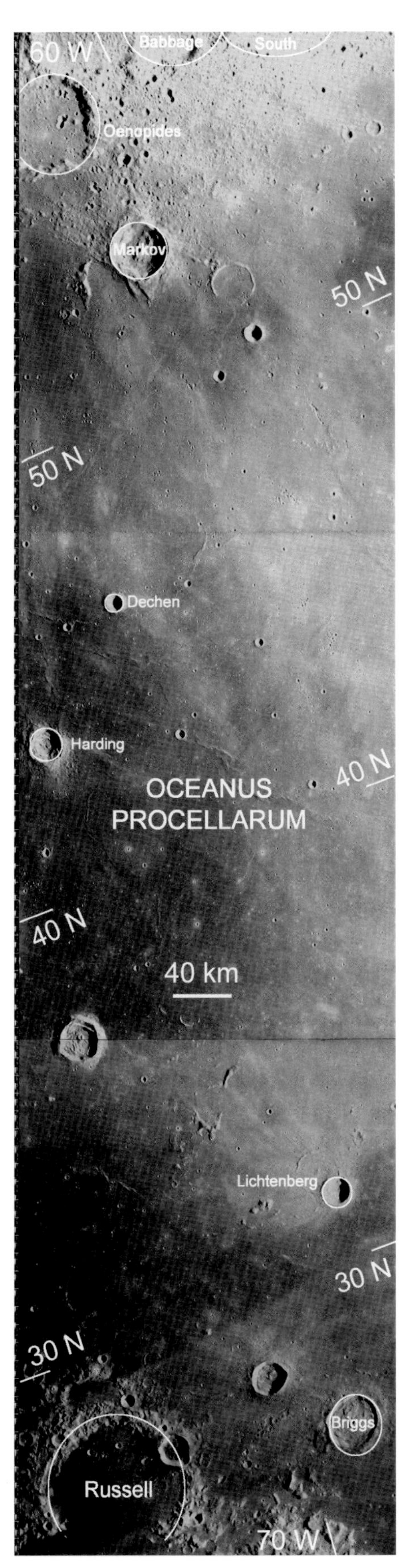

LO4-175H
Sun Elevation: 19.2°
Altitude: 2871.76 km

The field of sharp secondaries south of Oenopides and Babbage was deposited from the Eratosthenian crater Pythagoras. Later in the Eratosthenian Period a fresh lava flow south and west of Markov submerged parts of the secondary field. A deep trough and ridge pattern runs about 15 km south from Markov. It may be ejecta from the possible basin under northern Oceanus Procellarum.

Rays from a crater on the north rim of Oenopides intersect with rays from Lichtenberg; this cloudy appearance helped give Oceanus Procellarum its name ("Ocean of Storms"). The dark streak running east to west, passing just south of Lichtenberg, is one of the latest flows of lava in the maria. It covers rays and ejecta of the bright-rayed crater Lichtenberg. Although very few lava flows have taken place after the Eratosthenian Period, this particular flow has taken place in the Copernican Period, which began on the order of 1 billion years ago. Nearly all mare surfaces formed during the earlier Late Imbrian Epoch and the Eratosthenian Period.

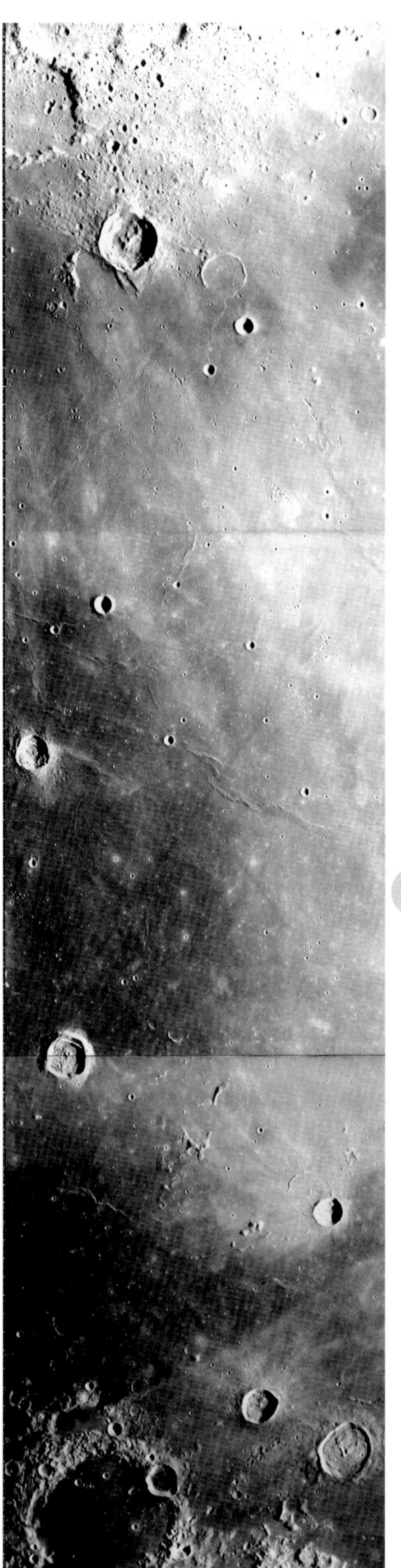

LO4-170H1 Sun Elevation: 19.8° Altitude: 2870.64 km

The dark young lava flow that covers the rays and ejecta from the Copernican crater Lichtenberg (see note for LO4-175H) continues to the east, where it is in turn covered by rays from the small crater near Dorsum Scilla. Neither Dorsum Scilla nor the wrinkle ridge near Lichtenberg impeded the dark flow, which suggests that they may have formed after the flow, perhaps as it cooled.

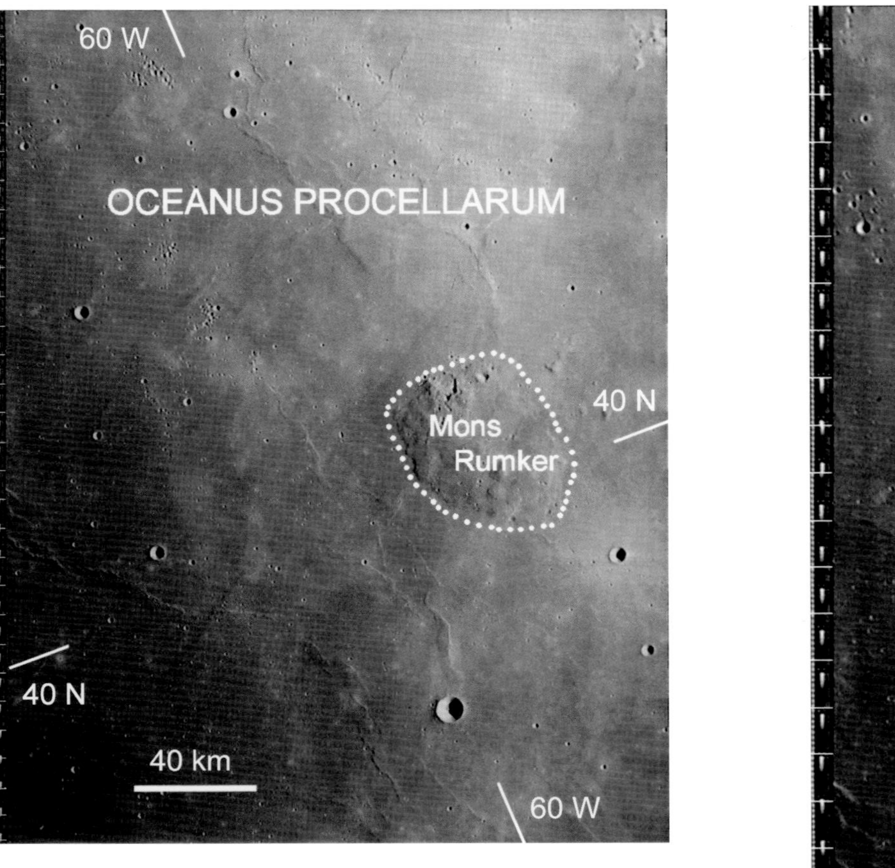

LO4-170H2 Sun Elevation: 19.8° Altitude: 2870.64 km

Several wrinkle ridges meet at Mons Rumker, which is formed of multiple round domes. Perhaps contraction of the cooling, solidifying mare surface caused lava to be extruded in this area a multiple number of times, ultimately forming Mons Rumker. If the lava here was of a different chemical composition or cooler than the usual mare material, its viscosity may have been greater, allowing the prominence to build up.

LO4-163H2 Sun Elevation: 20.1° Altitude: 2866.72 km

Rima Sharp is sinuous, unlike other rimae that are more linear. Although linear rimae are interpreted as surface manifestations of fractures that relieve tension, sinuous rimae (rilles) are interpreted as the result of flows of fluid along low slopes. On Earth, the most familiar analogy is that of meandering streams and rivers flowing in alluvial valleys. The curving and recurving meanders are dynamically caused by lateral oscillations of the momentum of a moving fluid, which is probably low-viscosity lava on the Moon. Such meanders may have been very common during filling of the mare, but perhaps only those associated with the last flows have survived being submerged.

LO4-163H3 Sun Elevation: 20.1° Altitude: 2866.72 km

This is the widest part of Rima Sharp, and is probably the source of its flow (see LO4-163H2). At its northern extreme, it actually breaks into relatively smaller tributaries, apparently gathering lava before carrying it southward. Perhaps lava upwelling into the area southwest of Harpalus was blocked by the highland prominence in the lower right corner of the photo (part of the rim ring of the Imbrium Basin) and formed Rima Sharp as it found a route to a lower elevation. The highly eroded crater South is covered with Fra Mauro Formation, ejecta from the Imbrium Basin. Harpalus is a Copernican crater with a broad ray pattern.

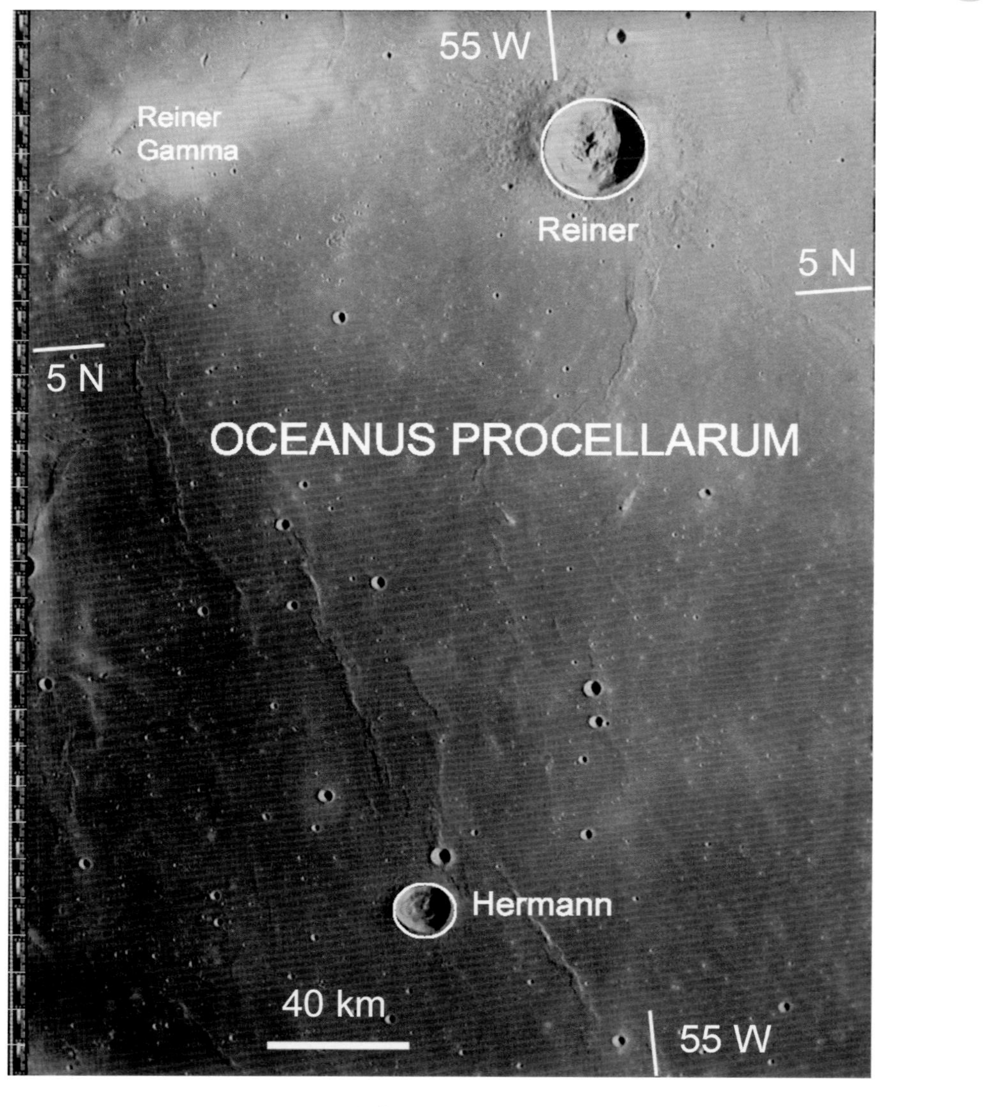

LO4-157H1 Sun Elevation: 16.9° Altitude: 2668.85 km

Reiner Gamma is a very unusual feature. The IAU classifies it as an albedo feature, to finesse the question of its origin as an impact or internal process. Although rare, there are a few other features with similar bright swirls. They occur in areas with magnetic anomalies. It has been suggested that the magnetic fields deflect the ionized solar wind. Without such protection, ionized hydrogen (protons) reduces oxides in the surface grains, freeing elemental iron. The iron absorbs light, darkening the surface. What causes the magnetic anomalies is unknown.

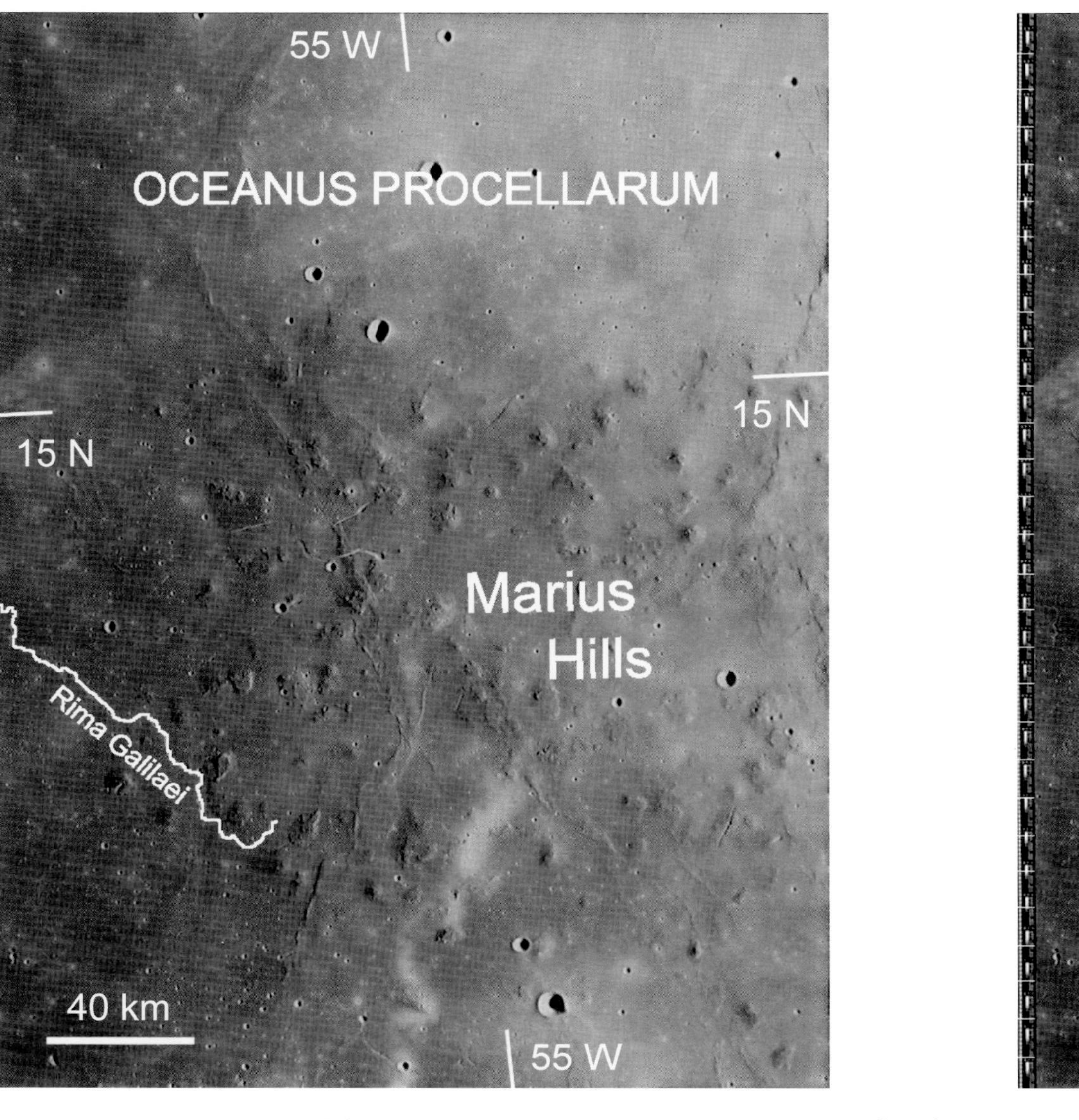

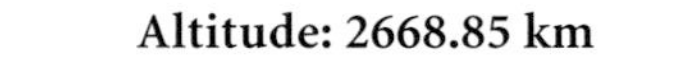

LO4-157H2 | Sun Elevation: 16.9° | Altitude: 2668.85 km

The Marius Hills have been interpreted as a volcanically active region. Rima Galilaei seems to have collected and transported lava from two tributaries near its southeast end. Other rimae in the area seem to start at craters that may have formed around an underground source of lava. The northern boundary of the elevated hilly terrain occurs along the ring of the proposed Lavoissier-Mairan Basin, so it may be an ejecta blanket of that basin. If so, the ejecta would have been modified by flooding with lava from below, filling any radial troughs. As in many cases where thin lava comes in contact with highland material, the area may have been covered with dark mantling material, generated by fountains of glass beads.

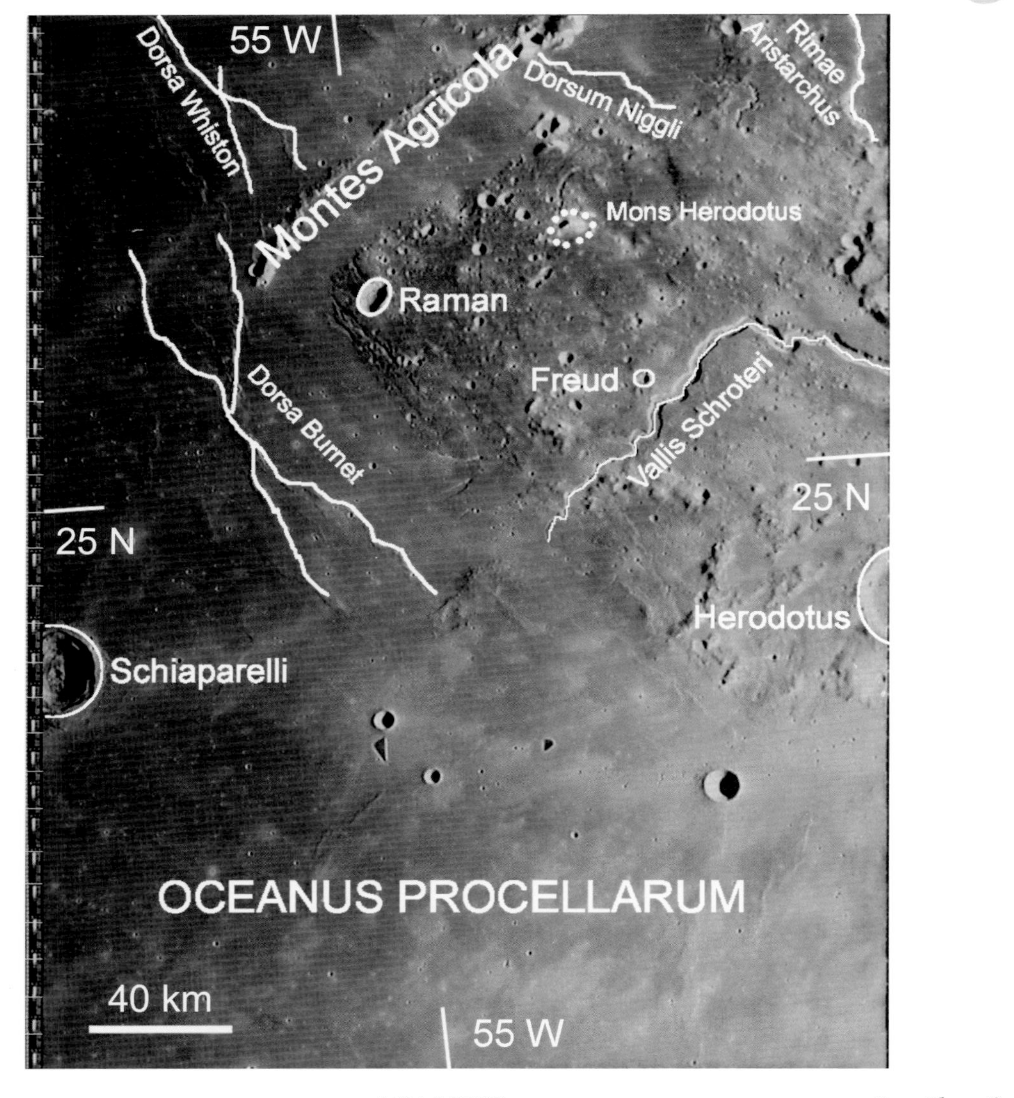

LO4-157H3 **Sun Elevation: 16.9°** **Altitude: 2668.85 km**

The highland area in the upper right quadrant of this photo is the eastern half of a large island in Oceanus Procellarum known as the Aristarchus Plateau. The eastern half can be seen in LO4-150H3. This plateau may have been formed by a triple intersection of rings; the first external elevated ring of the Imbrium Basin, the southeastern rim of the Lavoissier-Mairan basin, and the northeastern rim of the possible Cardanus-Herodotus Basin. Vallis Schroteri seems to be carrying lava down to the mare surface. See LO4-158H1 and its notes for more on this feature.

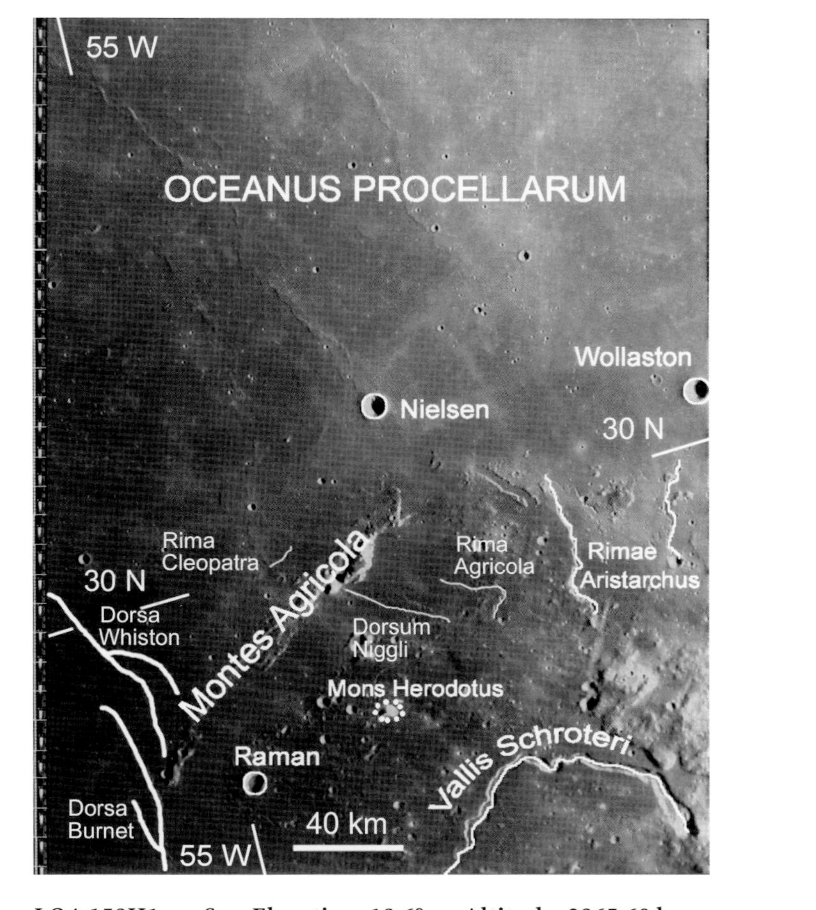

LO4-158H1 **Sun Elevation: 19.6°** **Altitude: 2865.60 km**

The northeastern sector of the Aristarchus plateau (lower right in this photo) is rich in evidence of subsurface thermal (endothermic) activity. Multiple sinuous rilles carry lava down to the mare surface from source craters. Vallis Schroteri shows evidence of a massive outpouring of lava that excavated the valley, falling nearly 4 km from Cobra Head, the source crater, to the mare surface 140 km away (LO157H3). Presumably the massive Aristarchus Plateau blocked upwelling of magma, which was finally flowing through Vallis Schroteri, Rima Agricola, and Rimae Aristarchus.

LO4-158H2 Sun Elevation: 19.6° Altitude: 2865.60 km

This curved shoreline of Oceanus Procellarum could be the rim and ejecta pattern of the possible Lavoissier-Mairan Basin. Some striations and crater chains in this area are in the appropriate radial direction, northwest to southeast, while others radiate from the Imbrium Basin. This area may be covered with two layers of basin ejecta blankets, the Fra Mauro Formation being the upper layer.

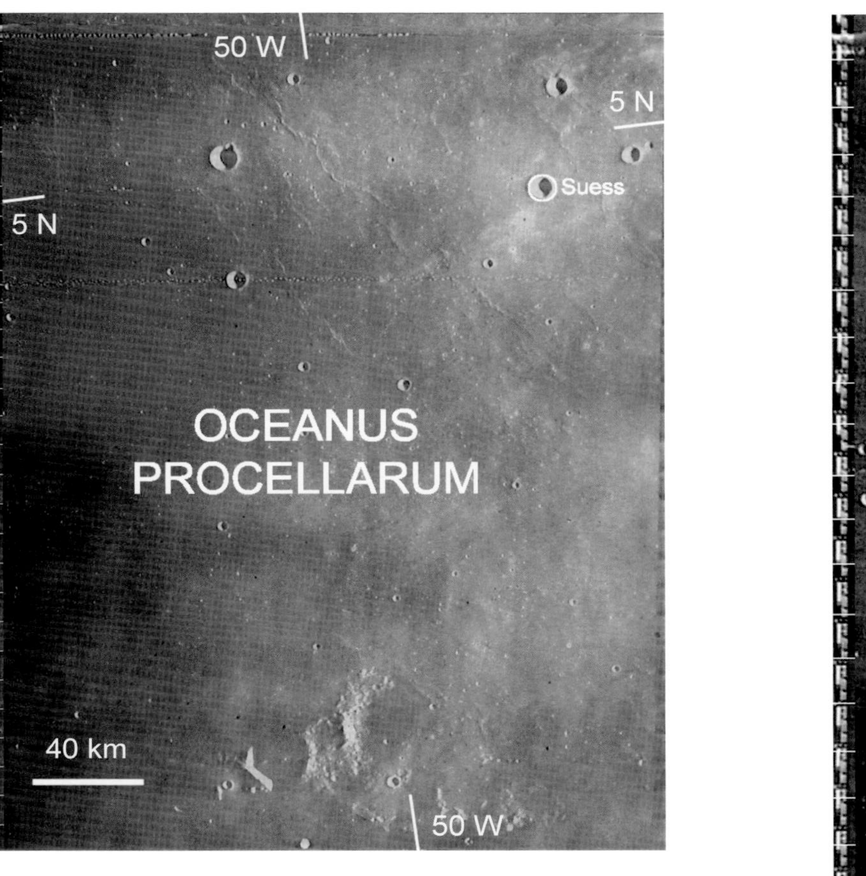

LO4-150H1 **Sun Elevation: 17.8°** **Altitude: 2668.40 km**

The ray passing to the south of Suess is from Kepler, to the east-northeast. Actually, all but the upper and lower left sides of the area in this image are lightly covered with Kepler ray material. The nearly submerged craters to the south indicate that the mare layer is relatively thin there, near the southern edge of Oceanus Procellarum.

LO4-150H2 Sun Elevation: 17.8° Altitude: 2668.40 km

This is the eastern edge of the Marius Hills (see LO4-157H2 for the western edge). Luna 7 failed due to premature ignition of its retro-rocket in October 1965 and crashed at the indicated site. Rima Marius flows north into the mare floor from the hills while Rima Suess flows down to the south. The chains of small craters to the east and southeast of Marius radiate from Kepler.

LO4-150H3 Sun Elevation: 17.8° Altitude: 2668.40 km

This is the eastern part of the Aristarchus Plateau: the western part can be seen in LO4-157H3. Copernican crater Aristarchus, its ejecta, and its ray pattern dominate this photo. Counts of craters on its floor show Aristarchus to be very young, nearly as young as Tycho. Clearly, there has been a great deal of upwelling of lava here, to generate all the ridges and lava streams. Remote sensing of the surface of this area shows a high percentage of REE (rare earth elements), which indicate radioactivity, because of high chemical affinity to the uranium-thorium series. The Lunar Prospector spacecraft confirmed a high level of thorium. Radioactivity generates heat and the heat causes the melting rock to expand and rise.

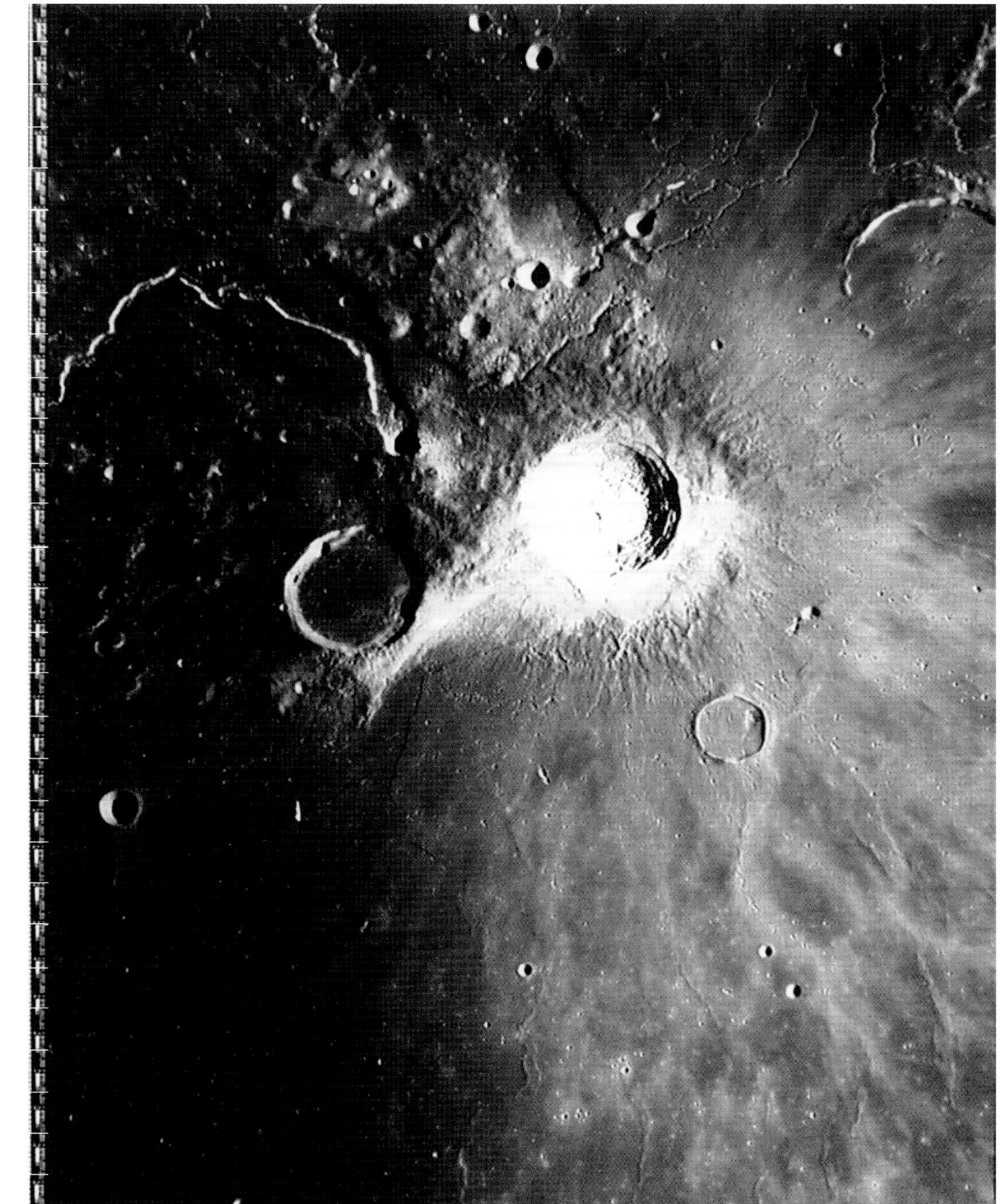

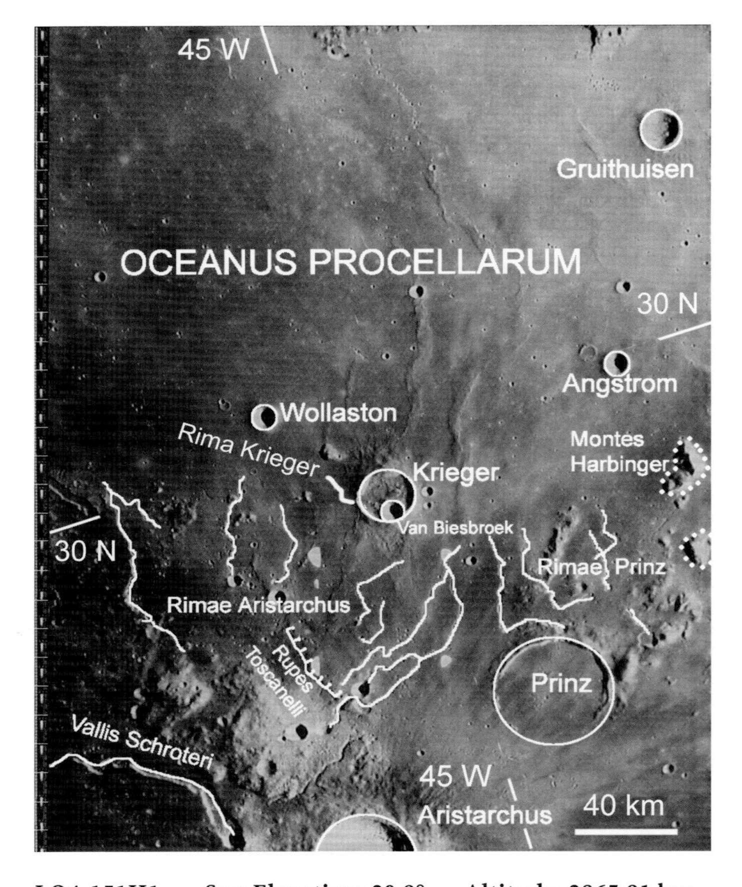

LO4-151H1 Sun Elevation: 20.9° Altitude: 2865.91 km

The northern part of the Aristarchus Plateau is shown in the lower half of this photo. Oceanus Procellarum seems thin here, leaving remnants of older features such as Prinz and Montes Harbinger projecting above the mare floor. However, there are extensive signs of lava flowing out from various sources on the Aristarchus Plateau and running through the rimae down to the mare surface. Note that each rima seems to start from a crater. The Aristarchus impactor probably arrived long after the eruptions ended.

LO4-151H2 Sun Elevation: 20.9° Altitude: 2865.91 km

Sinus Iridum, a bay off of Mare Imbrium, results from a circular impact feature whose rim diameter of 260 km results in it being characterized as a crater rather than as a basin. Montes Jura constitutes the rim of the Iridum crater. The chains of secondary craters and other lineations in this picture tend to be radial to Sinus Iridum, rather than to Imbrium, indicating that the Iridium crater occurred later than the Imbrium Basin. Mons Gruithuisen Gamma and Delta, with a crater on top of each of them, have been considered to be volcanic, but they may simply be relics of the Imbrium Basin, modified by the Iridum ejecta and coincidental primary or secondary craters at their peaks. There are several such features on the floor of Mairan.

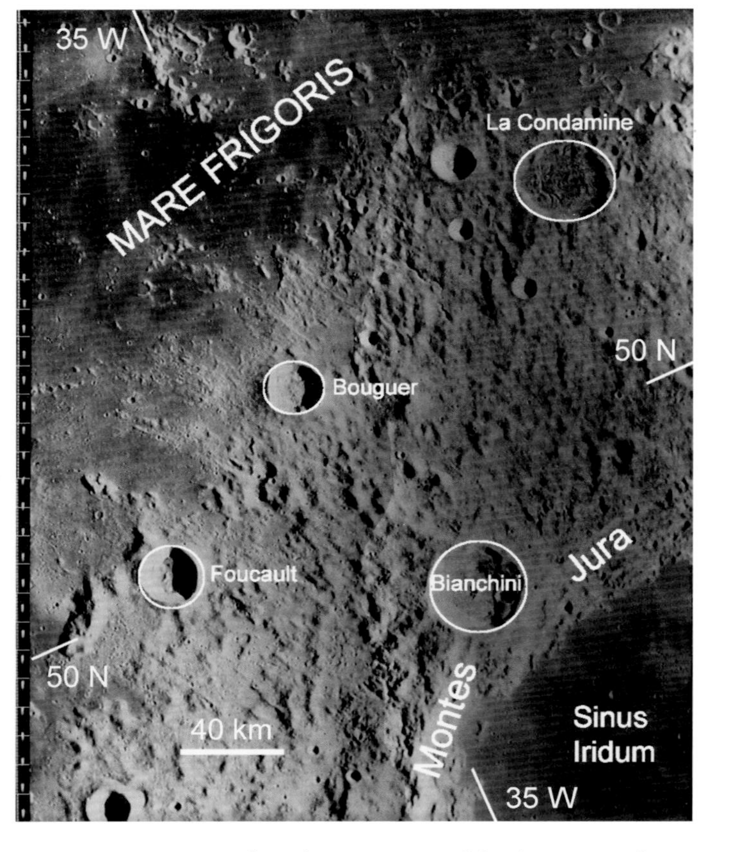

LO4-151H3 **Sun Elevation: 20.9°** **Altitude: 2865.91 km**

The northwestern rim of the Imbrium Basin crosses diagonally across this picture. Mare Frigoris has flooded the outer trough of the Imbrium Basin.The area between Sinus Iridum and Mare Frigoris has been crisscrossed with ejecta from different directions. Strong southeast to northwest lineations from the direction of the Imbrium Basin are clear, as are lineations radial to the Iridum crater.

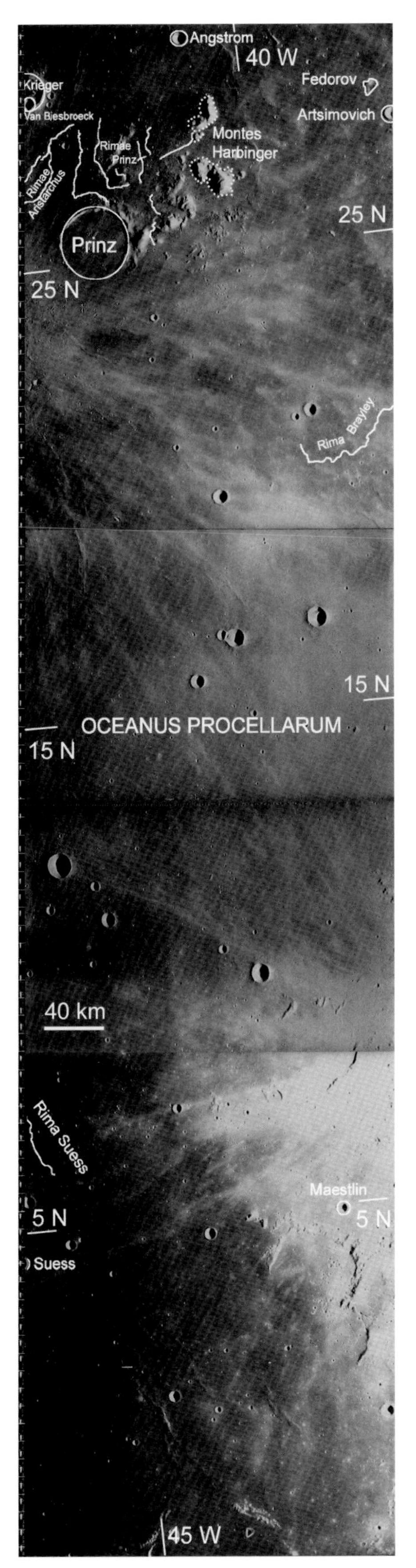

LO4-144H
Sun Elevation: 19.1°
Altitude: 2668.99 km

This area just east of the Aristarchus plateau has relatively thin mare coverage; it is higher than the central mare surfaces of either Oceanus Procellarum or Mare Imbrium. Numerous rilles (and the lakes that fed them) carried lava from the vicinity of Prinz to the north. Montes Harbinger is probably a remnant of the topographic ring of the Imbrium Basin.

Some of the valleys of Rimae Prinz are quite wide. Like meandering streams on Earth, some sinuous rilles of lava on the Moon are dynamic. In time, their individual bends move in a downstream direction and sweep out a wide valley.

The rays from Kepler, coming from the southeast, meet rays from Aristarchus, coming from the northwest. Kepler is larger than Aristarchus but is further away from the center of the picture. Note the pair of overlapping craters near 7° N, 43° W (the larger of the overlapping craters is called Bessarion B). The unusually sharp septum between the two suggests a simultaneous impact of the two craters; neither has pushed wall material into the other.

The flooded crater rims below Maestlin show striations from the Imbrium Basin.

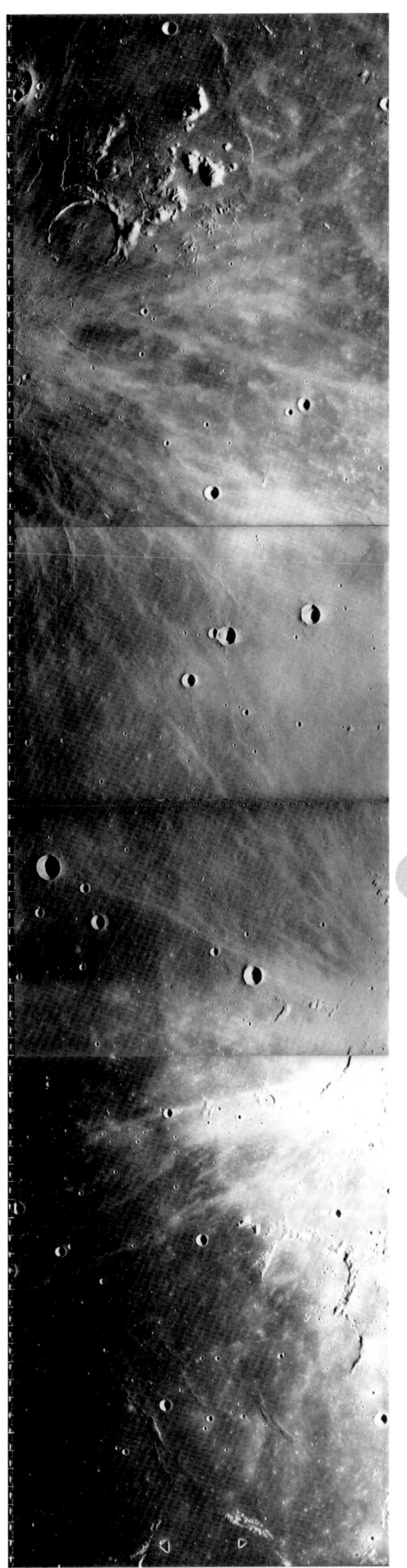

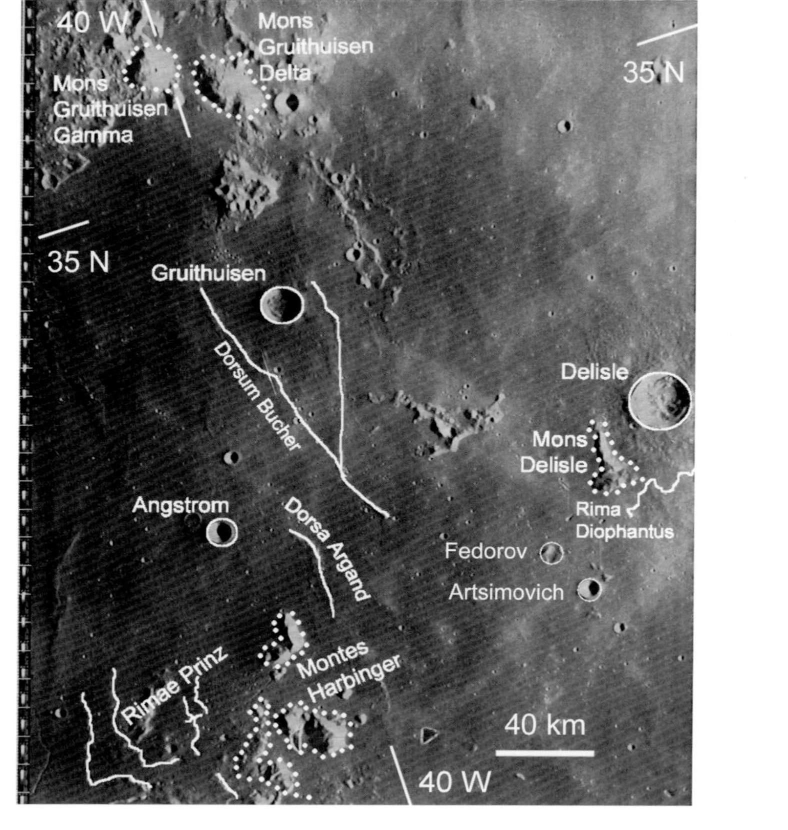

LO4-145H1 Sun Elevation: 20.7° Altitude: 2868.12 km

Mons Delisle and the archipelago of "islands" in the mare that lead to Mons Gruithuisen Gamma and Mons Gruithuisen Delta are remnants of the mostly submerged topographic ring of the Imbrium Basin.

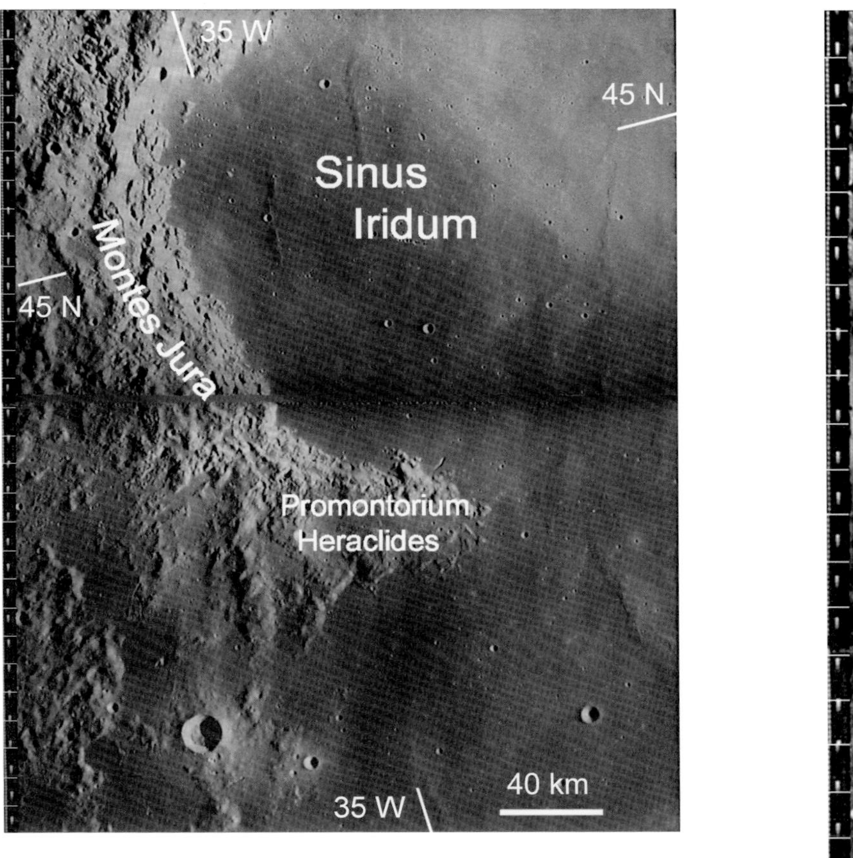

LO4-145H2 Sun Elevation: 20.7° Altitude: 2868.12 km

Promontorium Heraclides is a result of the intersection of the rim of the crater underlying Sinus Iridum (Montes Jura is part of that rim) and the highest ring of the Imbrium Basin. The curve at the southern border of the promontorium (peninsula) is not the curve of the Imbrium rim, but is formed by the Iridum crater pushing the Imbrium ring and floor material outward. The large vertical ridge at the base of the peninsula is more representative of the Imbrium ring in this area. Most of the heavy ejecta terrain on the left of the picture is from the Iridum event, but traces of striations from the Imbrium Fra Mauro Formation (probably lightly covered by Iridum ejecta as well) can be seen in the lower left and in the upper left of the photo.

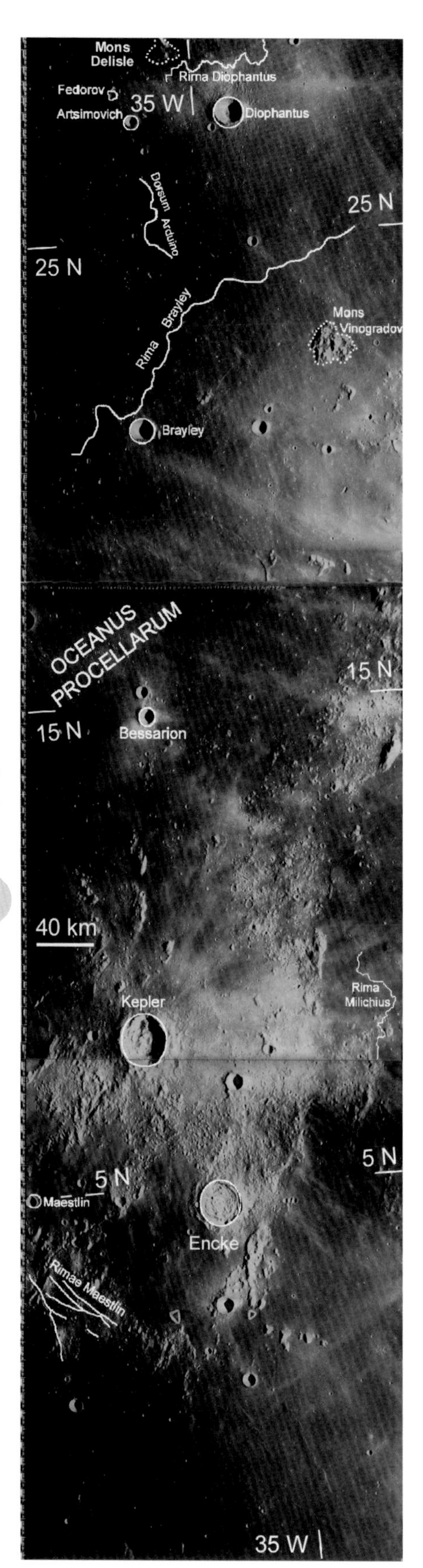

LO4-138H
Sun Elevation: 18.3°
Altitude: 2670.70 km

Oceanus Procellarum meets Mare Imbrium (upper right) near Mons Vinogradov, which is part of the highest ring of the Imbrium Basin. The darkness of the mare surface, plus the patterns of Dorsum Aduino and the long, narrow Rima Brayley suggest a young lava flow of low viscosity. Rima Brayley appears to guide lava in each direction away from a low rise near crater Brayley.

The knobby terrain curving along the right side of this picture is part of a circular feature that bounds Mare Insularum. The shapes of the interior of Kepler and its wall have been disrupted by the crater near its northern rim. This impact, being younger than Kepler, must have added its ray pattern on top of the one from Kepler itself. Southwest of Kepler, projecting from the mare surface of Oceanus Procellarum, are ridges and valleys radial to the Imbrium Basin. They are probably exposed because they top the first circumferential ring of the Imbrium Basin outside of its highest ring, the ring that bounds the trough occupied by Mare Frigoris.

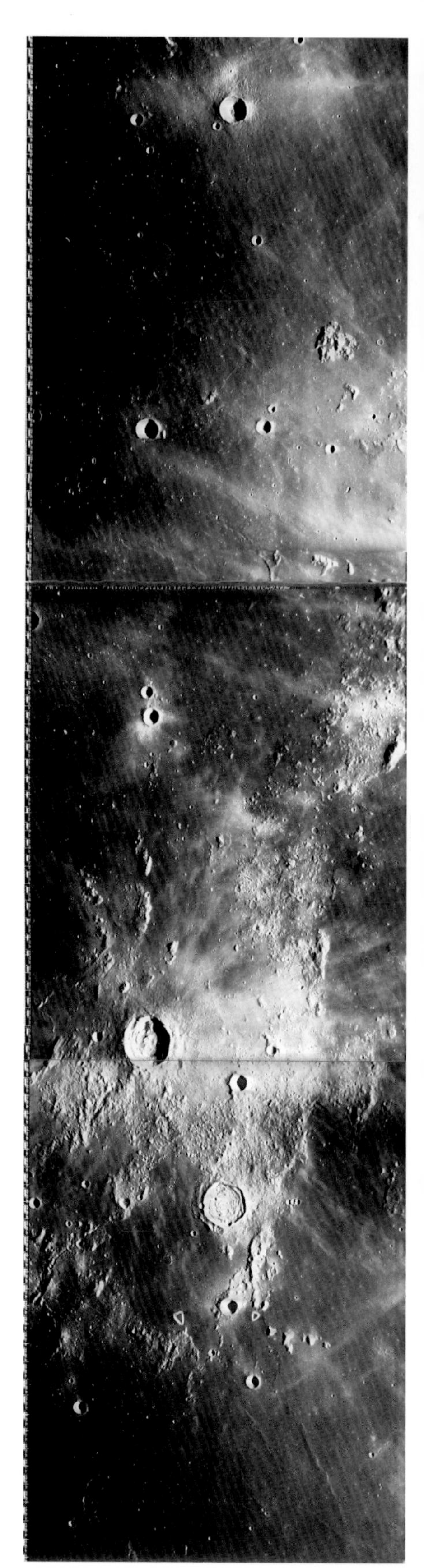

LO4-139H1 Sun Elevation: 20.2° Altitude: 2872.43 km

It is difficult to assign the age of Diophantus. Sharpness and crater counts are similar to those characteristics of older Copernican Period craters. However, the lack of a ray pattern has resulted in assignment to the previous Eratosthenian Period. The proliferation of dorsa (mare ridges), mostly circumferential to the Imbrium Basin, reflect the stresses of massive flow and subsequent cooling.

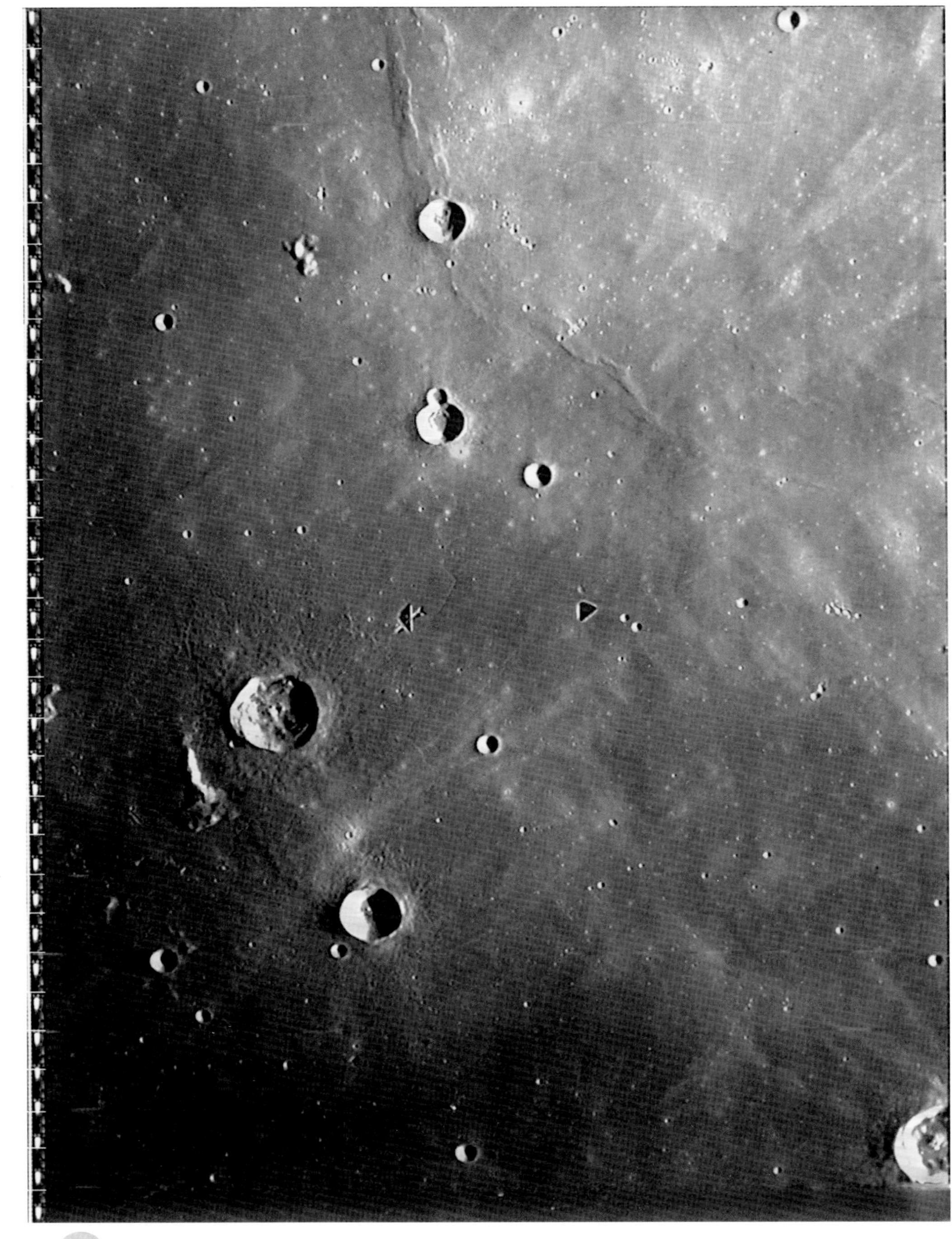

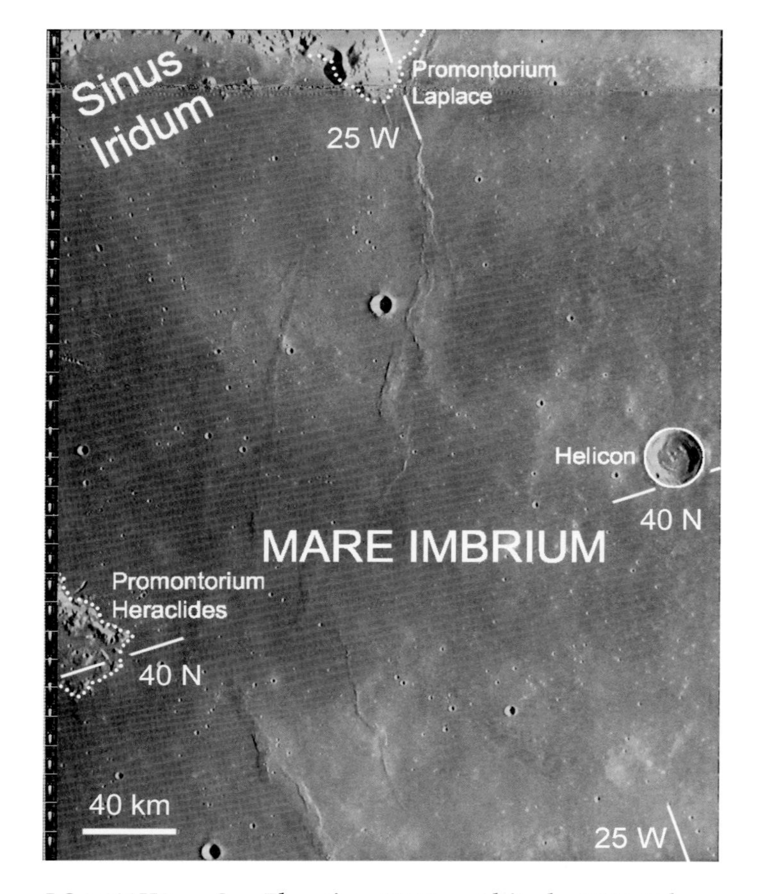

LO4-139H2 Sun Elevation: 20.2° Altitude: 2872.43 km

The presence of Sinus Iridum, more clearly seen in LO4-145H2, appears to have strongly affected the pattern of mare ridges in this area. Promontorium Heraclides and Promontorium Laplace mark the intersection of the rim of the crater that underlies Sinus Iridum with the highest ring of the Imbrium Basin. A large unnamed mare ridge curves between these two promontories and may mark the eastern and southern rims of the crater beneath Sinus Iridum.

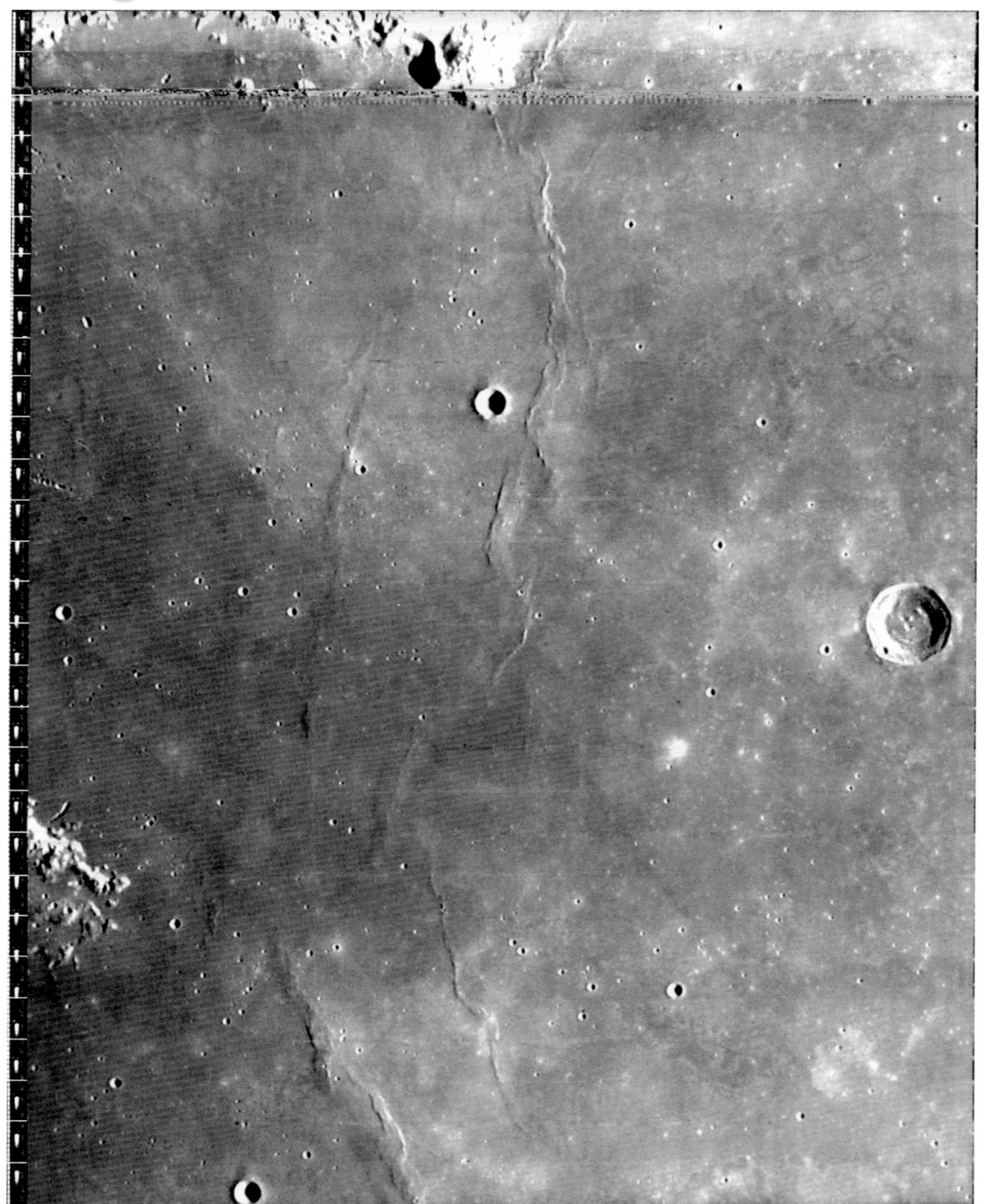

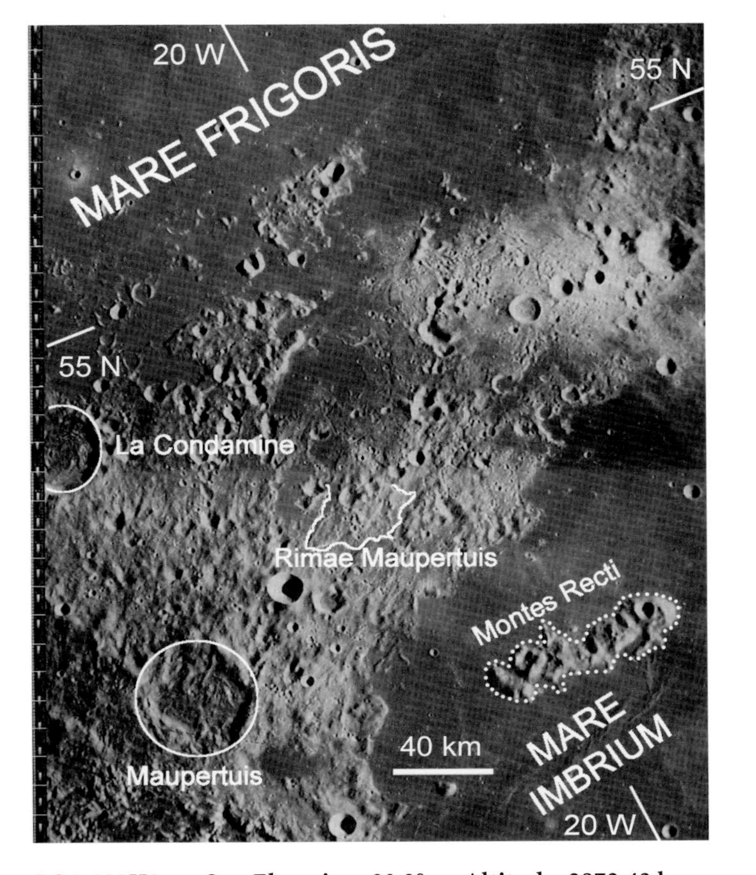

LO4-139H3 Sun Elevation: 20.2° Altitude: 2872.43 km

Montes Recti is a part of the main ring of the Imbrium Basin. Mare lava has penetrated the ring and flooded part of the Imbrium ejecta blanket (Fra Mauro Formation, covered by ejecta from the Iridum crater). This section of Mare Frigoris has flooded the trough beyond the main ring. The striations running from southwest to northeast come from the Iridum crater and have nearly (but not entirely) erased the ridges and valleys radiating from the interior of the Imbrium Basin. There are strong evidences of flow on the floor of Maupertuis and Rimae Maupertuis that may have resulted from a deposit of molten ejecta from the Sinus Iridum crater.

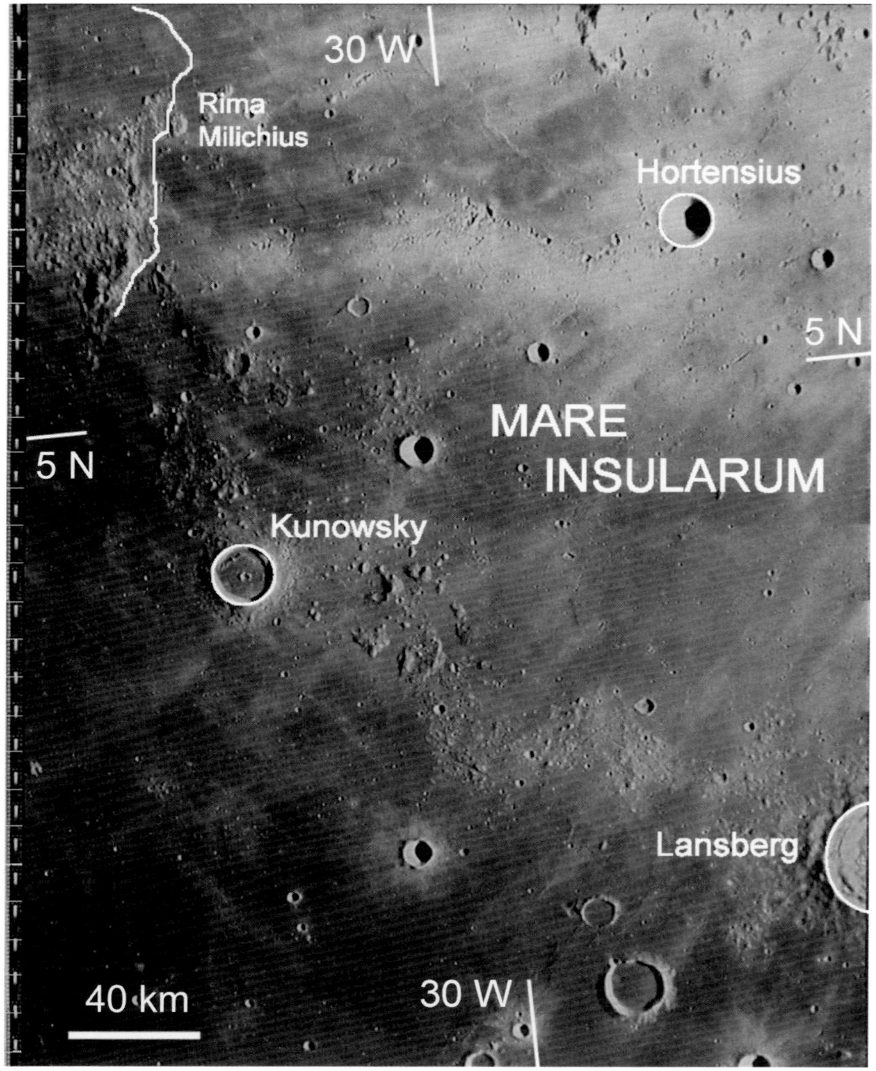

LO4-133H1 Sun Elevation: 18.5° Altitude: 2673.45 km

The arc of hills largely submerged by the mare surface, interpreted as islands, helped give Mare Insularum its name. This curved ridge of summits may represent the rim of the possible Insularum Basin underlying Mare Insularum. A cluster of low domes, each with a crater in its center can be seen northeast of Hortensius. These domes may be the result of lava extrusions. Retreating lava leaves a central crater in the top of the dome, a caldera.

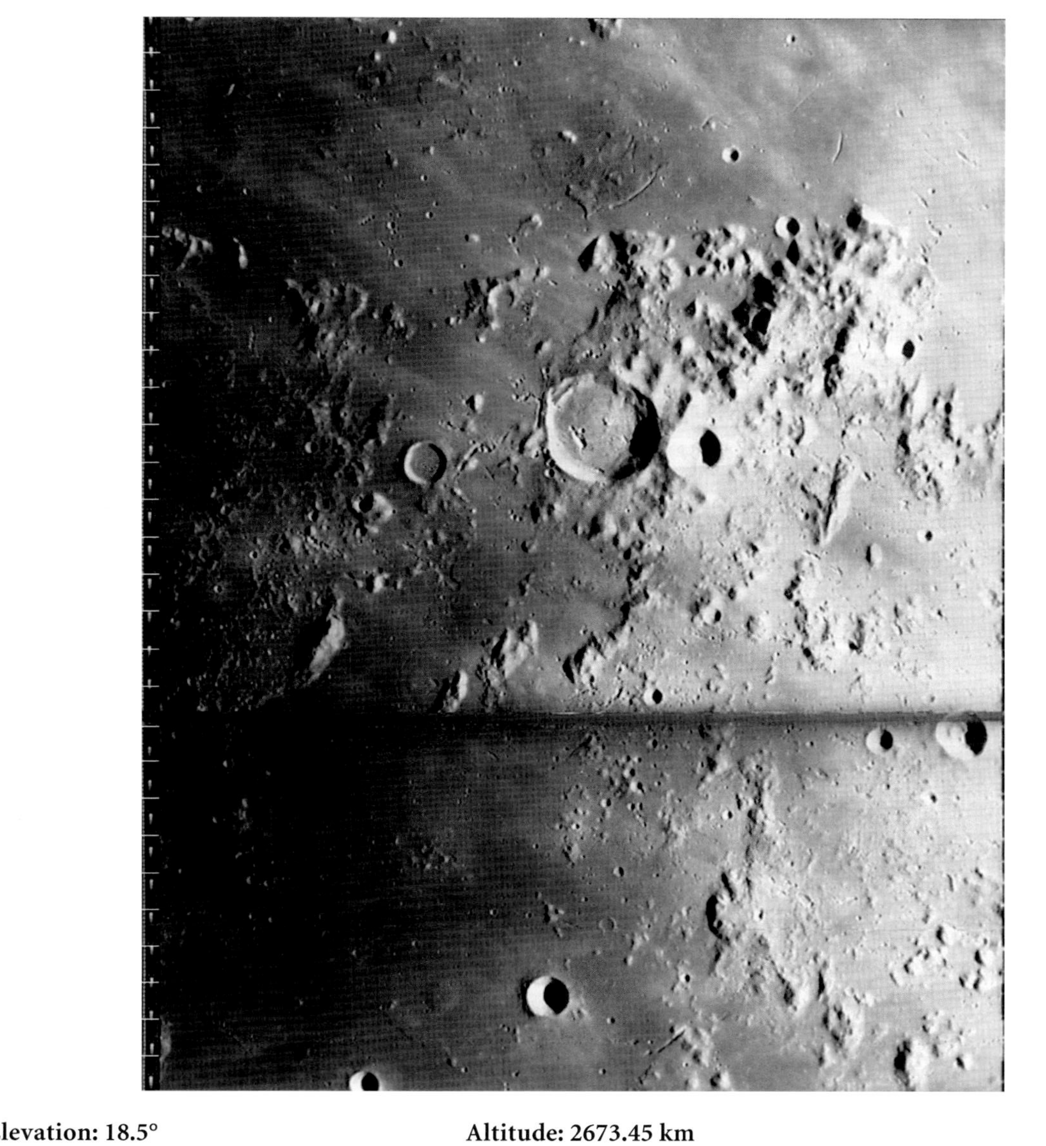

LO4-133H2 Sun Elevation: 18.5° Altitude: 2673.45 km

A low 8-km dome with a central crater or caldera can be seen about 20 km west of Milichius. This dome is surrounded by a lava flow with visible edges (scarps) that extend to the rim of Milichius. Rima T. Mayer has been deflected around another low dome with a central crater. Montes Carpatus is part of the rim of Mare Imbrium. The flooding of every possible small inlet along the northern shore of the Montes Carpatus illustrates how easily the lava flowed. The viscosity must have been much lower than typical terrestrial lava, low enough to overcome the effect of lower lunar gravity (one-sixth of Earth's gravity). The difference is attributed to a relative lack of volatile elements in lunar lava.

LO4-133H3 Sun Elevation: 18.5° Altitude: 2673.45 km

A very large lava flow extends from an area north of Euler through the gap between Mons La Hire and the hills 30 km to the northwest, all the way to Dorsum Zirkel. The edges of this flow are just barely visible in this image and in LO4-126H3, but can be seen more clearly in Apollo 15 frame M-1556, taken at a lower sun angle.

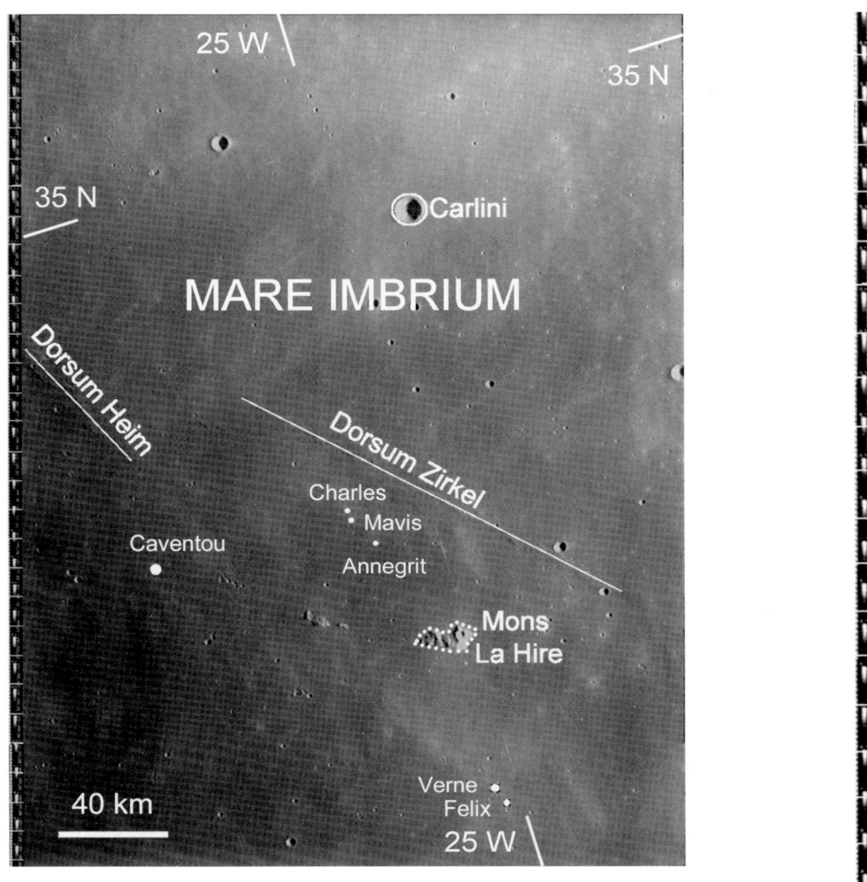

LO4-134H1 **Sun Elevation: 21.4°** **Altitude: 2878.51 km**

Craters Charles, Mavis, Annegrit, Verne, and Felix are small craters that were given special attention because they were photographed at high resolution by orbiting Apollo spacecraft. The official IAU names of such landmark craters are all "first names" (personal names from a variety of cultures) or in a few cases the names of gods and goddesses. This particular area, if chosen as a landing site, would have allowed surface exploration of distinctly different lava flows.

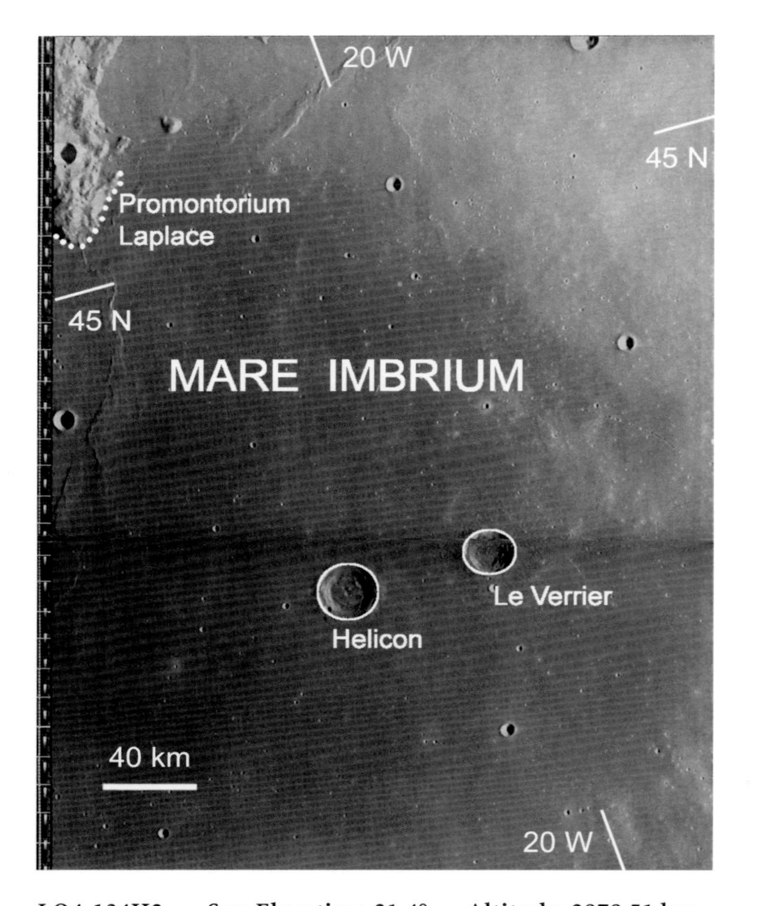

LO4-134H2 Sun Elevation: 21.4° Altitude: 2878.51 km

Pairs of impact craters with very similar sizes, ages, and other characteristics are found on the Moon and also on Earth. Helicon and Le Verrier are an example of such a pair on the Moon. Recent modeling of collisions of Kuiper belt objects indicates that there is a significant population of pairs of objects of similar size that mutually orbit each other. Pluto and Charon are the largest example of such a pair, if indeed they came from the Kuiper region. Comets come from the Kuiper region and are an important component of primary impactors on the Moon, so it is very reasonable that these pairs of craters come from paired comets. See the notes for LO4-127H1 for further discussion of Helicon and Le Verrier.

LO4-126H1 **Sun Elevation: 19.5°** **Altitude: 2677.28 km**

Massive deposits from Copernicus cover the eastern rim of the Insularum Basin, just northeast of Reinhold. Luna 5 (May 1965) suffered a bad midcourse correction and crashed in May 1965. Surveyor 3 (April 1967), seeking a soft landing on a western mare, bounced three times before finally touching down safely. Its cameras and soil scoop confirmed suitable terrain and flatness for an Apollo landing site. Matching features in Surveyor's photos with Earth-based maps and Lunar Orbiter 3 photography identified the precise landing location. This allowed Apollo 12 to land (November 1969) within 200 m of the Surveyor, a short walk away (in spacesuits!) for Alan Bean and Pete Conrad. This Apollo mission confirmed the ability to land at a precise location on the Moon, a capability that was critical for later Apollo planning.

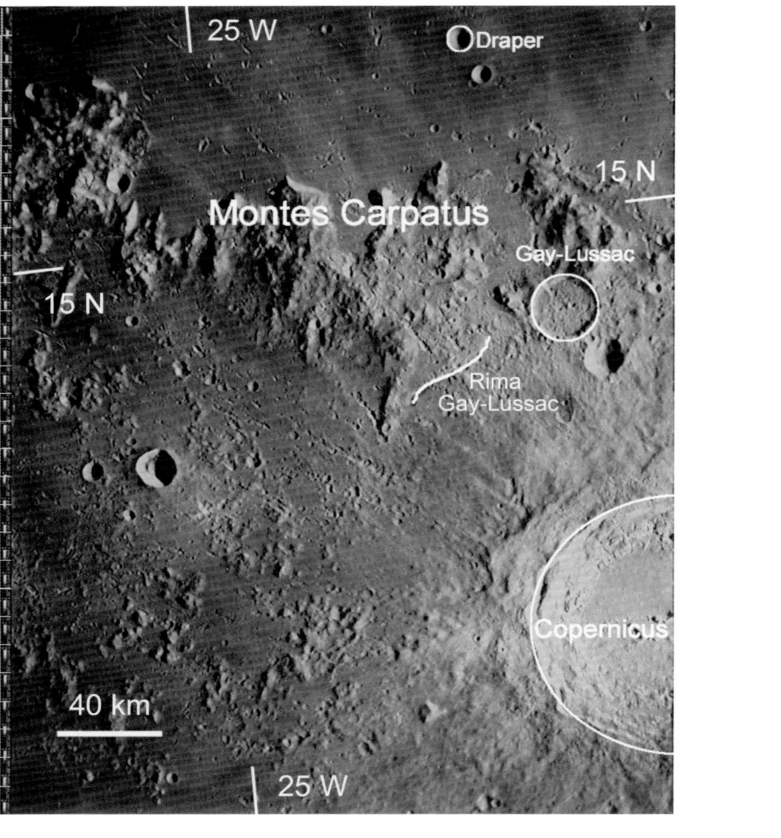

LO4-126H2 Sun Elevation: 19.5° Altitude: 2677.28 km

Copernicus dominates this dramatic scene of Montes Carpatus, the southern rim of the Imbrium Basin. Rima Gay-Lussac, like many other lava rilles, starts from a crater on the outside of the ring bounding a mare; however, Rima Gay-Lussac finds a valley leading across the ring into the basin floor. Rima Gay-Lussac appears more like a collapsed lava tube than a surface channel. A lava tube is formed when lava flows beneath the cooled surface and is kept open by the continuous flow of hot lava. The hardened roof of the tube insulates the lava from cooling by radiation into space. An intact lava tube is, of course, not visible from the surface but if the roof of the tube is less than a few meters in depth, primary and secondary impacts can collapse it in time.

LO4-126H3 **Sun Elevation: 19.5°** **Altitude: 2677.28 km**

Ejecta from Copernicus, landing along its ray near the right edge of this image, plowed a number of craters of unusual shapes (indicating a low angle of approach to the surface). South of Lambert, a 60-km buried crater (Lambert R) has left a faint image of its rim on the mare surface, like the fading smile of the Cheshire Cat of "Alice in Wonderland." Such patterns are sometimes called ghost craters. This crater has arrived after the Imbrium Basin event, but before Mare Imbrium covered the basin floor. The crater rim was probably completely covered with mare lava, but after the lava cooled and contracted, the rim reappeared as a mare ridge. Then Lambert arrived, covering the ridge over the northern rim of the ghost crater with ejecta. Copernicus then draped rays over the composition for further interest.

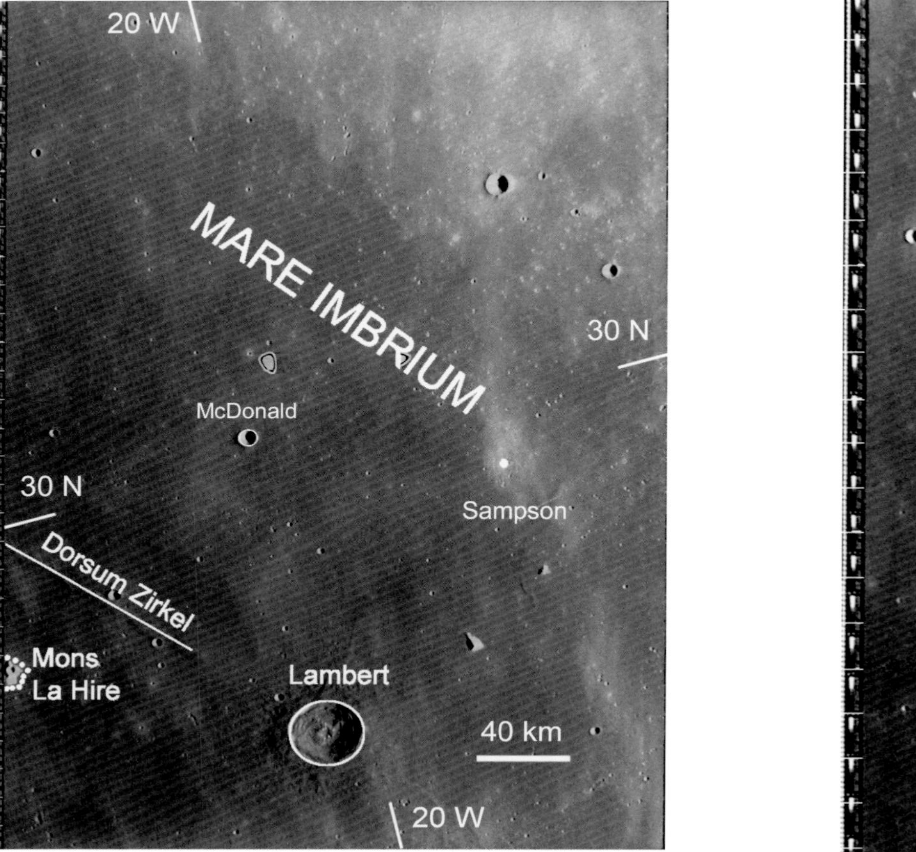

LO4-127H1 Sun Elevation: 21.8° Altitude: 2886.35 km

The 30-km crater Lambert has multiple circumferential rings in its floor, suggesting that it has penetrated the solid mare surface and hit something different below, semimolten lava or perhaps the melt sheet of the basin floor beneath the mare. Since Lambert left its ejecta blanket on the mare surface, a fresh flow (Apollo 15 frame M1009) has covered much of the eastern sector of Lambert's secondary field. The tiny crater Sampson is in the center of a shallow 30-km dome and could be a source of this flow. The pyramidal rocks northeast of Lambert could be peaks of an inner ring of the Imbrium Basin. The center of the basin (33° N, 18° W) is near the top of this photo.

15 W

Mons Pico

45 N

45 N

MARE IMBRIUM

Le Verrier

40 km

15 W

LO4-127H2 Sun Elevation: 21.8° Altitude: 2886.35 km

The center of Mare Imbrium, near the bottom of this picture and the top of LO4-127H1, is nearly devoid of mare ridges and rilles. Only impact craters and rays relieve the uniformity of surface topography.

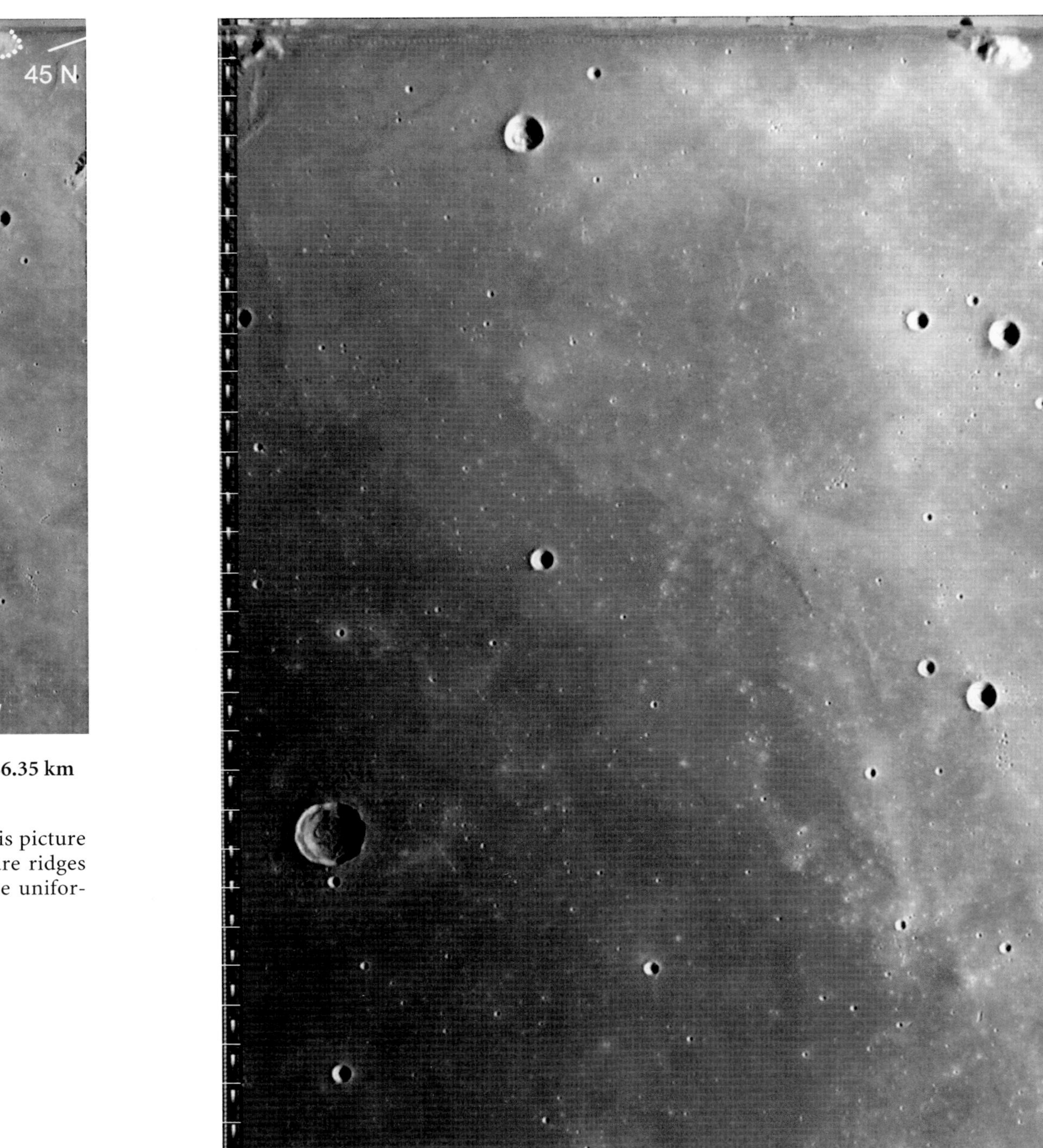

LO4-127H3 **Sun Elevation: 21.8°** **Altitude: 2886.35 km**

The Montes Alpes constitute the northern rim of the Imbrium Basin, which passes through the northern part of the magnificent Late Imbrian crater Plato. Plato falls on a "continental divide" in a sense: rima direct flows from its outer ramparts south toward the center of the Imbrium Basin and north toward the basin's outer trough, which underlies Mare Frigoris. The rille labeled Rimae Plato in this photo is a tributary of a larger system. It seems to flow beneath the surface near the edge of the photo, possibly into a lava tube that surfaces about 15 km downslope (LO4-115H3). Plato and Mare Frigoris are believed to have been flooded in the Eratosthenian age, well after the part of Mare Imbrium seen south of Plato in this picture. The sharp craters in the 10-km range are probably secondaries from Plato.

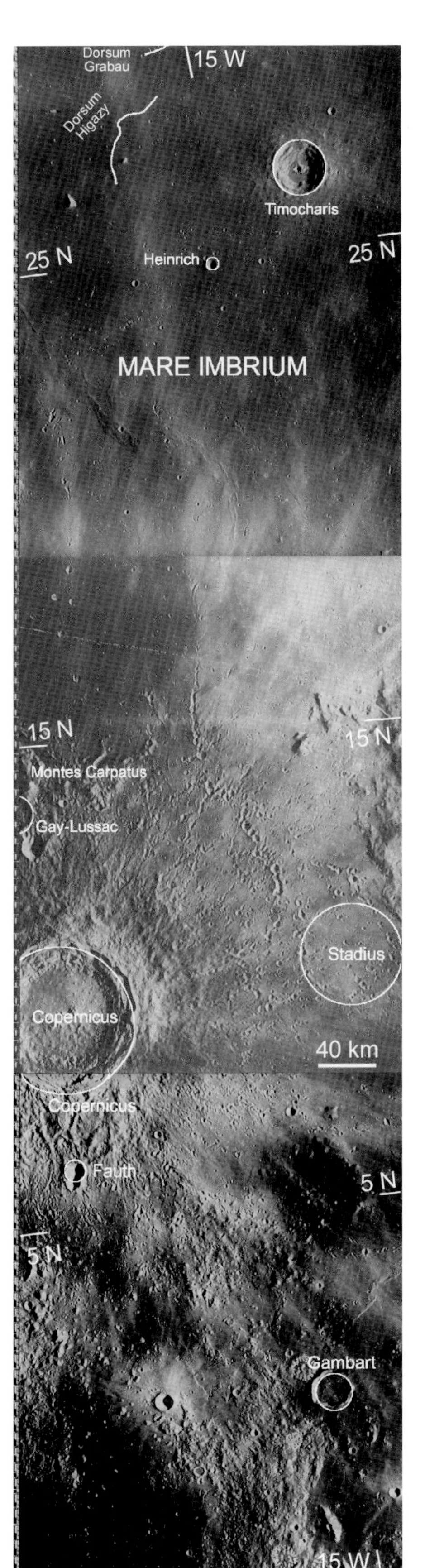

LO4-121H
Sun Elevation: 19.6°
Altitude: 2681.82 km

Timocharis is a crater with moderately sharp features but whose ray pattern has blended with the surrounding terrain due to exposure to the solar wind and gardening by meteorite bombardment. These characteristics, plus measurements of craters on its floor and ejecta, establish its age as Eratosthenian.

Copernicus is just south of Montes Carpatus, the local name of the highest ring of the Imbrium Basin. The ejecta pattern of this moderately young and spectacular crater is typical of both large craters and basins. A rampart of material is thrown outside of the rim. Further out, there is a heavy blanket of ejecta with radial striations. At about one radius away from the rim, the ejecta blanket thins, leaving uncovered areas, and then becomes a field of secondary craters for a distance of about one diameter away from the rim. Rays of fine, bright material extend for many diameters.

Well south of Copernicus, there are ridges and valleys from the Nubium Basin (overlain with Imbrium secondaries). A pair of simultaneous impactors produced Fauth and its smaller unnamed companion.

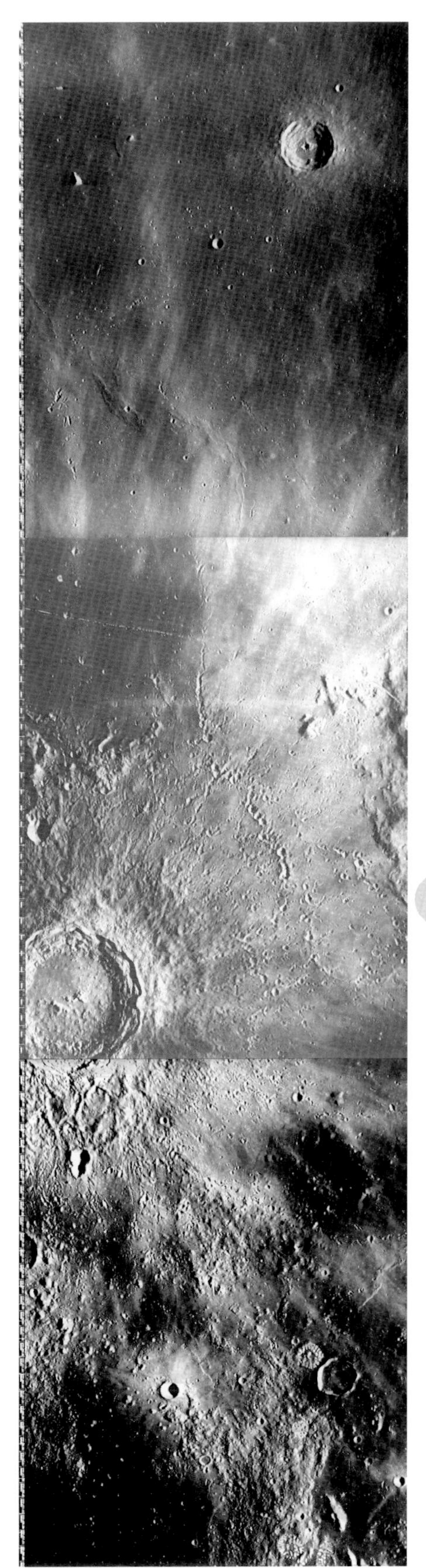

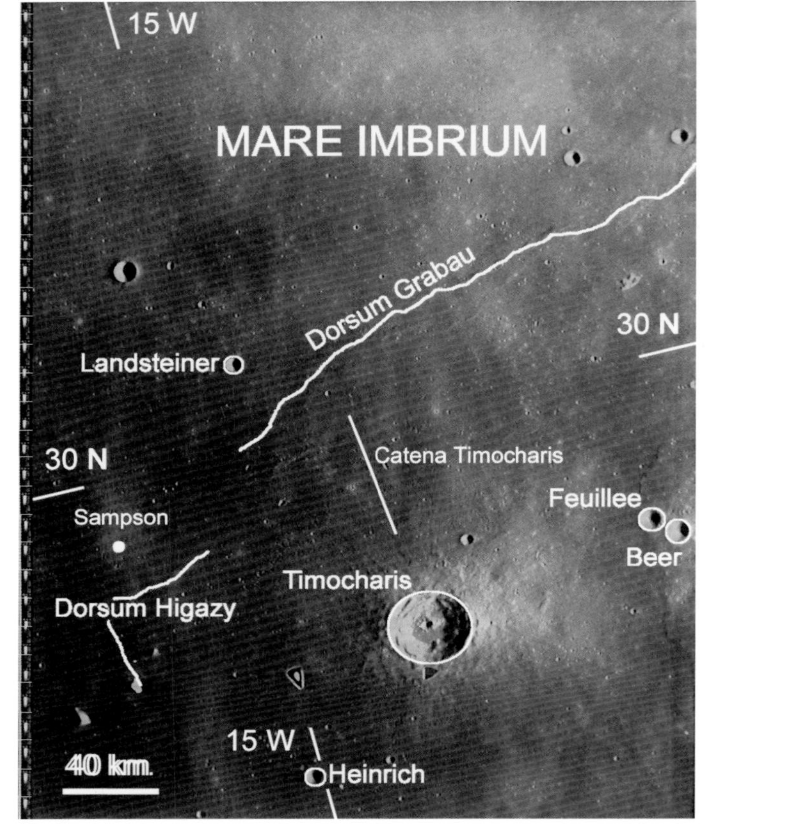

LO4-122H1 Sun Elevation: 20.7° Altitude: 2895.02 km

The surface of Mare Imbrium, like that of all large mare, is traversed by low ridges believed to result from compression stresses as the mare cools and contracts. Lava may find its way to the surface through faults in these ridges. There is a low 30-km dome topped by crater Sampson that may be a source for such a flow.

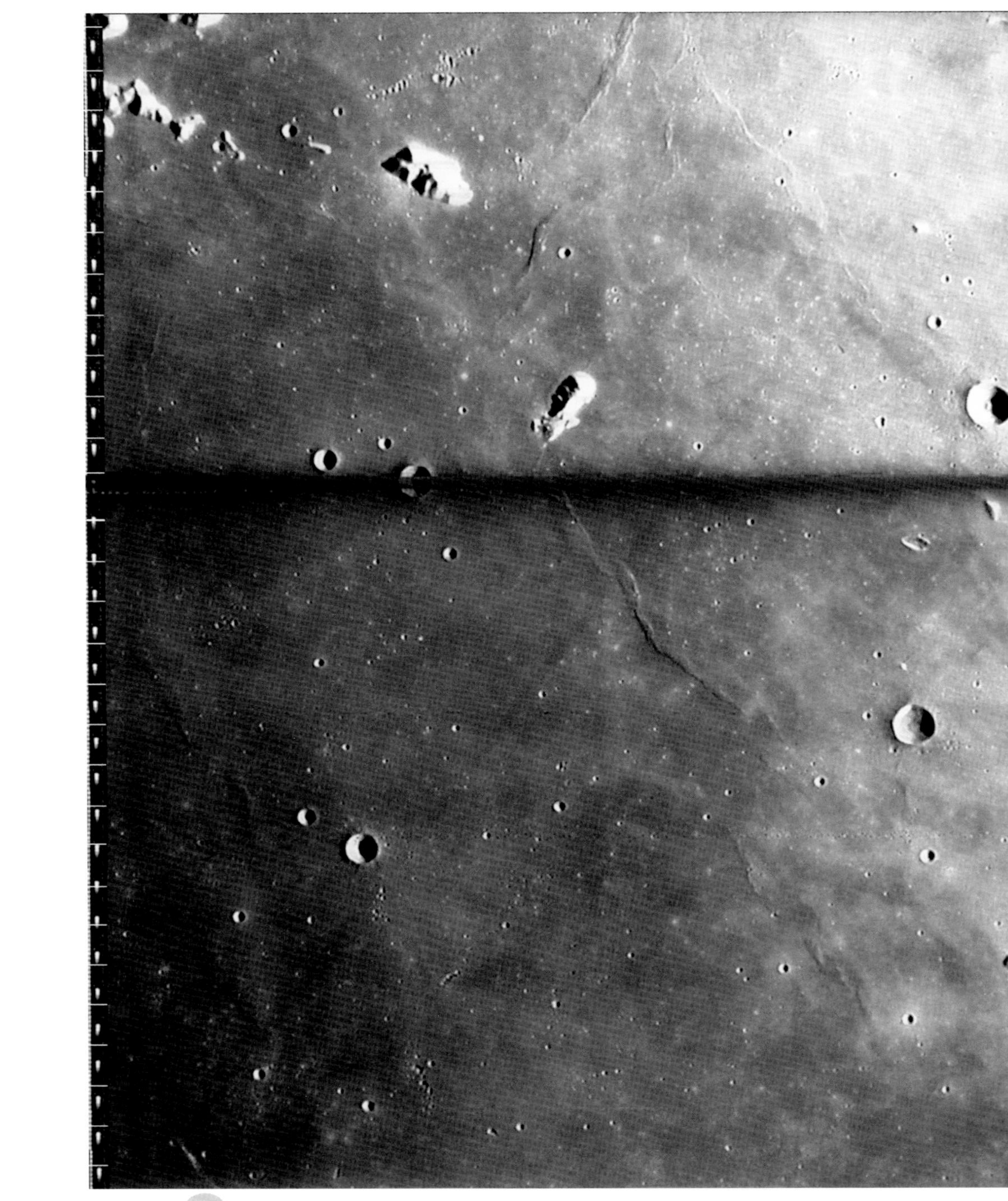

LO4-122H2 Sun Elevation: 20.7° Altitude: 2895.02 km

Basins often have rings inside their rim as well as outside. Montes Teneriffe, Mons Pico, and the peak to the south of Mons Pico may be signs of such a ring, most of which is covered by deep lava flows.

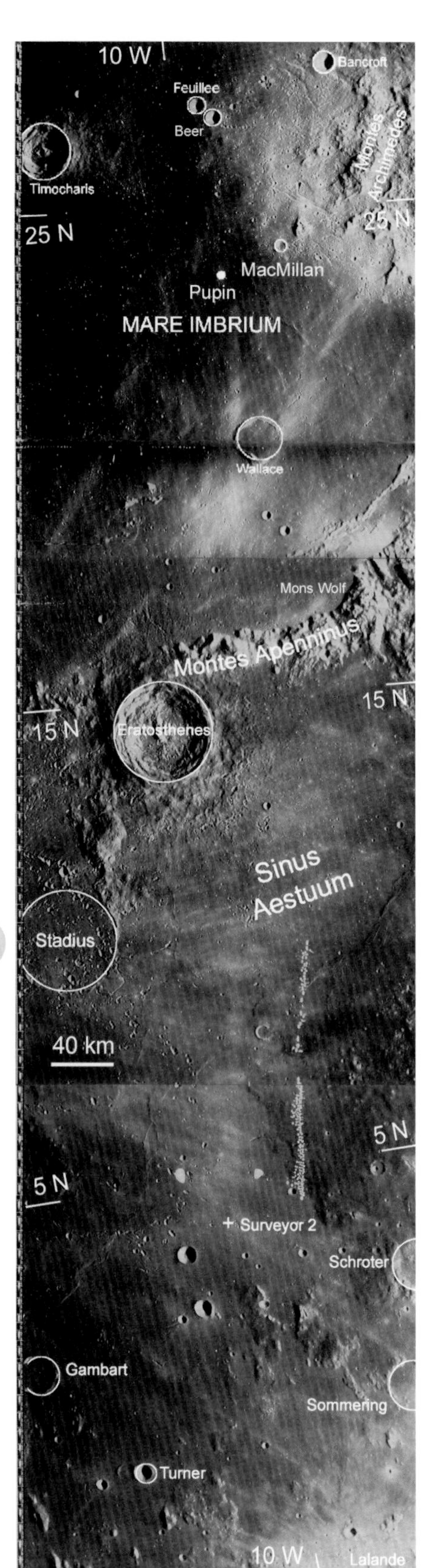

LO4-114H
Sun Elevation: 19.3°
Altitude: 2687.21 km

Montes Apenninus marks the rim of the Imbrium Basin and the edge of Mare Imbrium. As in other maria and other parts of Mare Imbrium, there are radial and circumferential ridges near this edge of the mare. Pupin is interesting because it is a bright mound with a central crater and no ray pattern, suggesting that it may be a volcanic feature. A pair of matched impactors may have caused Feuillee and Beer. Wallace has been nearly flooded, an indication of the depth of lava there.

Sinus Aestuum was formed by lava flowing into the valley south of the rim of Imbrium, as Mare Frigoris covers the valley north of the Montes Alpes. Eratosthenes is the type crater of the Eratosthenian Period; it formed after Sinus Aestuum was flooded. long after the Imbrium impact. However, rays have been nearly blended away and its rim structure has been slightly degraded.

Stadius, flooded by lava that has been covered by ejecta from Eratosthenes, must be from an earlier time. Dark mantling material surrounds Gambart and the nearby striations from Nubium. Surveyor 2 (September 1966) crashed after an unsuccessful midcourse correction.

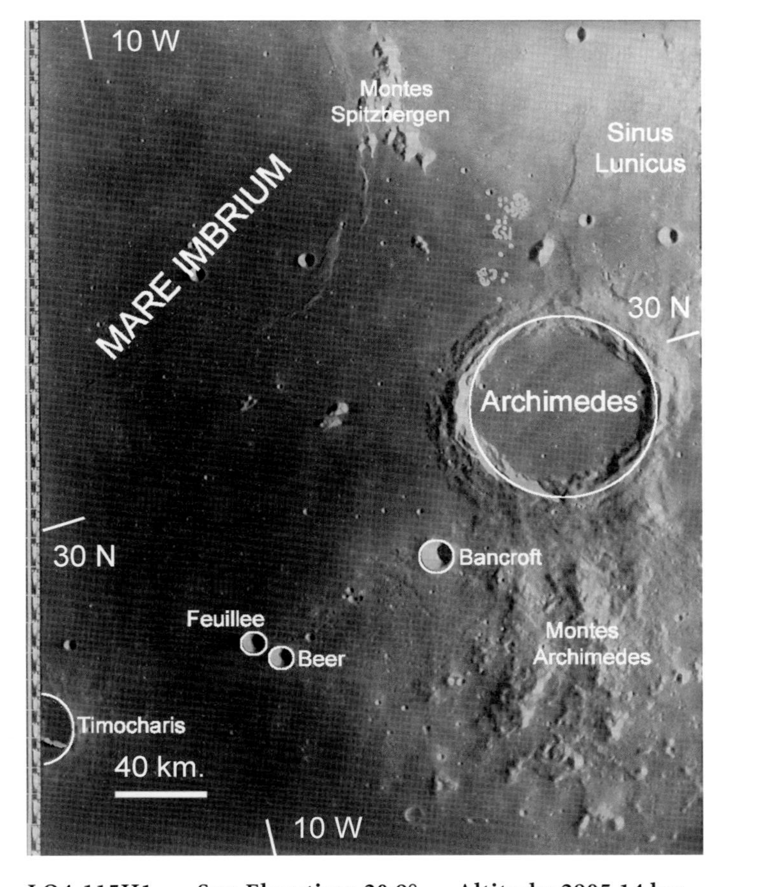

LO4-115H1 **Sun Elevation: 20.9°** **Altitude: 2905.14 km**

Montes Spizbergen and the nearby island arcs mark an inner ring of the Imbrium Basin, largely submerged under the mare surface; circumferential mare ridges mark its location in other sectors. The Montes Archimedes ranges, although overlain by ejecta from Archimedes, seem too massive to be caused only by that crater and probably were raised by one or more rings within the Imbrium Basin. The two rings may be analogous to the two ranges of Montes Rook. The plateau they rest upon is called the Apennine bench. An outer ring of the Insularum Basin to the southwest may also have helped to raise the elevation of the Apennine bench.

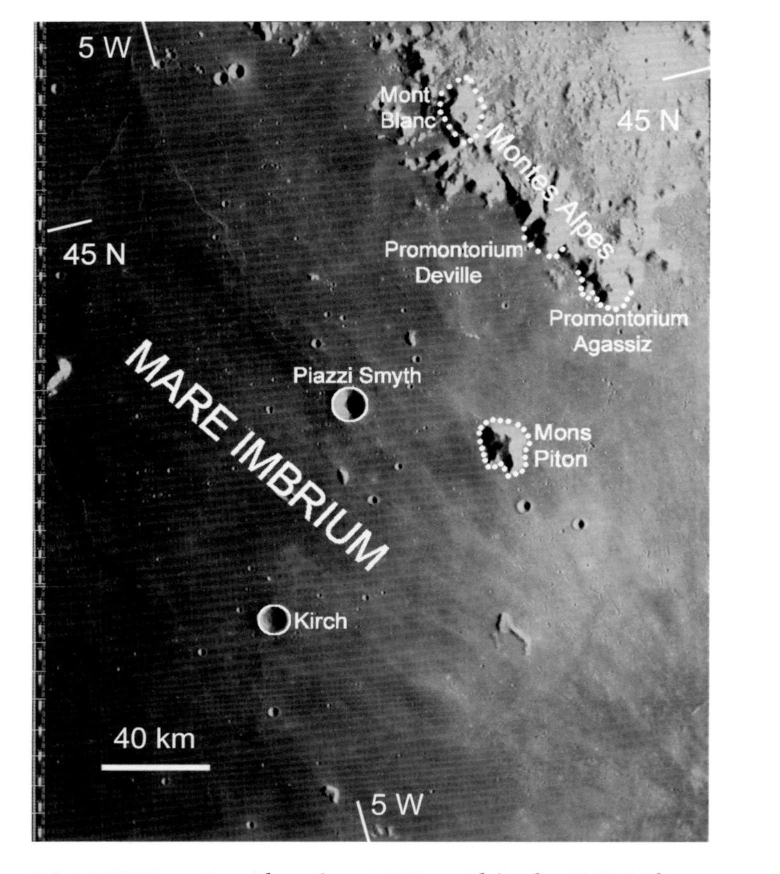

LO4-115H2   Sun Elevation: 20.9°   Altitude: 2905.14 km

Promontorium Agassiz marks the eastern end of Montes Alpes, at the northeastern sector of the rim of the Imbrium Basin. Mons Piton marks a ring within the basin (see LO4-122H2 for more peaks marking that ring). The large mare ridge that underlies Piazzi Smyth may also be related to that ring. The rays come from Aristillus to the southeast.

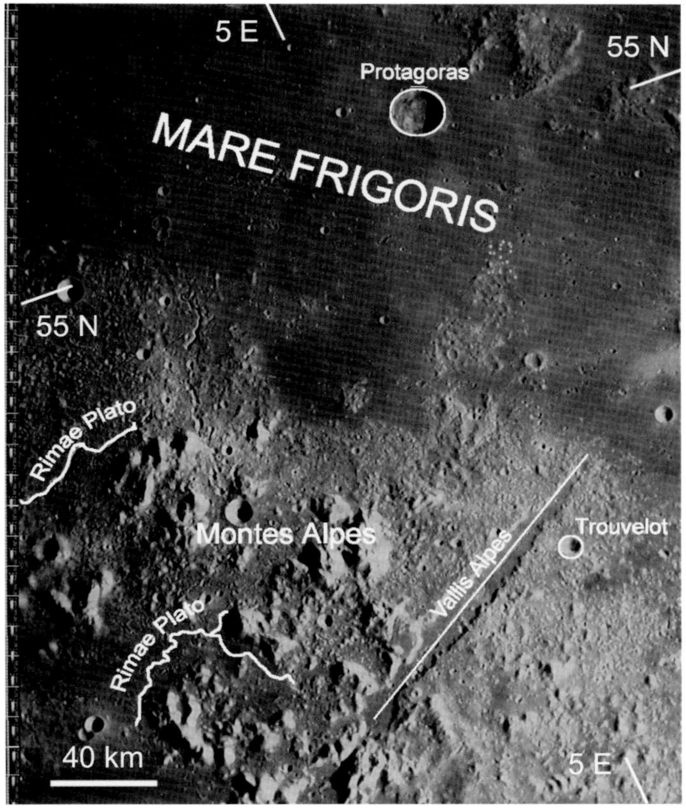

LO4-115H3 Sun Elevation: 20.9° Altitude: 2905.14 km

Vallis Alpes could be a valley formed in the rim of the Imbrium Basin during impact or could relate to a deep fracture opened by stresses during or shortly after the impact. It may have been flooded with lava from Mare Imbrium, but the lava may also have come from below, as with rimae in the area, flowing in this case both to Mare Imbrium and toward Mare Frigoris.

LO4-109H1 Sun Elevation: 20.5° Altitude: 2693.03 km

Sinus Medii and the unamed circular area of mare to the west of it may have been formed in craters. Bode has impacted the first outer ring of Imbrium; Pallas and Murchison have been degraded by the formation of that ring and then partly covered by Imbrium ejecta. The square bright area in the ridge between Sinus Medii and its companion to the west is an artifact due to a temporary loss of communication between Lunar Orbiter and the Deep Space Network.

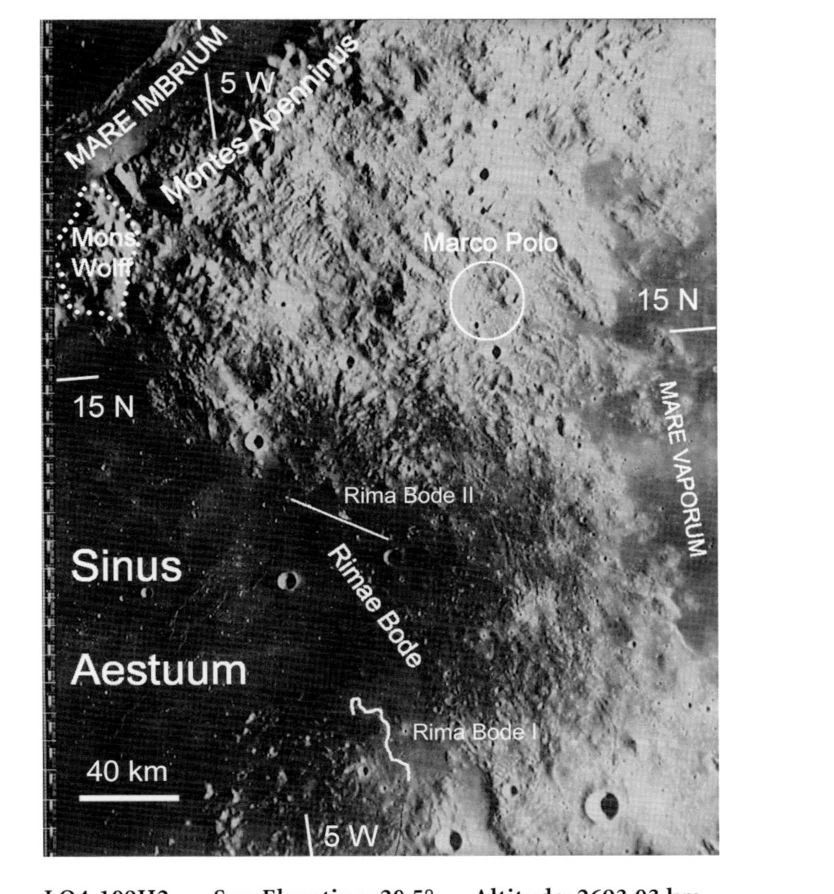

LO4-109H2 Sun Elevation: 20.5° Altitude: 2693.03 km

Sinus Aestuum and Mare Vaporum have flooded the valley outside of the rim of the Imbrium Basin, which is represented here by the Montes Apenninus and Mons Wolff. Here, large radial valleys formed by the Imbrium impact can be seen. Vallis Alpes (LO4-115H3) could be a similar (but deeper) valley. Marco Polo can just barely be perceived under its thick cover of Apenninus material, the inner Fra Mauro Formation, which is equivalent to the inner Hevelius Formation of Orientale. This material has been estimated to be 1 to 2 km thick near the crest of the Montes Apenninus. Dark mantling material covers both mare and highland surfaces in the vicinity of Rimae Bode.

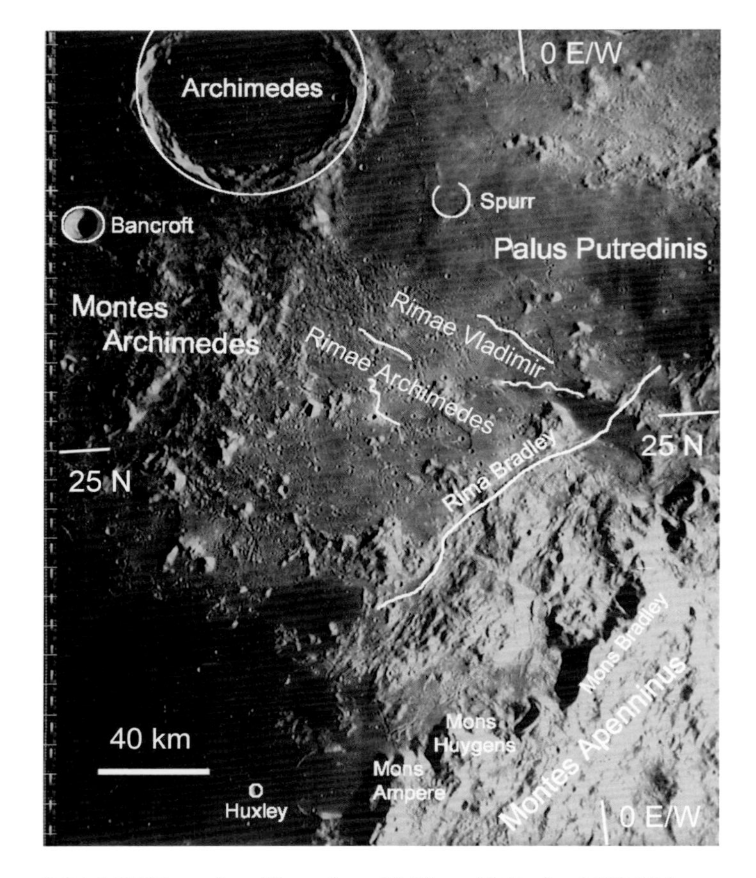

LO4-109H3 Sun Elevation: 20.5° Altitude: 2693.03 km

The Apennine bench stretches between Archimedes and Montes Apenninus (see the note for LO4-115H1). Rima Bradley is presumably a superficial sign of a circumferential fault caused by tensional stress in this area. It crosses material slumped from the rim of the Imbrium Basin (here called Mons Bradley) and mare surface with seeming indifference, an indication of a deep fracture. Note that the size of the slump blocks seem to fit the corresponding edges of Mons Bradley. This photo and LO4-110H1 are as close to the eastern edge of the Imbrium Basin as is covered in this chapter on the Imbrium Basin Region. The eastern edge and ejecta to the east are covered in the chapter on the Serenitatis Basin Region.

LO4-110H1 Sun Elevation: 20.8° Altitude: 2915.82 km

The Apollo 15 landing site was boldly chosen to be near Rima Hadley and Mons Hadley beyond the eastern edge of this picture. Autolycus has impacted a plains unit in the Imbrium Basin that appears to be rich in KREEP minerals (potassium, rare earth elements, and phosphorus). It is the probable source of a boulder containing KREEP that was sampled (15405) by the Apollo 15 crew. The age of the sample (as it was presumably restructured by the shock of the Autolycus impact) was found to be 1.29 billion years, the youngest age of all dated lunar rock samples. If Autolycus is not the source of this sample, then it may have come from Aristillus. For additional discussion of the Apollo 15 site, see the note for LO4-103H1 in the chapter on the Serenitatis Basin Region (Chapter 9).

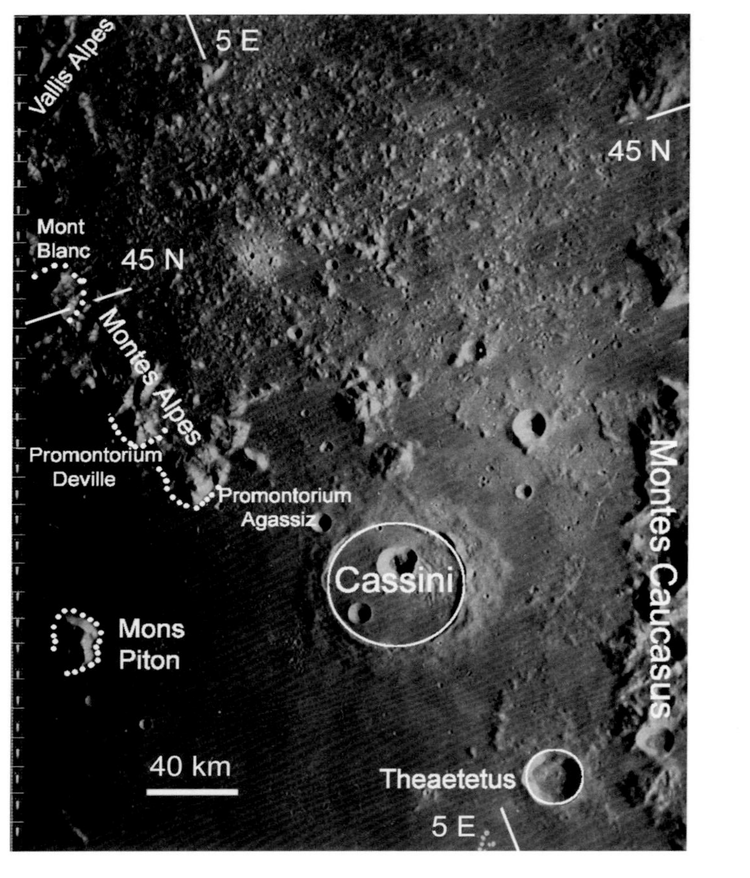

LO4-110H2 Sun Elevation: 20.8° Altitude: 2915.82 km

Beyond Montes Alpes lies the Alpes Formation. Although similar in relationship to the Imbrium Basin, the Alpes Formation has a very different appearance than the Fra Mauro Formation (LO4-109H2). An oblique impact may have induced asymmetry of the ejecta blanket. The hummocky appearance of the Alpes Formation, which extends far beyond Mare Imbrium, is similar to that of the Montes Rook Formation of the Orientale Basin. Cassini and other nearby craters of similar size seem to have suppressed the rim of the Imbrium Basin. The ejecta blanket of Cassini has a molten appearance, as if the impactor landed in mare material that had not fully solidified or that was sufficiently hot that it was liquefied by the energy of impact.

Chapter 8

Nectaris Basin Region

8.1. Overview

Basins, Maria, and Highlands

The Nectaris region (Figure 8.1) has extensive highlands, intensively cratered terrain whose large-scale topography has been established in very ancient times (in the Pre-Nectarian Period).

From the time of the formation of the Nectaris Basin, the highlands have been affected by ejecta (including secondary impactors) from basins such as Nectaris and Imbrium and also from craters such as Tycho, whose rays can be seen streaking across the southeast portion of the photo in Figure 8.1.

Figure 8.1. LO4-096M. This picture is centered on the southeastern part of the central highlands of the near side. The small mare east of the central highlands is Mare Nectaris. It occupies the middle of the Nectaris Basin, which is much larger than the mare. East of Mare Nectaris, at the edge of the picture, is the western portion of Mare Fecunditatis. North of Mare Nectaris and Mare Fecunditatis is Mare Tranquillitatis in the Crisium Basin Region.

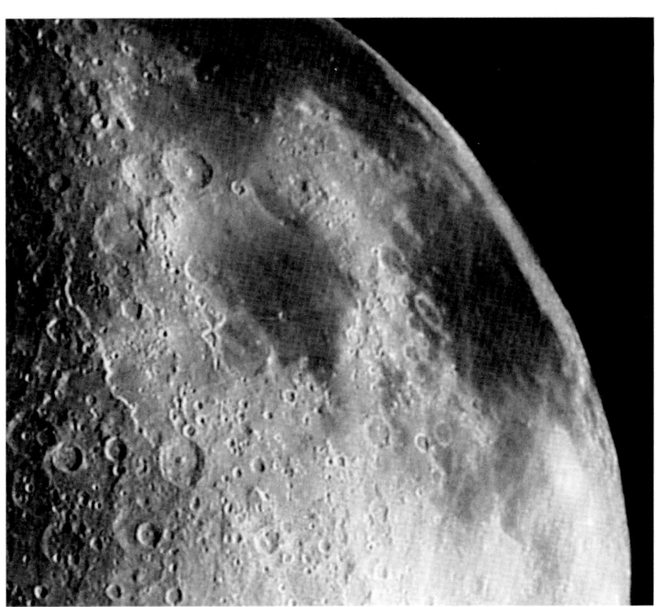

Figure 8.2. Part of LO4-083M. The arc of cliffs outlined by the morning sun is the Rupes Altai (the Altai scarp). It outlines the major ring of the Nectaris Basin.

An especially interesting part of this region is the area where the Nectaris Basin intersects with the Fecunditatis Basin to the west and the Tranquillitatis Basin to the north.

Figure 8.2 illustrates the size of the Nectaris Basin, much larger than Mare Nectaris.

Figure 8.3 shows three of the rings of the Nectaris Basin. These rings have been drawn so that the second ring from the center is the square root of 2 greater than the inner ring and the third ring from the center is the square root of 2 greater than the second ring. With this constraint, the size and center of the set of rings have been adjusted to match topographic highs.

Like other major basins, Nectaris has thrown a heavy blanket of ejecta that forms radial ridges and troughs outside of its major ring, the rim. In the case of Nectaris, this ejecta blanket has been degraded by further basin impacts to its north and east. However, the formation is clearly seen to the south (Figure 8.4).

Figure 8.3. LO4-084M. The ridge marked by the innermost ring, whose eastern sector passes over the Montes Pyrenaeus (LO4-065H2), bounds the central area of mare flooding (partly blocked or covered by ejecta from later craters). The outer ring is the main ring; aligned with Rupes Altai. Troughs between the rings are partly filled with mare.

Apollo Landing

Apollo Mission 16 landed in the midst of the south-central highlands (LO4-089H3), near the boundary between the Cayley and Descartes Formations. This was the first mission with a Lunar Roving Vehicle, which permitted the astronauts to reach both formations. Before the mission, these terrain types were thought to be volcanic. In the course of the mission it became clear to both the astronauts, who were particularly well trained in geology, and the support geologists at Mission Control that ejecta from Imbrium had covered both areas. The astronauts revised their sample collection strategy to improve characterization of the impact-dominated terrain. Analysis of the samples helped establish the age of the Imbrium impact event (3.85 billion years). In addition, some samples from the Descartes Formation, probably excavated from an underlying layer of ejecta from the Nectaris Basin, were dated at 3.92 billion years.

Figure 8.4. LO4-052M. These deep troughs and ridges south of the Nectaris Basin were produced by ejecta from that basin. The pattern is similar to the Fra Mauro Formation of the Imbrium Basin, to the Hevelius Formation of the Orientale Basin, and to the secondary-impact crater fields of those basins.

8.2. High-Resolution Images

Table 8.1 shows the high-resolution images of the Nectaris Basin Region in schematic form.

The following pages show the high-resolution subframes from south to north and west to east. That is, they are in the order LO4-100H1, LO4-100H2, LO4-100H3, LO4-101H1, LO4-101H2, LO4-101H3, LO4-095H1... LO4-053H3.

Photos LO-095H1, LO-083H1, and LO4-071H1 are redundant and have not been printed, although the cleaned images are in the enclosed CD.

LO096H and LO077H are printed as full frame.

Latitude Range	Photo Number										
0–27 N	109	102	097	090	085	078	073	66	061	054	
0–27 S	108	101	096	089	084	077	072	65	060	053	046
27 S–56 S	107	100	095	088	083	076	071	64	059	052	
56 S–90 S	094		082		070		058		044		
Longitude at Equator	3 W	4 E	10 E	10 E	16 E	24 E	30 E	38 E	49 E	57 E	63 E

Table 8.1. The cells shown in white represent the high-resolution photos of the Nectaris Basin Region (LO4-XXX H1, -H2, and -H3, where XXX is the Photo Number). The Humorum Basin Region is to the west, the Imbrium Basin Region is to the northwest, the Serenitatis Basin Region is to the north, the Eastern Basins Region is to the east, and the South Polar Region is to the south.

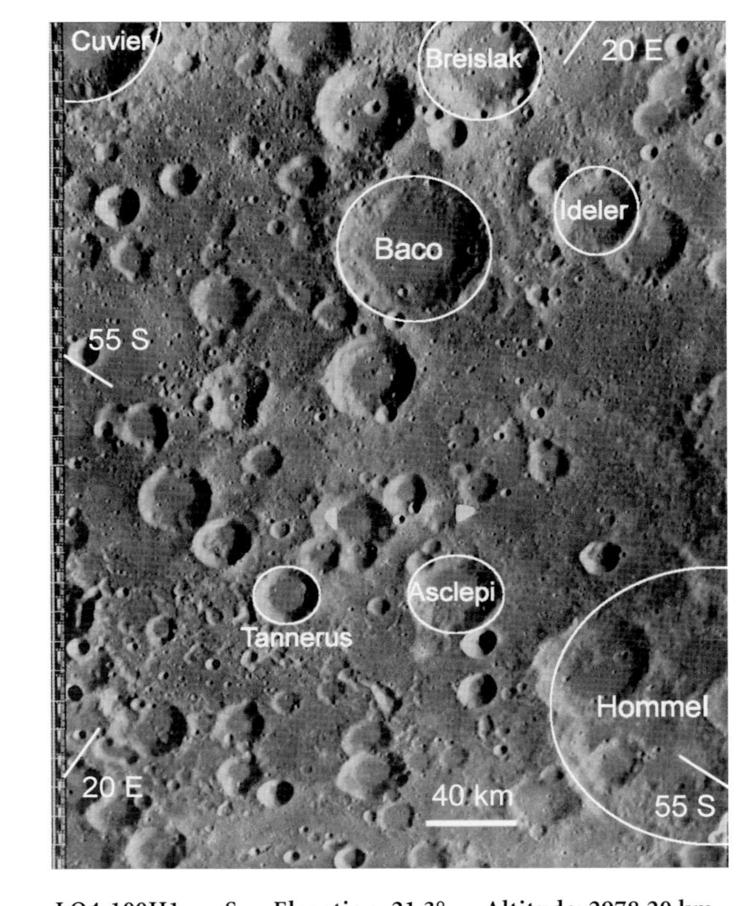

LO4-100H1 Sun Elevation: 21.3° Altitude: 2978.20 km

In this highlands region, the relative ages of craters are judged by the amount of degradation of their rims and the density of small craters on their floors. Although the large craters here are assigned to the Pre-Nectarian Period, Baco seems much fresher than Hommel, but Tannerus may be younger than Baco. Fresh-looking craters in the 5-km range in this area are believed to be Orientale Basin secondaries, even though the center of that basin is fully 90° around the Moon (about 1500 km to the west) away. Light rays in the east-west direction and the strings of craters south of Tannerus are from Tycho.

Stofler
Faraday
Maurolycus
Licetus
Barocius
Clairaut
10 E
45 S
40 km

LO4-100H2 Sun Elevation: 21.3° Altitude: 2978.20 km

The cluster of craters near the intersection of Stofler and Faraday is aligned in the direction of the Imbrium Basin, whose rim is about 2000 km away. The clusters along the 10° east meridian are aligned with the direction of the South Pole–Aiken Basin, about 1300 km away, but may come from the younger Serenitatis Basin, about 1800 km to the north-northeast.

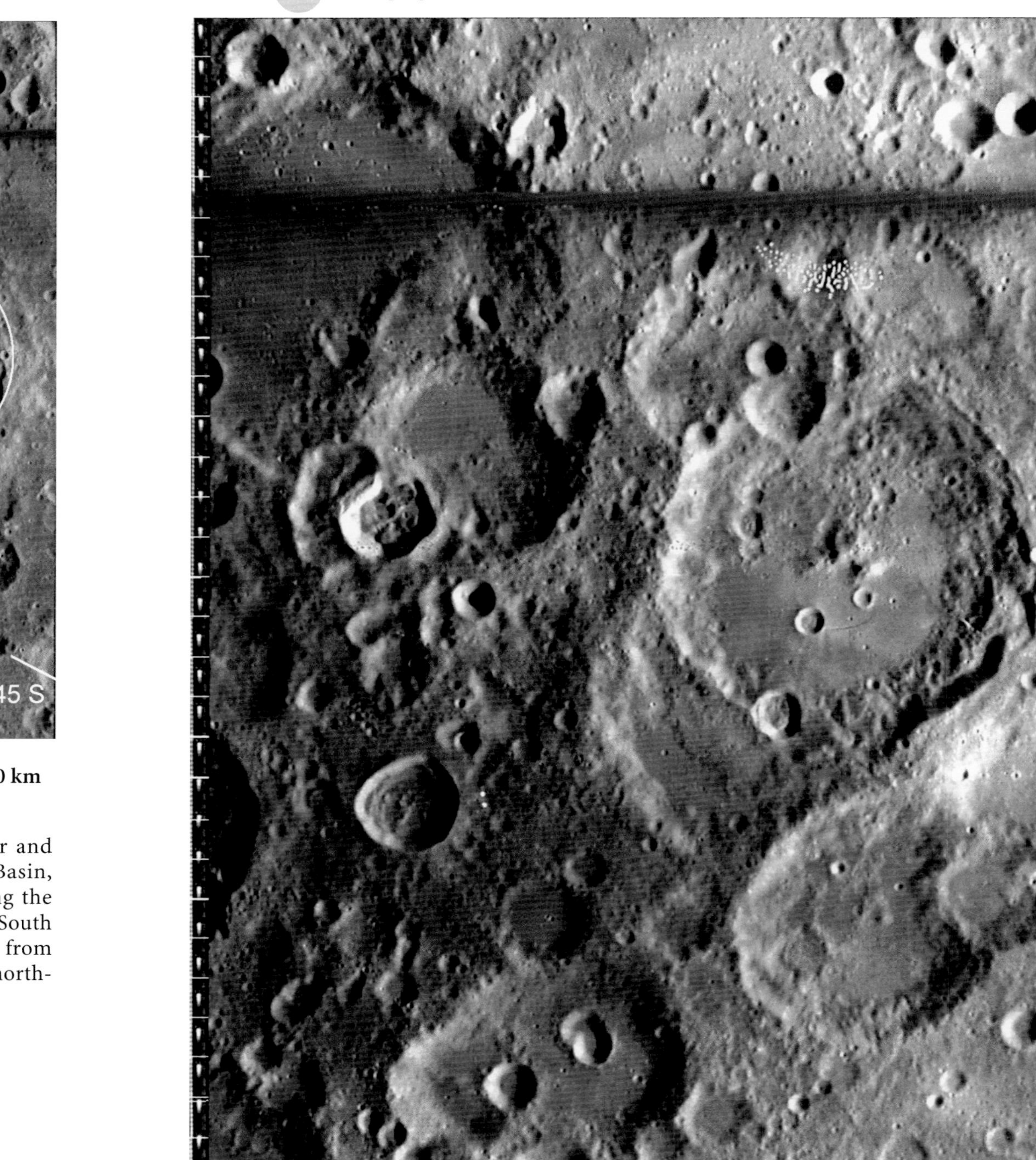

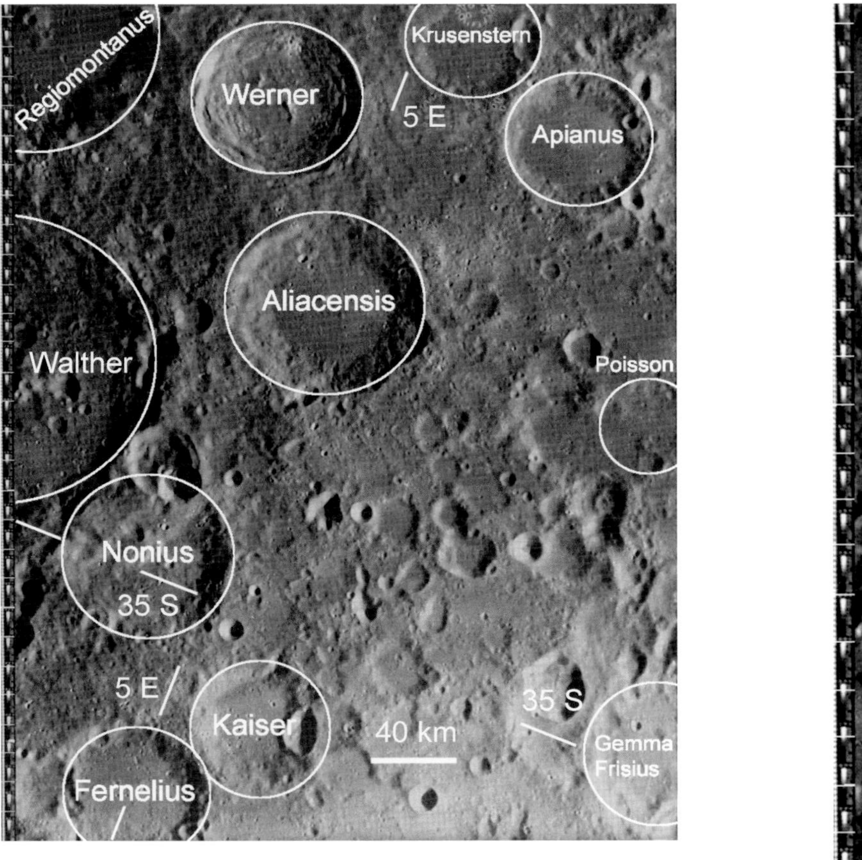

LO4-100H3 Sun Elevation: 21.3° Altitude: 2978.20 km

The neighboring craters Werner and Aliacensis reveal much of the mechanics of crater formation and subsequent erosion in this highland area. The ejecta from Werner has been deposited on the rim of Aliacensis; the force of the deposition, perhaps aided by the shock wave of the impact, has pushed about 20 km of the rim of Aliacensis onto its floor. Because these two craters are similar in size and have impacted similar terrain, they probably had similar topography when they were formed. The older Aliacensis has had the texture of its ejecta blanket eroded away by subsequent impacts. Many more small craters (less than 5 km) can be seen in the area of its former ejecta blanket than in that of Werner.

LO4-101H1 Sun Elevation: 21.1° Altitude: 2720.44 km

Radial ejecta features from Eratosthenian crater Werner can be seen up to two radii from its rim. This ejecta blanket is superimposed on irregular craters, ridges, and valleys that are radial to the Imbrium Basin, part of the Fra Mauro Formation. Secondary craters from the Imbrium Basin also pervade this area, except where they are covered by the Werner ejecta blanket. Beneath the Fra Mauro Formation, Playfair, Apianus, and Krusenstern have redistributed the rim material from an older unnamed 115-km crater. The crater cluster near the right edge of this photo is from the Nectaris Basin.

LO4-101H2 **Sun Elevation: 21.1°** **Altitude: 2720.44 km**

Deep grooves formed by chains of craters illustrate this area of the Fra Mauro Formation. Traces of such grooves remain on the floor of Albategnius, but appear to have been buried by a relatively smooth layer of plains material. In some areas such material may have been deposited by basin ejecta, but the abrupt transition here between grooves and plains suggests a volcanic origin from within the crater. Vogel and its companion to the north seem to be secondaries from Imbrium associated with a large groove. Airy, on the other hand, has been nearly buried by Imbrium ejecta.

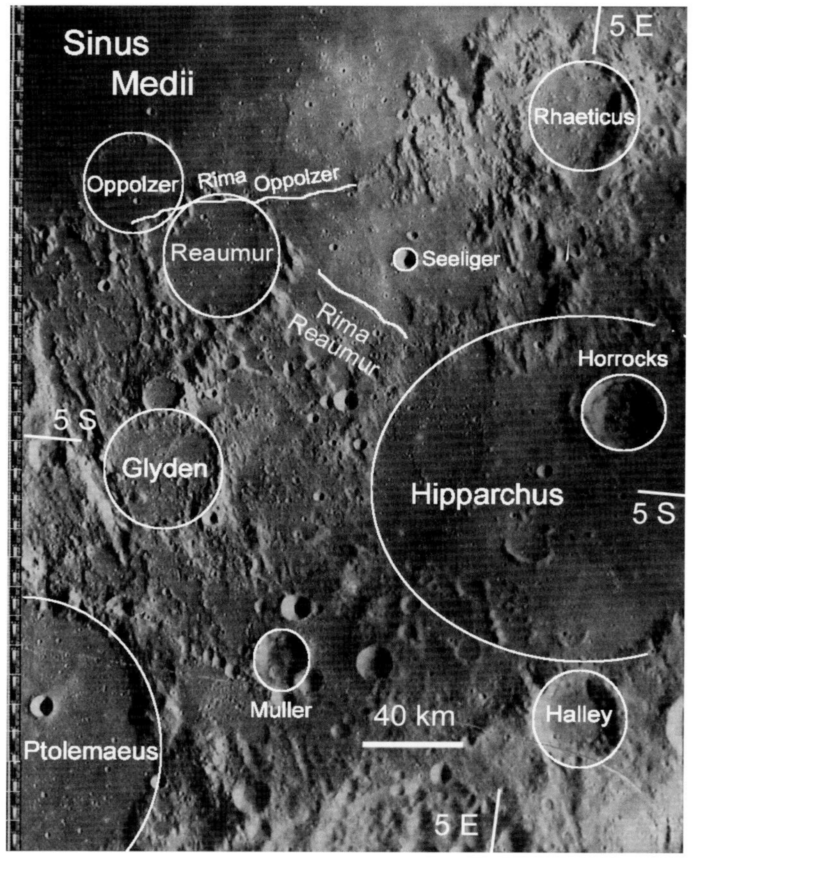

LO4-101H3 Sun Elevation: 21.1° Altitude: 2720.44 km

This area is dominated by the Fra Mauro Formation, emerging from the mare surface of Sinus Medii and the floor of the Hipparchus crater. The floor of Ptolemaeus, like that of Albetegnius (LO4-101H2), is covered with light plains material that has obscured the Fra Mauro Formation.

LO4-095H2 **Sun Elevation: 21.6°** **Altitude: 2975.51 km**

The northwest floors of Maurolycus and Barocius are covered with heavy deposits of ejecta thought to be the margin of the outer deposits from the Imbrium Basin. Some chains of small secondaries from the north-northeast are from Nectaris, whose rim is about 1.5 radii away. Fresher clusters of craters are probably from the Imbrium Basin, even though this area is two basin radii away from the Imbrium rim.

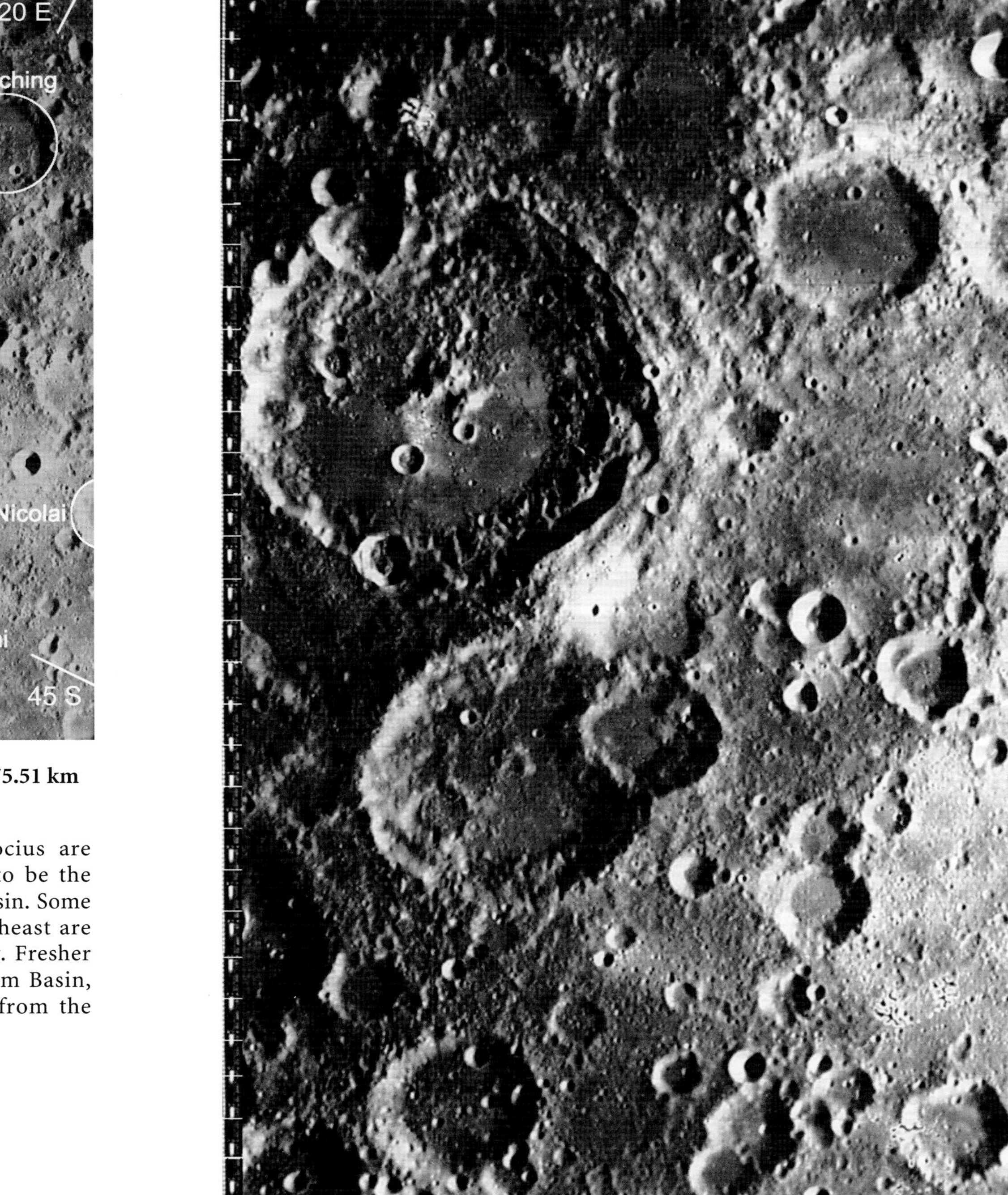

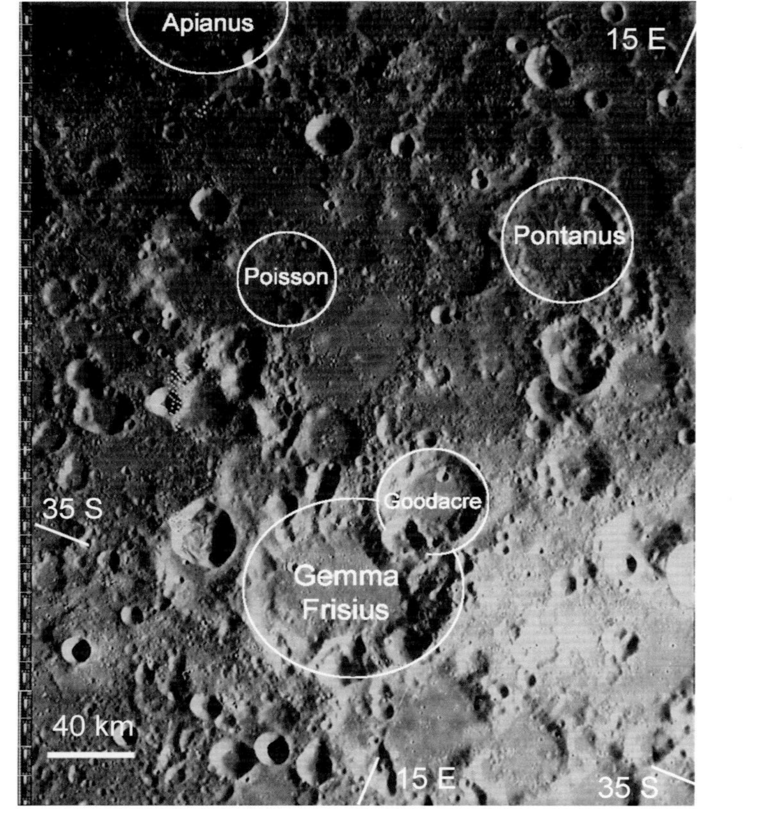

LO4-095H3 Sun Elevation: 21.6° Altitude: 2975.51 km

This interesting picture shows crossing patterns of crater chains and striations from Imbrium to the north-northwest and Nectaris to the east-northeast. The floors of Gemma Frisius, Goodacre, and the craters to the southeast show resurfacing by light plains materials.

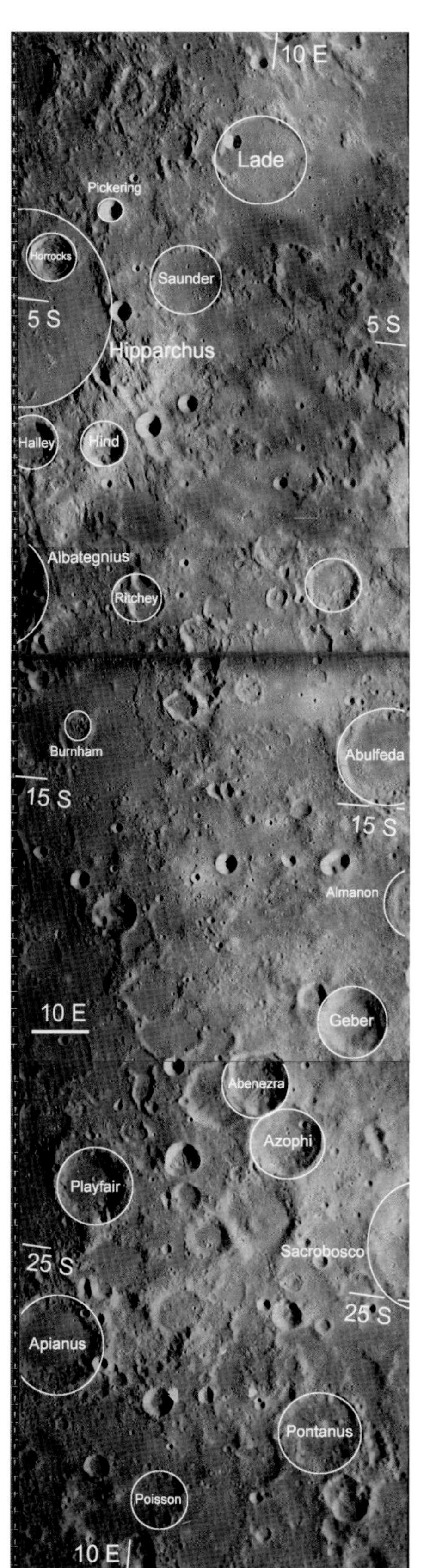

LO4-096H1
Sun Elevation: 21.8°
Altitude: 2722.26 km

The fresh, sharp craters in the cluster southeast of Hipparchus are all bright in Clementine albedo data. The rays and bright ejecta come mostly from the crater to the northeast of Hind. The mottled dark areas are concentrated in low-lying terrain, with the exception of a dark plateau south of Lade with sharp edges that may be a flow. Clementine gravity data show a positive anomaly here that is similar to those under mare areas, but weaker. Could there have been an early stage of mare formation here that did not go to completion?

Burnham, in the midst of light plains, has a roughly circular raised rim, but the floor is not depressed. Further, the rim is breached with grooves that suggest flow to the outside. Between Burnham and Abulfeda there is another irregular rim, surrounded by a flow with a lobate edge. These features may have been produced by molten ejecta from the Nectaris Basin.

The ridge and the crater chain near Playfair are from the direction of the Nectaris Basin to the east-northeast, as are many other striations in this area. Many of the craters show signs of being covered by Nectaris ejecta and so are considered Pre-Nectarian. However, Playfair is relatively unmarked, so it has been classified as Nectarian.

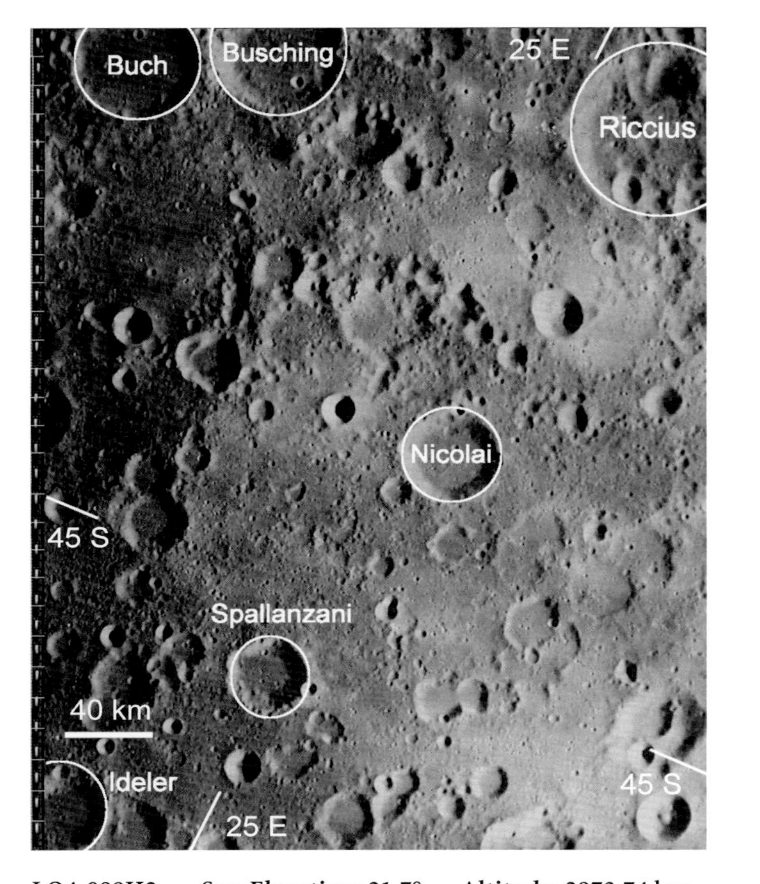

LO4-088H2 Sun Elevation: 21.7° Altitude: 2973.74 km

Note how flat this area seems, relative to surrounding areas. It would be the northeast quadrant of the floor of a possible large Pre-Nectarian basin named Mutus-Vlaq. The basin floor would have obliterated any earlier cratering and provided the basic flatness. Then, secondaries and other ejecta from the Nectaris Basin would have arrived, followed by secondaries from Imbrium.

LO4-088H3 Sun Elevation: 21.7° Altitude: 2973.74 km

Pons, Wilkins, Zagut, Lindemau, and Rabbi Levi lie on a massive highlands ridge that has the highest elevation on the entire near side, about 4000 m above the floor of the nearby Nectaris Basin. The highly eroded craters Zagut and Rabbi Levi are Pre-Nectarian in age and have been covered in ejecta from the Nectaris Basin. Grooves and ridges from Nectaris can be seen in the northeast rim of Zagut and on the floor of Rabbi Levi. Chains of secondaries from Imbrium lie between Pontanus and Wilkins.

LO4-089H1 Sun Elevation: 21.9° Altitude: 2724.41 km

Clementine elevation data show a depression that may mark a very old crater about 225 km in diameter, centered near 17° east, 26° south (dashed circle). The floor of this possible crater would be about 1500 m below the rim. The area to the east is heavily covered with the Nectaris ejecta blanket.

LO4-089H2 **Sun Elevation: 21.9°** **Altitude: 2724.41 km**

This unusual terrain has been called the Descartes Formation; the area is heavily covered by ejecta from the Nectaris Basin, whose main ring is only one basin radius to the east. Descartes itself has been nearly buried by Nectaris ejecta. Abulfeda, however, shows no sign of ejecta, and must be of the Nectarian Period or younger. Catena Abulfeda is not quite radial to the center of the Imbrium Basin, although it aligns with the southwestern rim of that basin. The catena's impactors could be a set of primaries from a disrupted comet like Shoemaker-Levy 9.

LO4-089H3 Sun Elevation: 21.9° Altitude: 2724.41 km

Apollo 16 landed in this site to explore both the light plains (Cayley Formation) and the nearby hummocky Descartes Formation. A major objective was to confirm that the two formations were volcanic. Instead, analysis of the rock samples showed that the formations were ejecta deposits. Both Descartes and Cayley Formations are probably local aspects of ejecta from the Imbrium Basin. Most of the samples were of course of the upper layer, from the younger Imbrium Basin, but some samples may be from Nectaris. If so, the ages of these samples establish the age of the Nectaris Basin (and thus the boundary between the Pre-Nectarian and Nectarian Periods) as 3.92 aeons (billions of years before the present).

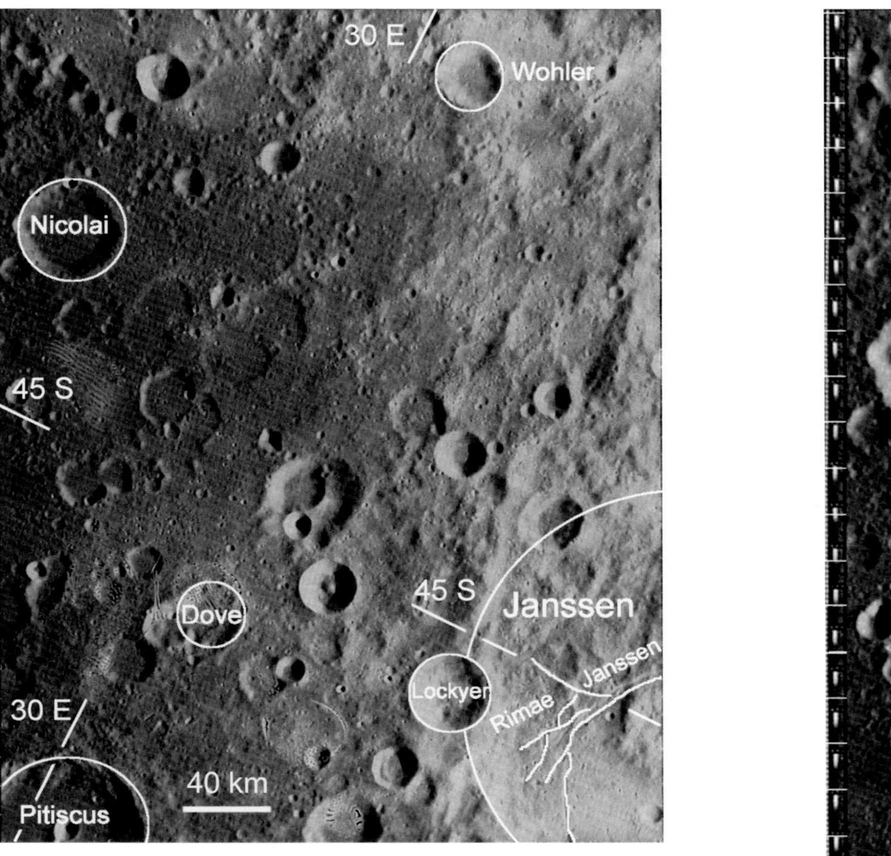

LO4-083H2 **Sun Elevation: 22.8°** **Altitude: 2971.86 km**

The region covered by heavy, hummocky deposits in the vicinity of Janssen is the type area of the Janssen Formation, an ejecta blanket from the Nectaris Basin. It is comparable to the Inner Hevelius Formation of the Orientale Basin and the Fra Mauro Formation of the Imbrium Basin. This formation has been obscured by Imbrium ejecta to the west of the basin (note the chain of secondaries in the upper left corner). The floor and ejecta of Janssen (LO4-076H2) are covered with Nectaris ejecta.

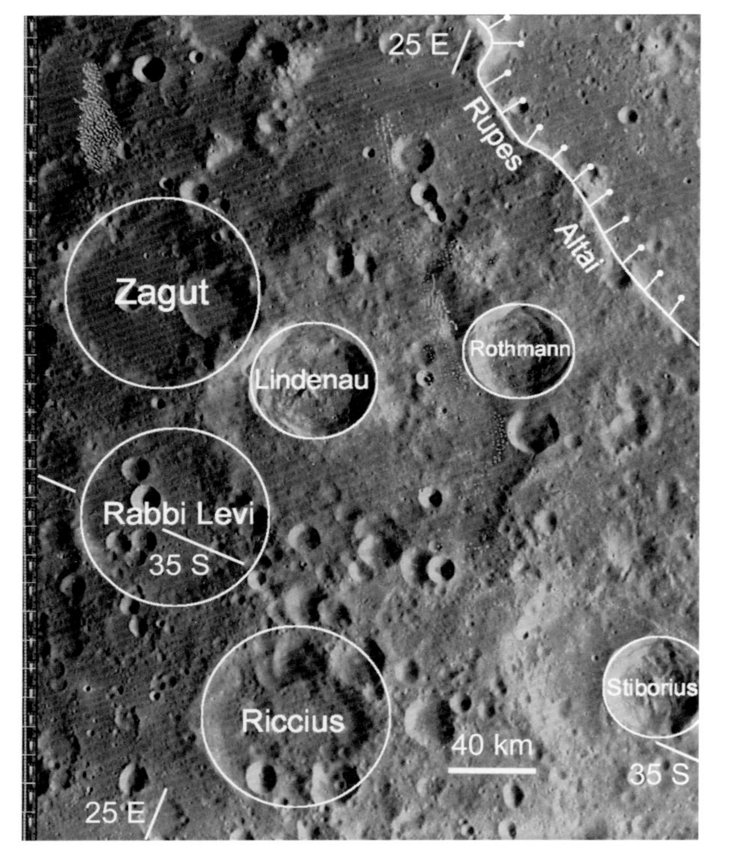

LO4-083H3 Sun Elevation: 22.8° Altitude: 2971.86 km

Rupes Altai (LO4-084H1) marks the main ring of the Nectaris Basin. This area illustrates the methods of assigning ages to large craters. Zagut, Rabbi Levi, and Riccius have been completely or partly covered by ejecta from Nectaris and are therefore assigned to the Pre-Nectarian Period. Lindenau and Stiborius are punched into the Nectarian ejecta and in addition are free of Imbrium secondaries that are common here. The sharpness of their rims, central peaks, and ejecta are similar to those of Schluter, which overlies Orientale (LO4-181H3); therefore, they have been assigned to the Late Imbrian Epoch. Rothmann, sharper yet and relatively uncratered, is considered Eratosthenian.

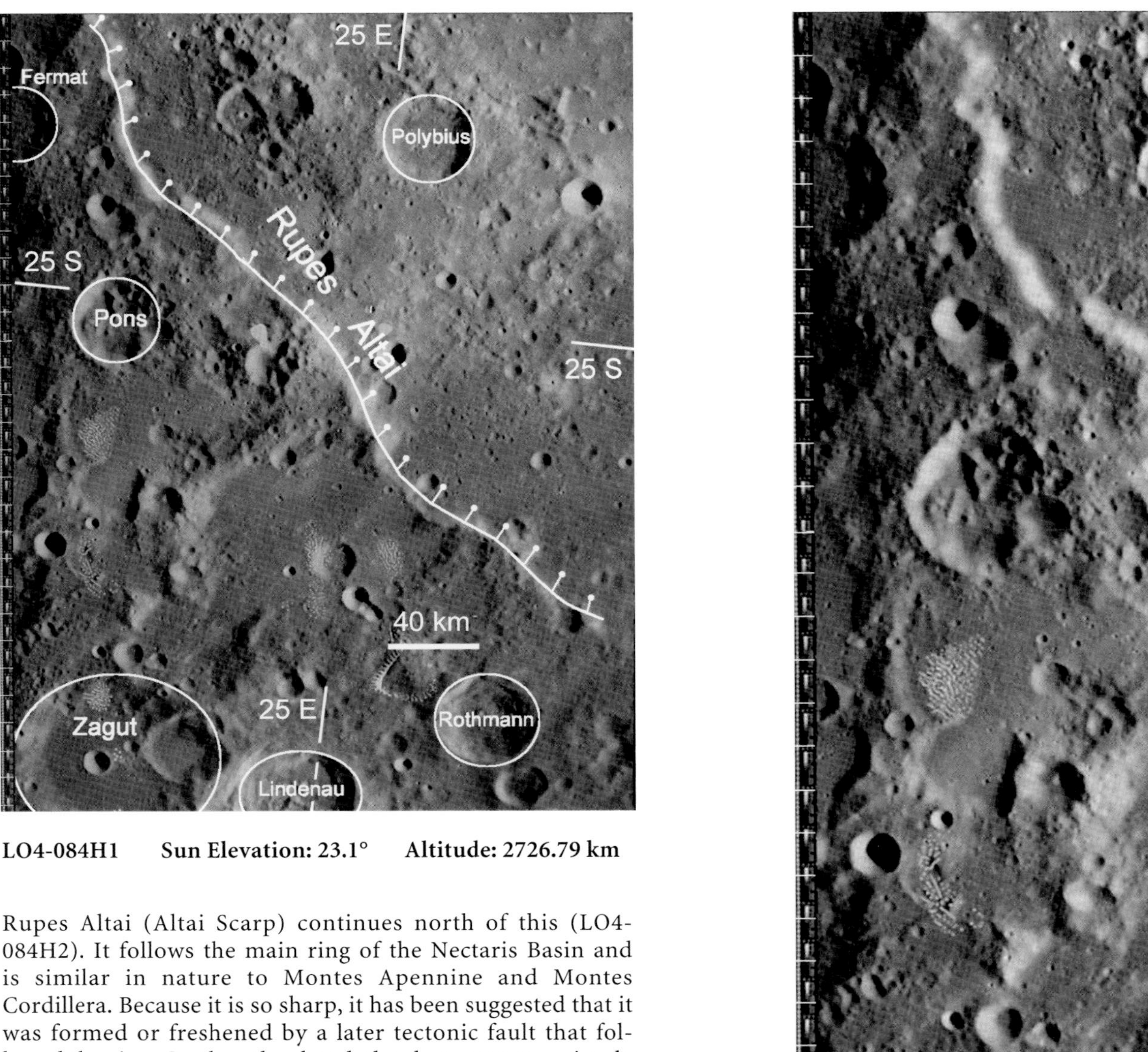

LO4-084H1 **Sun Elevation: 23.1°** **Altitude: 2726.79 km**

Rupes Altai (Altai Scarp) continues north of this (LO4-084H2). It follows the main ring of the Nectaris Basin and is similar in nature to Montes Apennine and Montes Cordillera. Because it is so sharp, it has been suggested that it was formed or freshened by a later tectonic fault that followed the ring. On the other hand, the sharpness may simply have been maintained by slumping of the wall.

Kant
Ibn-Rushd
25 E
Theophilus
Cyrillus
15 S
Tacitus
15 S
Catharina
40 km
25 E

LO4-084H2 Sun Elevation: 23.1° Altitude: 2726.79 km

The northern end of Rupes Altai is in the lower left-hand corner of this photo. First Cyrillus and then Theophilus have landed in the area between the main ring and one of the inner rings of the Nectaris Basin. Cyrillus is considered to be of the Nectarian Period and Theophilus, which appears to have pushed rim material of Cyrillus onto its floor, is free of Imbrium striations, and has a low incidence of craters on its floor, has been assigned to the Eratosthenian Period. Theophilus is believed to have thrown some fragments of Mare Nectaris to the highland landing site of Apollo 16 (LO4-89H3). Samples from such rocks were aged at 3.74 aeons (billions of years ago). The Nectarian impact would be earlier, estimated at 3.92 aeons.

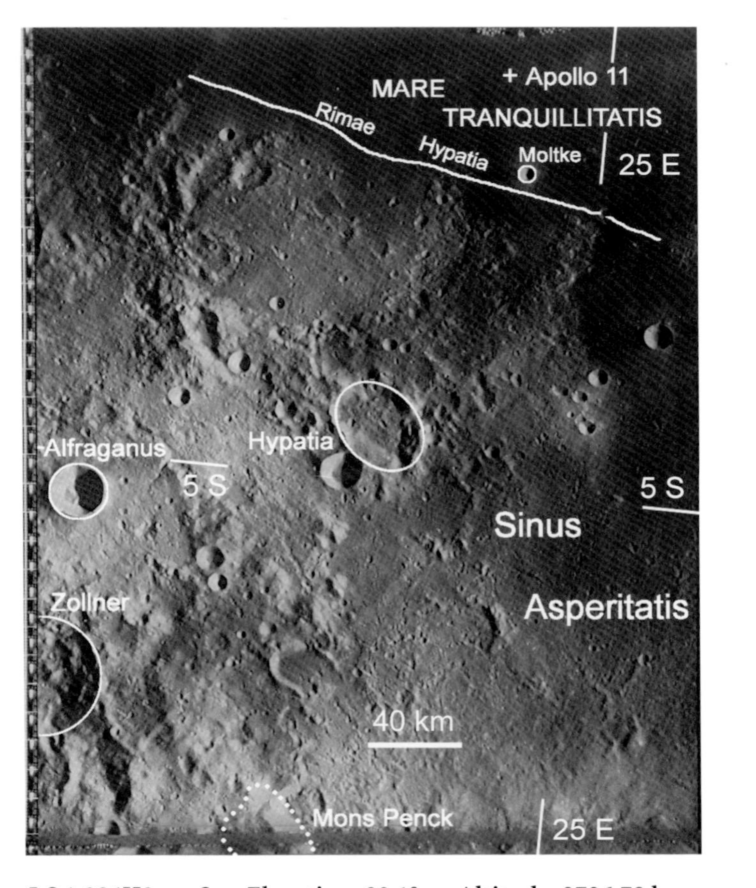

LO4-084H3 Sun Elevation: 23.1° Altitude: 2726.79 km

Mare Tranquillitatis and the Apollo 11 landing site are discussed in the chapter on the Serenitatis Basin Region (Chapter 9). The Kant Plateau, from Mons Penck to east of Zolner (LO4-089H3), rises 2 km above the surrounding plains. Uplift from a triple intersection of rings from the Nectaris, Imbrium, and Tranquillitatis basins may have raised it. Sinus Asperitatis has a small gravity anomaly (mascon) that indicates that this bay has its own source of lava. Much of Sinus Asperitatis has been covered with Theophilus ejecta, the lighter material in the lower right corner of the photo. Beyond the Theophilus ejecta is a heavy ejecta blanket from the Imbrium Basin, whose rim is about 1.5 basin radii away.

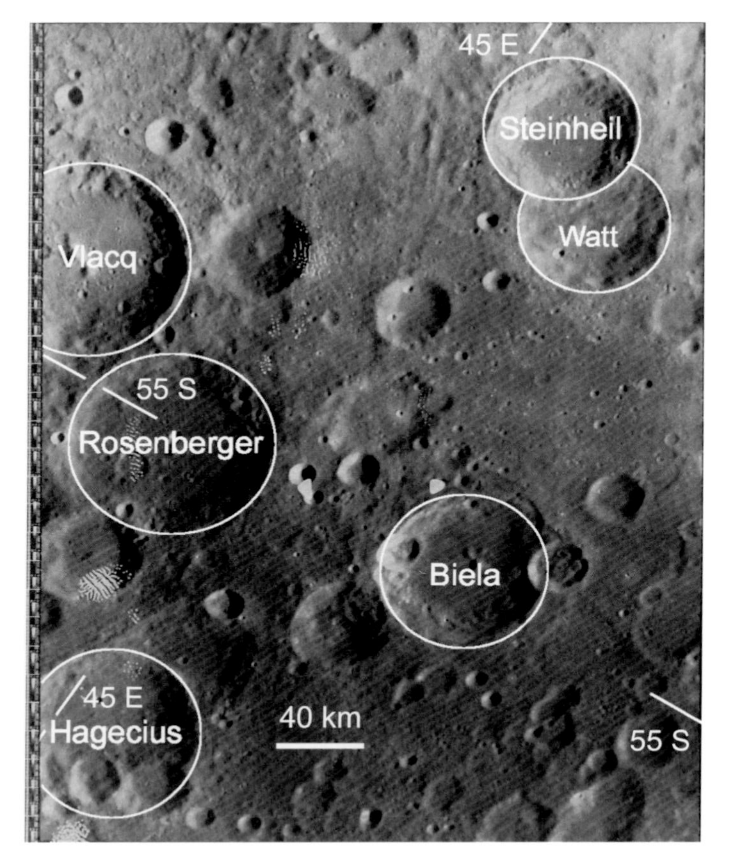

LO4-076H1 Sun Elevation: 22.1° Altitude: 2971.56 km

Watt is clearly of the Pre-Nectarian Period because the striations of Nectarian ejecta traverse its wall and floor. Steinheil and Biela have similar levels of degradation and are therefore likely to be of the same period. Because Steinheil is in an area of heavy Nectarian ejecta but is free of it, they must both be Nectarian or younger. The craters that have impacted Biela are Imbrium secondaries, so both craters must be older than the Early Imbrian Epoch. Therefore, both Biela and Steinheil are classified as of the Nectarian Period. Light plains are north and east of Biela.

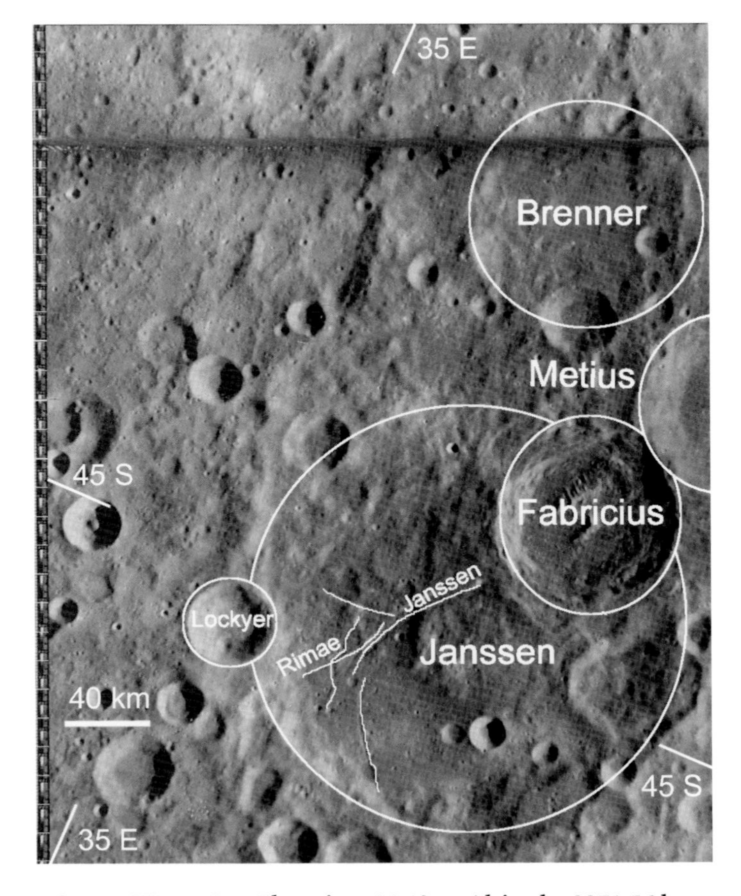

LO4-076H2 Sun Elevation: 22.1° Altitude: 2971.56 km

Fabricius and Janssen are a study in contrasting age. Janssen, of the Pre-Nectarian Period, has its rim obscured by ejecta from the Nectaris Basin and both rim and floor are impacted by Imbrium secondaries. Fabricius, on the other hand, shows sharp texture in its ejecta, central peak, and terraced wall. The Fabricius impact could be as late as the Eratosthenian Period. Rimae Janssen could be the surface manifestation of faults caused by stress from the heavy Nectarian ejecta or could be faults induced by the impact of Fabricius. An even older, more degraded, unnamed crater underlies Janssen, with its center near the northeastern rim of Janssen.

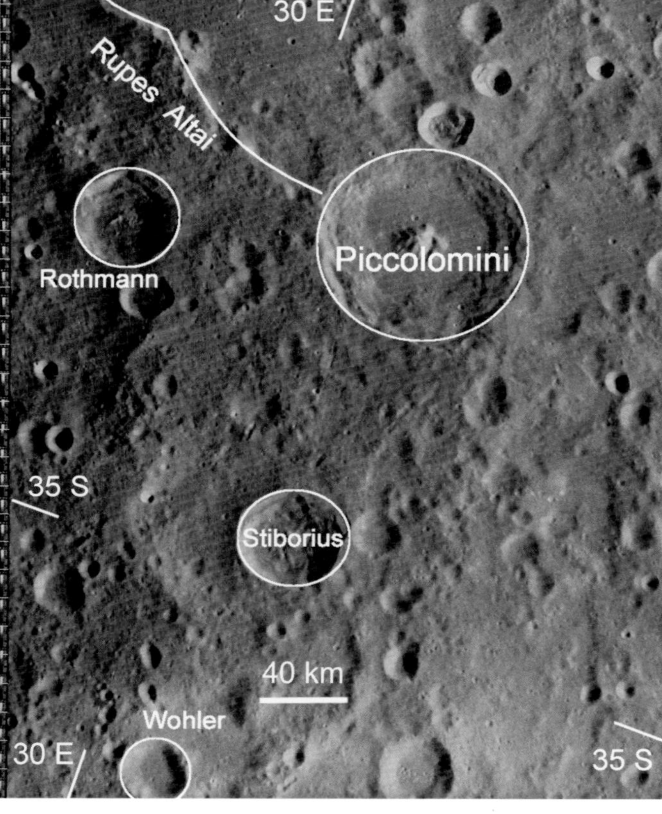

LO4-076H3 Sun Elevation: 22.1° Altitude: 2971.56 km

Piccolomini has obliterated the southeastern end of Rupes Altai The Piccolomini ejecta blanket has spread into craters thought to be secondaries of Imbrium (to the northwest). Piccolomini is dated in the Late Imbrian Epoch, as is Stiborius. Rothmann, which has sharper detail, is dated in the Eratosthenian Period.

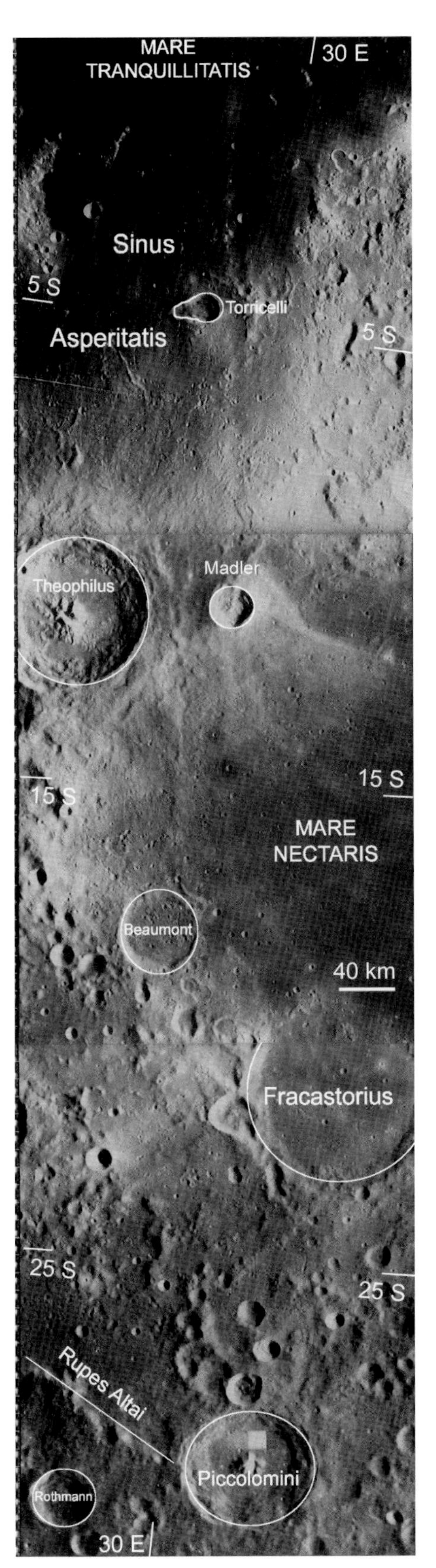

LO4-077H
Sun Elevation: 22.9°
Altitude: 2729.55 km

Torricelli may have been formed by a pair of impactors. The rim of a flooded 88-km crater surrounding Torricelli can be seen just above the mare surface. There is a positive gravity anomaly at Sinus Asperitatis, suggesting that it had its own source of lava. The last (surface) flow of southern Mare Tranquillitatis did not enter the sinus; its edge can be seen at the northeast sector of Sinus Asperitatis.

The northwestern part of Mare Nectaris has been covered with ejecta from Theophilus, covered in turn with rays from Madler. A couple of small (about 2 km) craters have penetrated the Theophilus ejecta to produce their own dark halos of mare material. It appears that a landslide from a hill to the north has cascaded over the rim of Madler onto its floor, as well as over the ejecta blanket of Madler, exposing very light material.

The next ring inside of Rupes Altai can be seen arcing across this image, passing halfway between Piccolomini and Fracastorius. The next inner ring is tangent to the southern rim of Fracastorius. Imbrium secondaries, crater chains, and other ejecta cover this part of the floor of the Nectaris Basin.

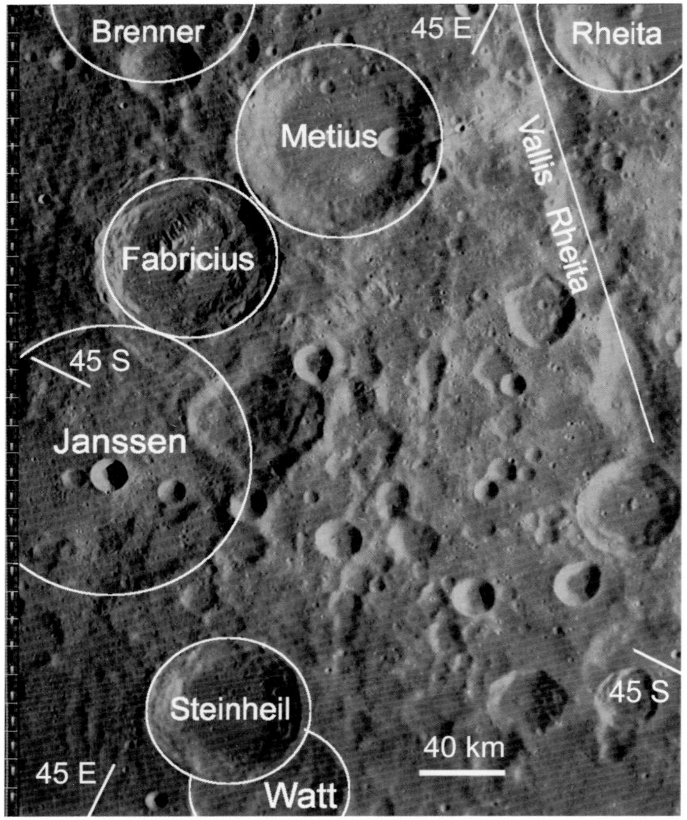

LO4-071H2 Sun Elevation: 23.5° Altitude: 29.72.29 km

Vallis Rheita and crater chains to its west are radiating from the Nectaris Basin. This area is the best preserved part of the Janssen Formation, the ejecta blanket from the Nectaris Basin. The visibility of Janssen itself, although clearly covered by the deposit from Nectaris, indicates that the blanket is less than 1 km thick. The Steinheil impact or some other activity seems to have thrown an unusual layered ejecta onto the southeast floor of Janssen.

LO4-071H3 Sun Elevation: 23.5° Altitude: 29.72.29 km

Neander has impacted the southern sector of the main ring of the Nectaris Basin. It appears that lava has flowed from the outside of the main ring south to the crater Brenner. The flat-floored valley has subsequently been covered with ejecta, probably from Piccolomini or Imbrium. Similar flows from a main ring outward (but with narrower valleys) can be seen at Rima Plato and Rima Archytas in the Imbrium Basin Region (LO-122H3).

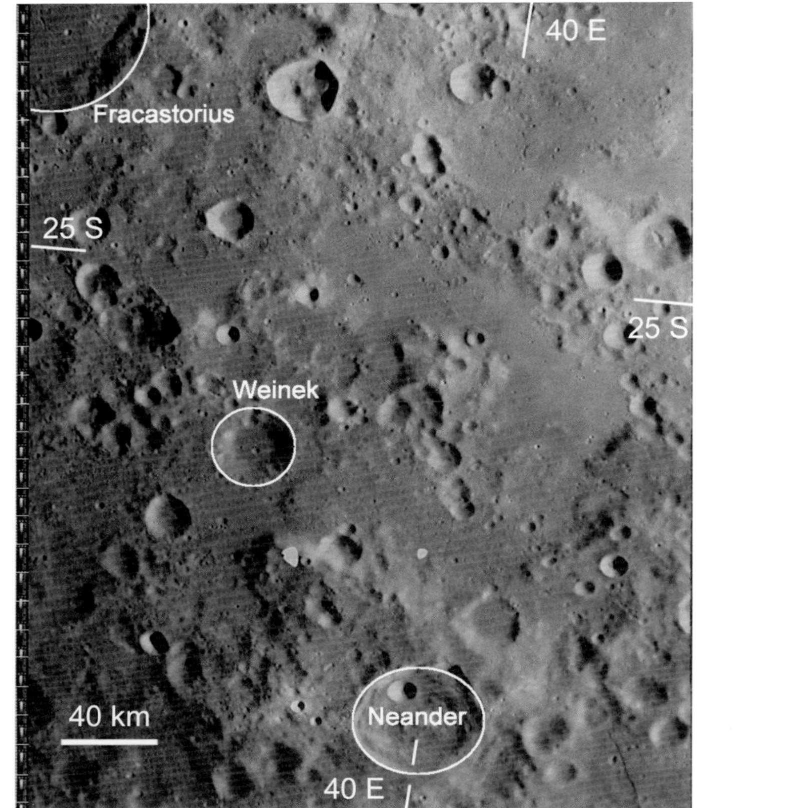

LO4-072H1 **Sun Elevation: 24.3°** **Altitude: 2742.43 km**

This southern sector of the Nectaris Basin covers three rings. The inner ring grazes the southern rim of Fracastorius, the next ring (somewhat intermittent) passes to the north of Weinek, and the main ring passes through Neander. Ejecta from Fracastorius and Imbrium obscure whether the floor of the basin has been flooded with mare and then covered, but the flatness suggests that it has been flooded. The smoothness and flatness of the plains area to the northeast suggest that unconsolidated material has been strongly shaken down. This is only one sign that parts of the floor of the Nectaris Basin have subsided in one or more sudden episodes.

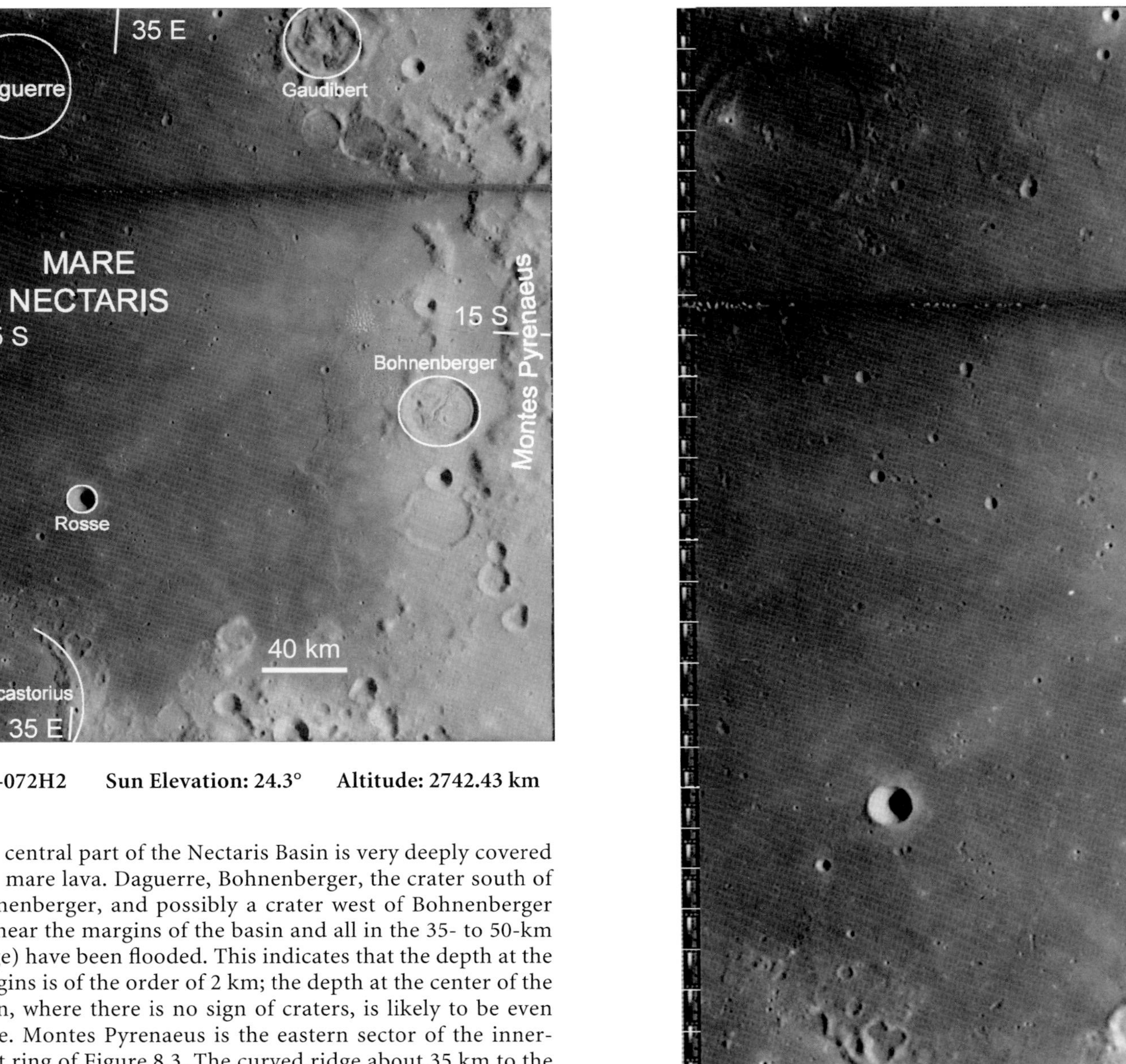

LO4-072H2　　Sun Elevation: 24.3°　　Altitude: 2742.43 km

This central part of the Nectaris Basin is very deeply covered with mare lava. Daguerre, Bohnenberger, the crater south of Bohnenberger, and possibly a crater west of Bohnenberger (all near the margins of the basin and all in the 35- to 50-km range) have been flooded. This indicates that the depth at the margins is of the order of 2 km; the depth at the center of the basin, where there is no sign of craters, is likely to be even more. Montes Pyrenaeus is the eastern sector of the innermost ring of Figure 8.3. The curved ridge about 35 km to the west of Montes Pyrenaeus is likely to be simply composed of vestigial sectors of submerged crater rims. The floor of Gaudibert is an extreme example of uplift in craters near the margins of maria.

LO4-072H3 Sun Elevation: 24.3° Altitude: 2742.43 km

Imbrium ejecta has covered all but the larger features of the Nectaris Basin, such as Vallis Capella, Rimae Gutenberg, and the main ring. The plateau north of Leakey has been raised at the intersection of rings of three basins (LO4-065H3). Vallis Capella was probably formed by ejecta from Nectaris. Rimae Gutenberg appear to mark radial fractures, often found near mare edges (both Mare Nectaris and Mare Fecunditatis are nearby). Censorinus is a small 4.5-km Copernican crater whose much larger ray system is about 30 km across. This crater and its ray pattern have been photographed at very high resolution (LO5-063H1, -H2, and -H3; 2-m resolution) and display detailed mechanisms of ejecta and ray formation. Gutenberg has produced a large ejecta blanket and possibly smoothed out preexisting Imbrium ejecta by its impact.

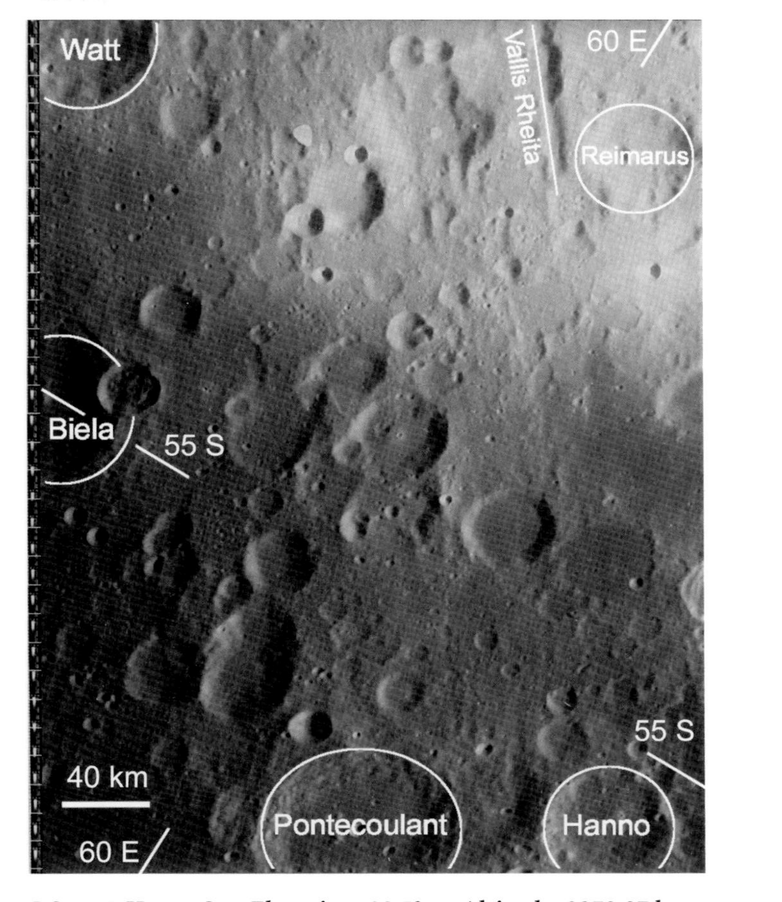

LO4-064H1 Sun Elevation: 22.5° Altitude: 2972.87 km

The heavy Nectaris ejecta blanket formed Vallis Rheita and covers Reimanus thoroughly. The ejecta thins out to the south and southwest, becoming patchy toward Pontecoulant and Hanno. The flooded craters between Vallis Rheita and crater Hanno are part of the Australe Basin. Ejecta from that basin probably underlies the region of this photo but has been overlain with Nectaris ejecta. The high brightness in the upper right corner of this image is an artifact.

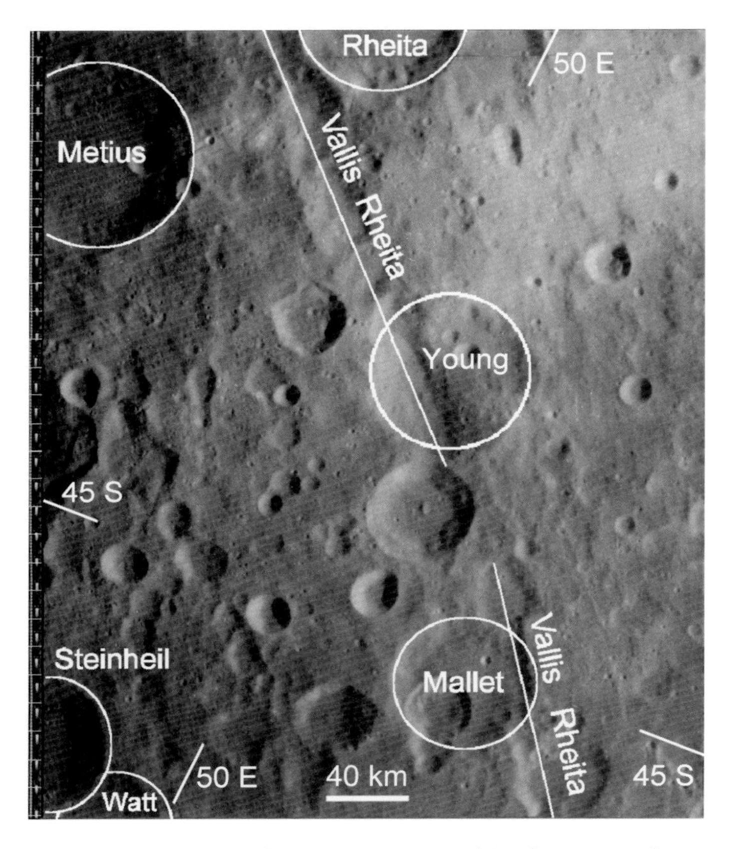

LO4-064H2 **Sun Elevation: 22.5°** **Altitude: 2972.87 km**

This area is covered with the Janssen Formation, ejecta from Nectaris. Vallis Rheita, formed by the impact of a chain of secondary impactors from the Nectaris Basin, has excavated two sectors of the rim and the floor of Pre-Nectarian crater Young, but has been interrupted in turn by the younger crater (Young D). Young D has plowed material from the wall of Vallis Rheita across its floor and over the other side of the valley. The high brightness in the upper right corner of this image is an artifact.

LO4-064H3 Sun Elevation: 22.5° Altitude: 2972.87 km

The much-cratered main ring of the Nectaris Basin passes from Neander to the upper right corner of this image. Between Neander and Reichenbach there is a light plains unit with a fault running through it in a direction that is radial to the nearby Mare Nectaris. Reichenbach itself has a light plains unit on its floor. The greater brightness on the right side of this image, and the fine crater chains southwest of Reichenbach, come from nearby Stevinius to the east (LO4-059 in the Australe Basin Region). The ray pattern of Stevinius may be accompanied with a fine ejecta deposit that forms light plains units in this area.

LO4-065H1 Sun Elevation: 23.8° Altitude: 2735.07 km

The Nectarian crater Borda is on the border between the main ring of the Nectaris Basin and the trough within it. Santbech lies on the middle ring shown in Figure 8.4 (discontinuous here). The trough may have been flooded with mare in this sector (the surface has been lightened by rays from Stevinius). Imbrium deposits have landed in the trough and piled up against the main ring, blocking surface lava flow in this area.

LO4-065H2 Sun Elevation: 23.8° Altitude: 2735.07 km

The Montes Pyrenaeus range forms the eastern sector of the innermost ring of the Nectaris Basin and the hills south of Colombo mark the middle ring of Figure 8.3. Rimae Goclenius mark typical radial and circumferential fractures near the edge of a mare. In this case, the faults may relieve stresses due to both Mare Nectaris and Mare Fecunditatis, which meet near the right edge of this picture. Mare Fecunditatis is covered in the chapter on the Eastern Basins Region (Chapter 10). The floors of Goclenius and Magelhaens are flooded with lava, but probably not from the surrounding mare, because these floors are at a much lower level than the surrounding mare.

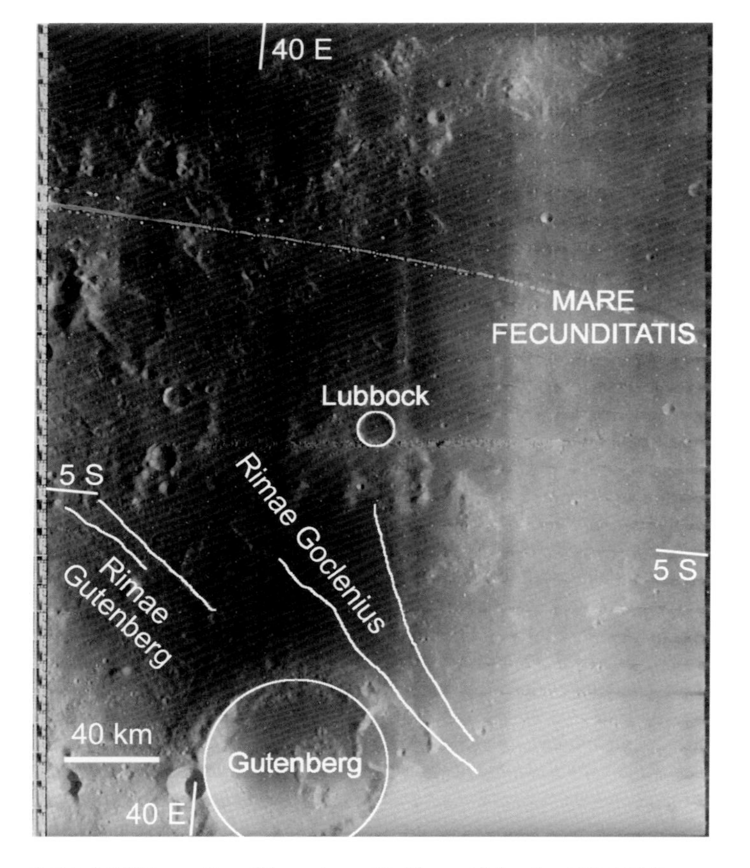

LO4-065H3 Sun Elevation: 23.8° Altitude: 2735.07 km

This image is partly fogged, apparently by ground processing, as LO4-065H2 was not affected. Mare Fecunditatis intrudes on the Nectaris Basin near Gutenberg, and part of Mare Tranquillitatis appears near the northwestern corner of the photo. A triple intersection of the main rings of the Nectaris, Fecunditatis, and Tranquillitatis Basins lifts the Censorinus plateau, whose eastern edge can be seen about 100 km down from the top of this photo, along the left-hand edge. See LO4- 072H3 for a full view of this interesting plateau. The Tranquillitatis Basin is covered in the chapter on the Serenitatis Basin Region (Chapter 9).

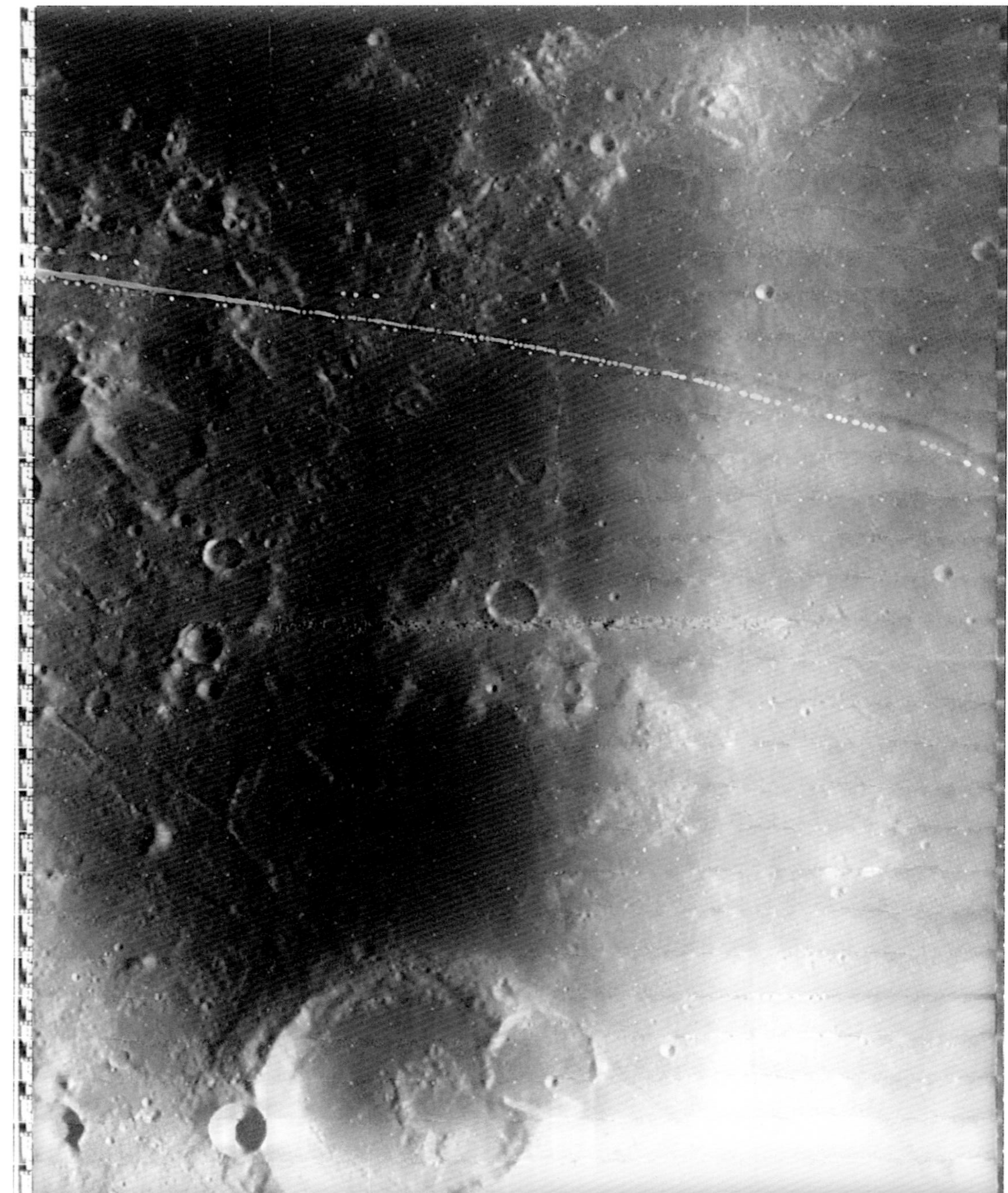

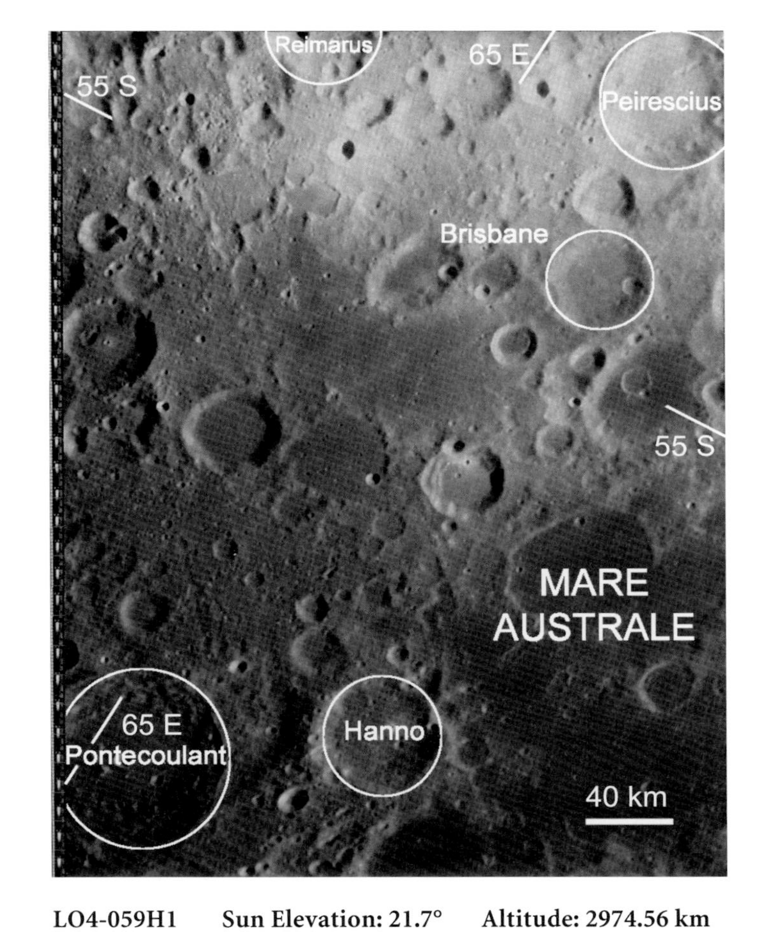

LO4-059H1 **Sun Elevation: 21.7°** **Altitude: 2974.56 km**

The main ring of the ancient, degraded Australe Basin passes through Hanno, bounding the eastern part of Mare Australe. Signs of east-west ridges and valleys radial to Australe have been erased by later deposits. The depression southwest of Brisbane and the Nectarian crater Pontecoulant lie in an outer trough of the Australe Basin. Crater Brisbane was named for astronomer Sir Thomas Brisbane, a Scot who served as governor of New South Wales in Australia before establishing observatories in Australia and Scotland.

LO4-059H2 **Sun Elevation: 21.7°** **Altitude: 2974.56 km**

Vallis Rheita and the other chains of large secondary craters radiate from the Nectaris Basin. In the area near Pre-Nectarian Vega, about one basin radius from the main ring of Nectaris, the heavy Janssen Formation gives way to fields of secondary craters.

LO4-059H3 Sun Elevation: 21.7° Altitude: 2974.56 km

The brightness in this area comes from a 20-km crater about 70 km southeast of the rim of Stevinus and a 5-km crater about 35 km northeast of the rim of Stevinus, as well as Stevinus itself. Stevinus has been assigned to the Copernican Period because of its fresh, sharp appearance. Compare its sharpness with the 35-km Eratosthenian crater north of Reichenbach (called Riechenbach A). The ejecta blanket from the smaller crater has lost its detail. The very long Vallis Snellius (radial to the Nectaris Basin) continues to the southeast toward the Australe Basin, becoming much less distinct. There may actually be two valleys (one from each basin) that approximately meet. The ridge and valley that comes from the north-northwest, continuing beyond Stevinus, is radial to the Fecunditatis Basin.

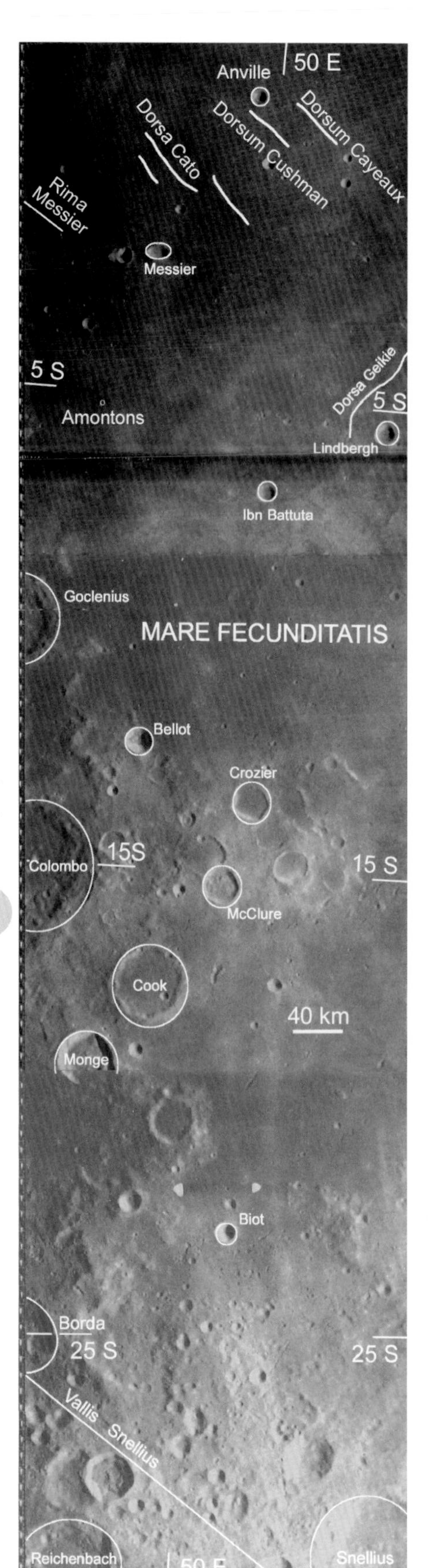

LO4-060H
Sun Elevation: 23.3°
Altitude: 2737.76 km

Messier and its companion to the west (called Messier A) are characteristic in shape to low-angle (about 5° from the horizontal) impacts of a lower-density material into a higher-density material. Messier A may actually have been caused by a pair of impactors. The ray pattern of these Copernican craters takes three directions: to the left and right of the impact axis and downstream. Except for the few surface features, this area shows a deep mare lava floor fed by a pluton of rising lava from the mantle.

The southern part of Mare Fecunditatis (east of Monge and Cook) has probably flooded an intersection of troughs (LO4-060H1). The main ring of the Fecunditatis Basin passes through the low ridge between Crozier and McClure.

This area of Mare Fecunditatis may have flooded a depression formed by the intersection of troughs from the Fecunditatis, Nectaris, and Balmer-Kapteyn Basins. The main ring of the Nectaris Basin passes through Borda and runs west of Biot. Vallis Snellius is radial to the Nectaris Basin.

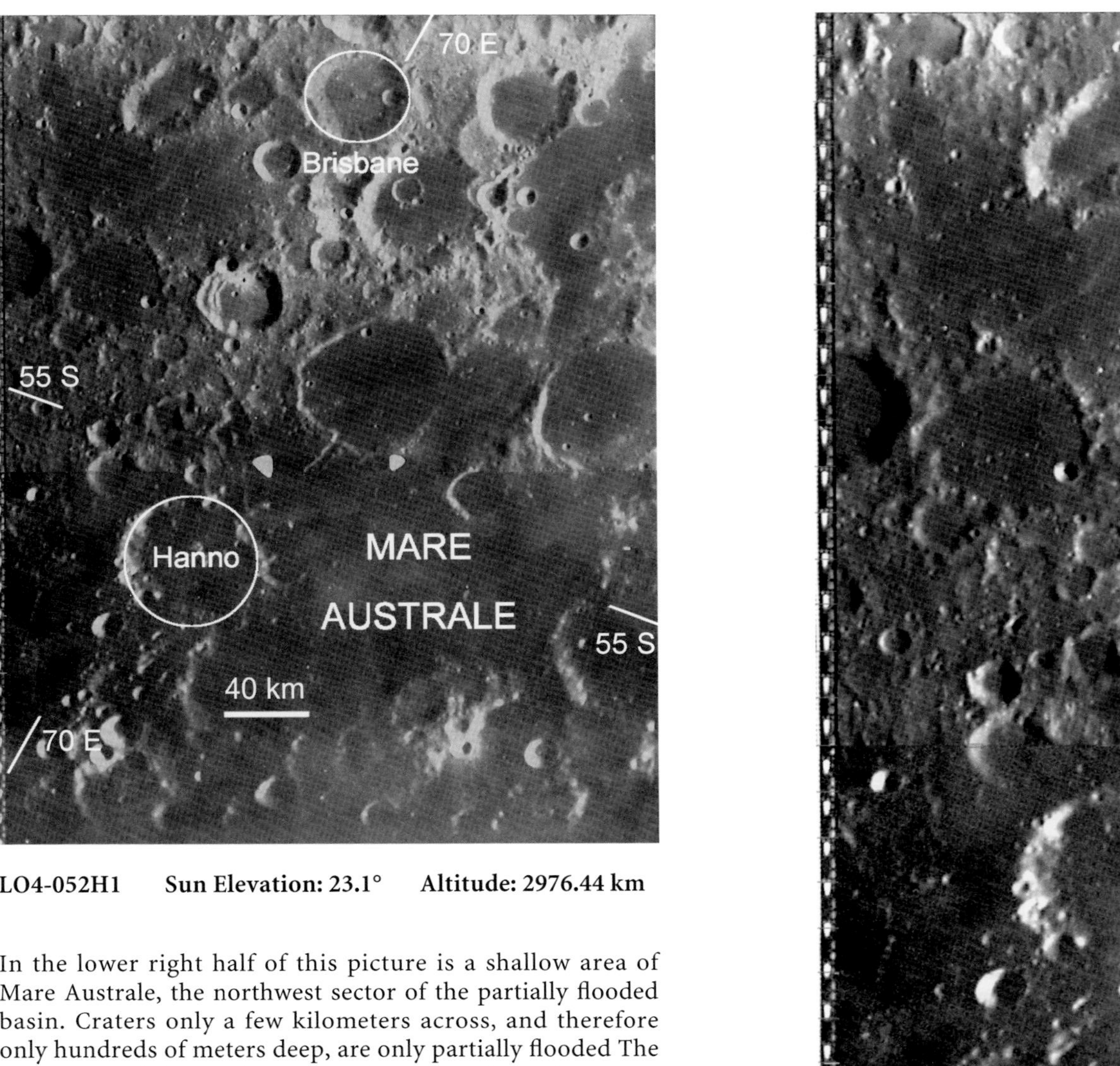

LO4-052H1 **Sun Elevation: 23.1°** **Altitude: 2976.44 km**

In the lower right half of this picture is a shallow area of Mare Australe, the northwest sector of the partially flooded basin. Craters only a few kilometers across, and therefore only hundreds of meters deep, are only partially flooded The main ring of the ancient, degraded Australe Basin passes through Hanno, bounding the eastern part of Mare Australe. Later deposits have erased most signs of Australe ejecta. The depression southwest of Brisbane and the Nectarian crater Pontecoulant lies in an outer trough of the Australe Basin.

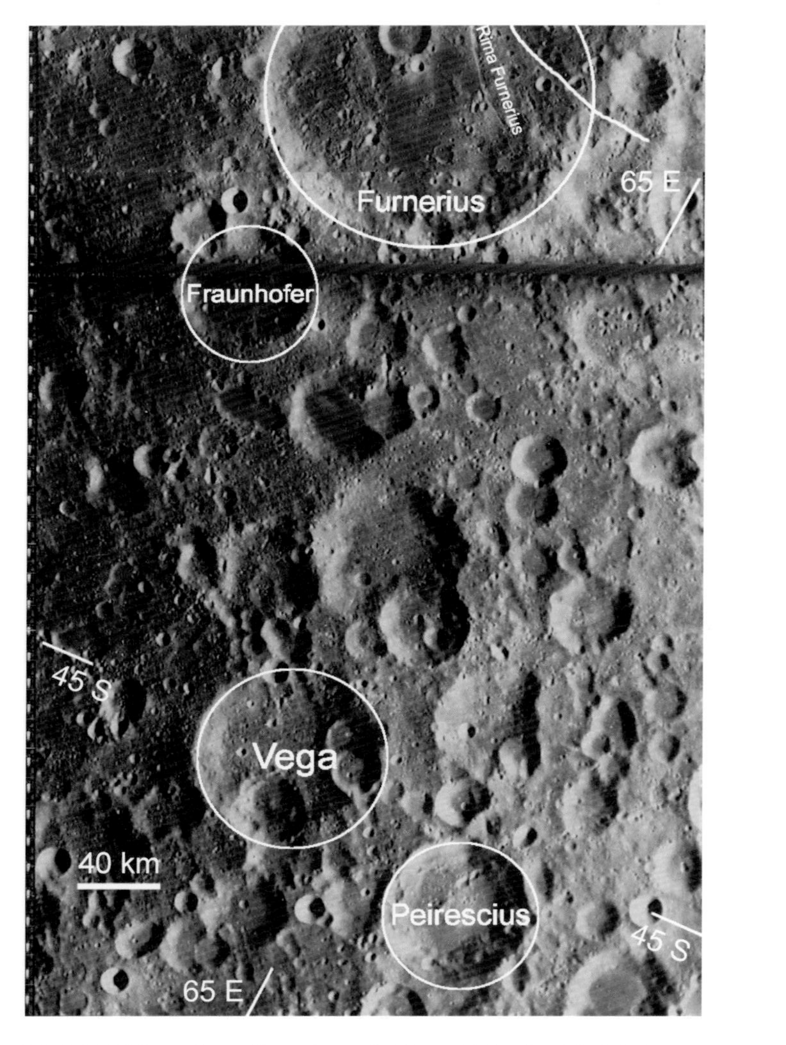

LO4-052H2 and -H3 **Sun Elevation: 23.1°** **Altitude: 2976.44 km**

An interruption of transmission between the spacecraft and Earth resulted in the loss of most of subframe LO4-052H3. The six framelets that have been received have been added to the top of LO4-052H2 in this mosaic. Chains of secondary craters in this area radiate from the Nectaris Basin. Vega and the craters within it are clearly Pre-Nectarian because the valleys aligned with the crater chains scar them all. Furnerius is a Pre-Nectarian crater; besides being fractured, it is over-

LO4-053H1 Sun Elevation: 23.3° Altitude: 2737.76 km

Petavius is a "poster child" of fractured-floor craters. It is large enough to have a prominent central peak of some complexity and a height of 3.5 km above the crater floor. In addition, Rimae Petavius reveal fractures that seem larger than can be explained by cooling and more likely to reflect strong uplift forces acting on a melt sheet. The upward pressure may be related to the lava flooding the nearby southern part of Mare Fecunditatis, to the upper left in the photo. The dark areas near the walls of Petavius suggest the release of dark glass particles (dark mantling materials) from fountains driven by the release of volatiles. Crater counts have resulted in Petavius being assigned to the Early Imbrian Epoch.

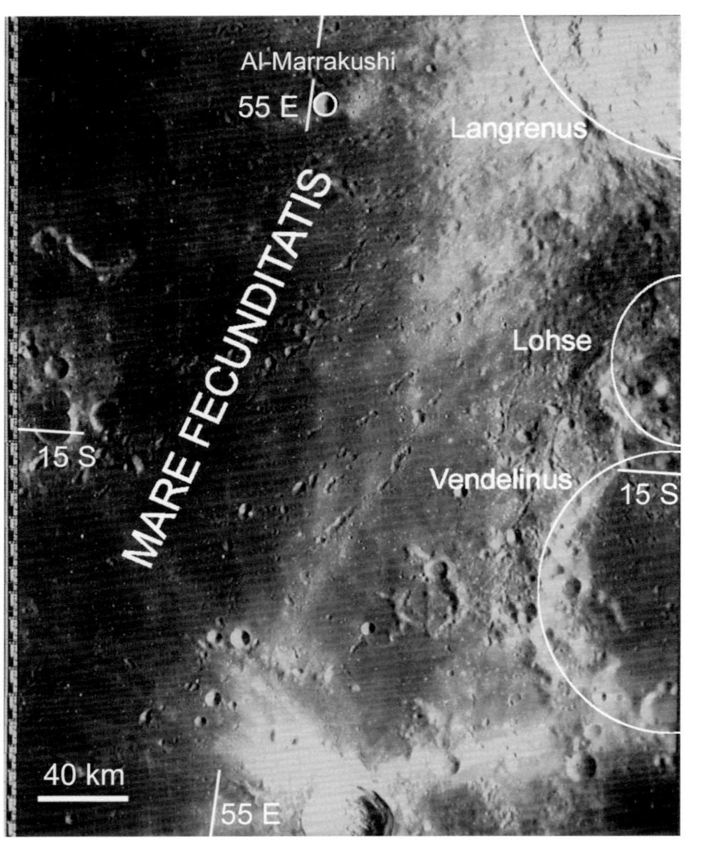

LO4-053H2 Sun Elevation: 23.3° Altitude: 2737.76 km

The main ring of the Fecunditatis Basin can be seen entering this image just north of 15° south latitude and curving toward Langrenus in the upper right corner. This ring may have been subsequently destroyed by the impact of one or more primary objects, allowing joint flooding by mare lava. Langrenus, variously assigned to the younger Eratosthenian Period or the older part of the Copernican Period, has a modest set of rays for such a large crater. The 35-km crater near the bottom of this image (and the top of LO4-053H1) is more clearly Copernican, with a three-pronged ray system typical of a primary crater that has approached from the north with a low angle. Vendelinus and Lohse may be in a trough of the Balmer-Kapteyn Basin (see Chapter 10), which would influence their being flooded with lava.

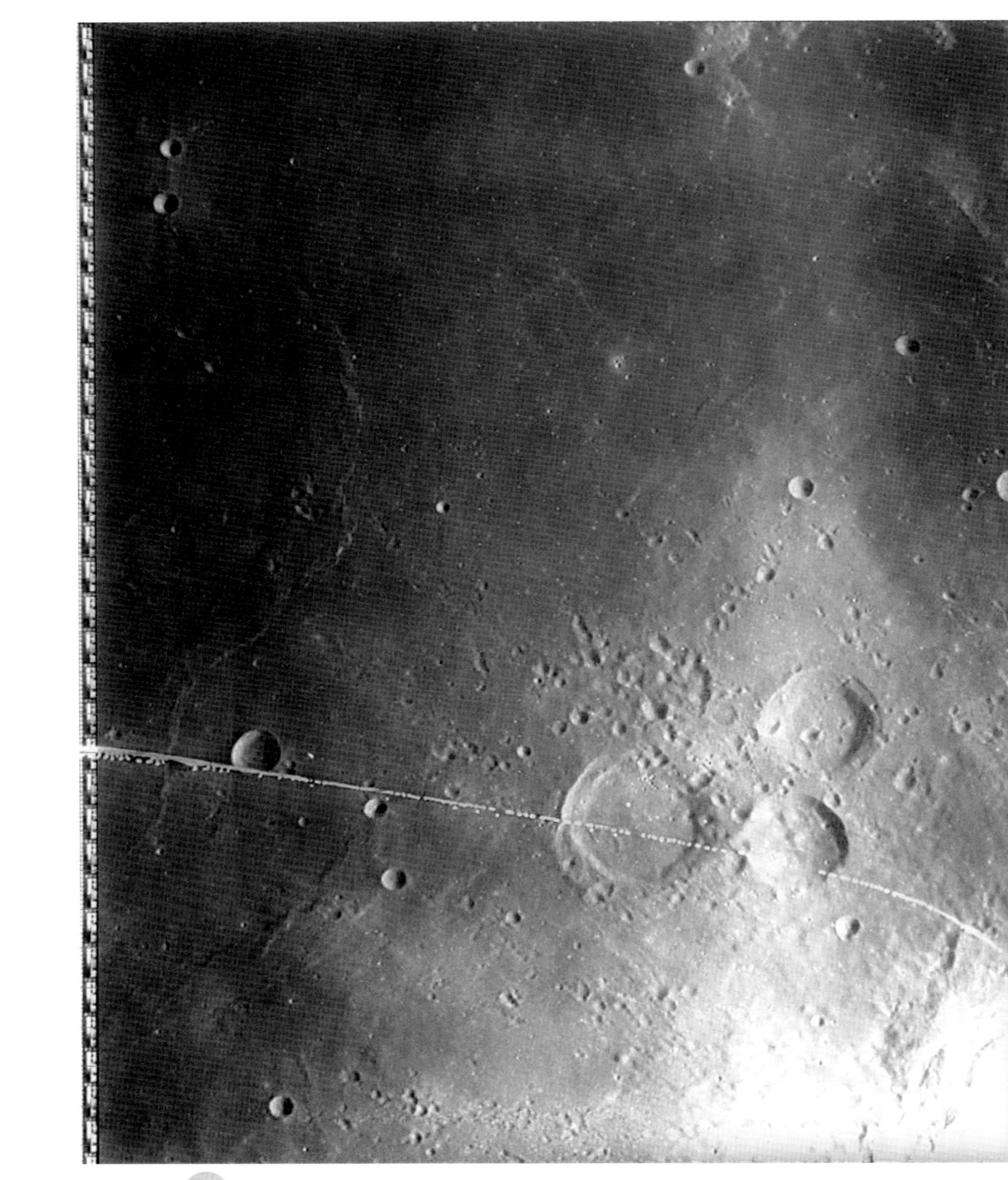

LO4-053H3 Sun Elevation: 23.3° Altitude: 2737.76 km

The successful landing of Luna 16 is discussed in the chapter on the Serenitatis Basin Region (LO4-054H1) (Chapter 9). This photo covers the deepest part of Mare Fecunditatis (the center of the Fecunditatis Basin is at 4° S and 52° E). A rich field of ejecta and secondaries from Langrenus has covered the mare floor, establishing Langrenus as either Eratosthenian or Copernican. The modest ray system (for such a large crater) places its age near the border between those two periods. Bilharz, Naonobu, and Atwood, formed by three impactors that may have landed in a salvo, are older than the mare floor but are clearly younger than Langrenus. As in other maria, a few wrinkle ridges (dorsa) appear on its surface. Northern Mare Fecunditatis continues into the Serenitatis Basin Region. The eastern shore can be found in the Eastern Basins Region.

Chapter 9

Serenitatis Basin Region

9.1. Overview

The Serenitatis Basin Region stretches from Mare Imbrium, whose eastern edge is near the 0° meridian, across Mare Serenitatis and Mare Tranquillitatis to the western edge of Mare Crisium.

A series of four basins, Imbrium, Serenitatis, Tranquillitatis, and Fecunditatis, cover an arc of 120°, one-third of the circumference of the Moon. Each member of this chain of basins has formed a circular depression that was subsequently (long after the impact event) filled with lava erupted from below, forming a mare. The Fecunditatis Basin, whose edge is visible in the lower right-hand corner of Figure 9.1, is covered in the Nectaris Basin Region (see Chapter 8).

The diameters of the main rings of these four basins (Spudis, 1993) are as follows:

- Imbrium: 1160 km
- Serenitatis: 920 km

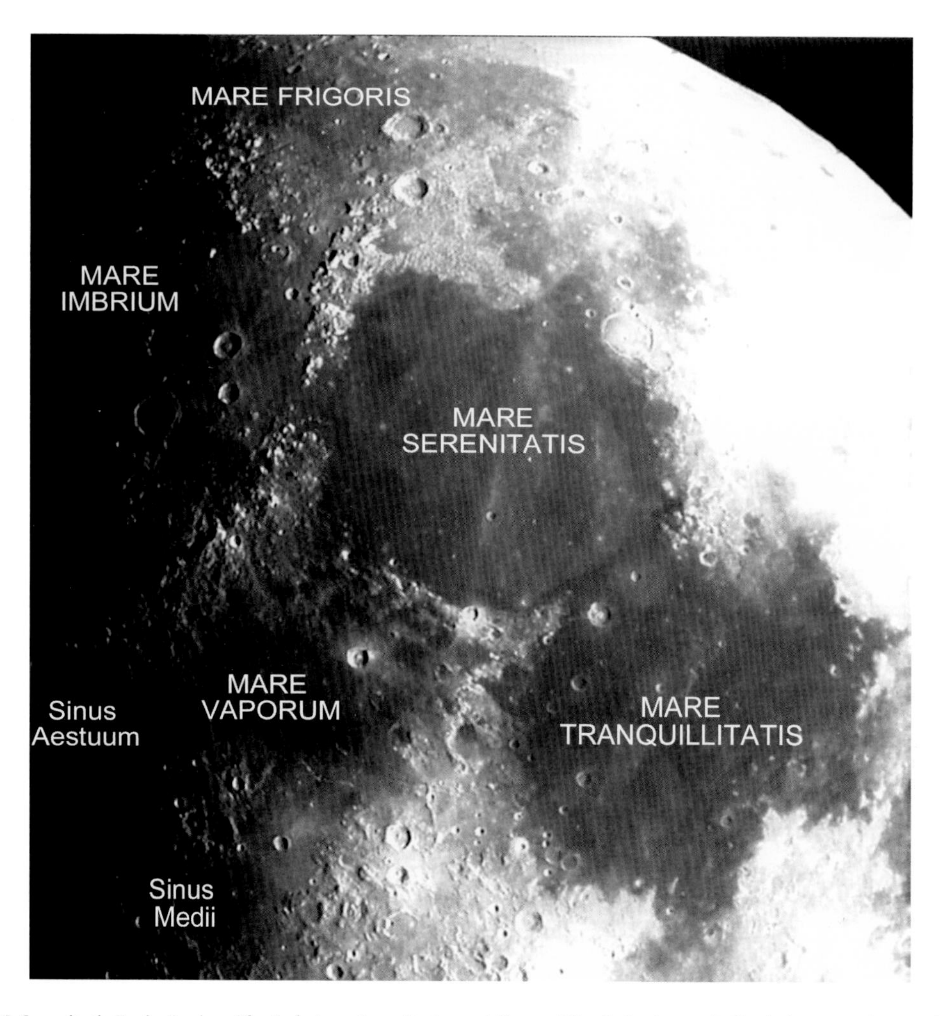

Figure 9.1. LO4-097M: Serenitatis Basin Region. The Imbrium, Serenitatis, and Tranquillitatis Basins underlie their respective maria. Sinus Aestuum lies in an outer trough of the Imbrium Basin, and Mare Vaporum lies in a depression formed by the intersection of outer troughs of the Imbrium and Serenitatis Basins. Mare Frigoris lies partly in the northern part of the same outer trough of the Imbrium Basin as Sinus Aestuum and Mare Vaporum.

- Tranquillitatis: 700 km
- Fecunditatis: 690 km

To provide a sense of scale, Spain and Portugal would be a close fit in Mare Serenitatis. Massive mountain ranges were thrown up at arcs of the main ring of these basins, especially where the main rings of two or more basins come together. For example, at the boundary between the Imbrium and Serenitatis Basins, Montes Apenninus (to the south) rise as much as 4,430 m (14,400 feet) above the nearby mare surface. Radial ridges formed by ejecta from the Imbrium event can be seen south of Mare Serenitatis, running for 500 km. Such ranges and ridges are formed by impact in a matter of minutes and hours.

Serenitatis Basin

Mare Serenitatis is unusual in that it has apparently covered two basins: the Serenitatis Basin and an older, smaller basin north-northeast of the main Serenitatis Basin called the Northern Serenitatis Basin. The main ring of the Serenitatis Basin is interrupted in the sector where it would have crossed the floor of the older basin, perhaps suppressed by a melt sheet within the main ring of that basin. Similarly, the melt sheet of the Serenitatis Basin may have suppressed the main ring of the younger Imbrium Basin where it would have crossed the floor of the Serenitatis Basin. In the same way, the melt sheet of the Tranquillitatis Basin may have suppressed the main ring of the Serenitatis Basin where they intersect.

Apollo Landings

Apollo 11 landed on Mare Tranquillitatis (LO4-085H1), the first landing of humans beyond Earth and the first mission to return rock samples from beyond Earth. The samples collected by Armstrong and Aldrin established many of the characteristics confirmed by subsequent missions. The rocks were composed of basalt, a complex material that hardens from lava flows when molten rock from the mantle (the region of relatively dense rock below a crust) rises to the surface. On Earth, basalt paves the floors of the oceans and erupts from volcanoes. The lunar soil and rocks are free of organic compounds and of volatile elements and compounds (hydrogen, sulfur, sodium, and water). The age of the basalt at Tranquillitatis Base was found to be 3.65 aeons (billion years).

A few rock samples were not basalt but had been thrown from highlands by impacts. These rocks contain plagioclase, a low-density mineral composed of calcium and aluminum silicates. This is the material, along with other minerals that combine easily with it, that rises from a body of molten rock (the magma ocean of the primitive Moon) to form the crust.

Apollo 15 landed near the sinuous canyon of Rima Hadley and Mons Hadley, a mountain in the Montes Apenninus range, part of the main ring of the Imbrium Basin (LO4-102H3). Equipped with a Lunar Roving Vehicle, Scott and Irwin explored a wide variety of terrain types. The walls of Rima Hadley showed layers of mare material from a series of flows; the deepest flow was 60 m in depth. "Genesis Rock," a crustal rock aged at 4.15 billion years, was collected from the slope below the mountain Hadley Delta. Some samples, rich in highland plagioclase, were probably thrown to this area from the crater Autolycus from 150 km away; the impact event was aged at 1.29 million years.

Apollo 17 landed in a valley (Taurus-Littrow Valley) of Montes Taurus, part of the main ring of the Serenitatis Basin (LO4-078H3). Schmitt and Cernan, their travels aided by a Lunar Roving Vehicle, sampled ejecta from the Serenitatis Basin. Radioactive aging of the samples established the age of the Serenitatis Basin to be 3.86 or 3.87 billion years. One-meter layers of orange and black glass beads were discovered, establishing the character of dark mantling material that is common near the edges of maria. Some samples of material from a ray of Tycho were dated at 109 million years.

9.2. High-Resolution Images

Table 9.1 shows the high-resolution images of the Serenitatis Basin Region in schematic form.

The following pages show the high-resolution subframes from south to north and west to east. That is, they are in the order LO4-102H1, LO4-102H2, LO4-102H3, LO4-103H1, LO4-103H2 ... LO4-054H1, LO4-054H2... LO4-055H3.

Subframes LO4-098H3, LO4-086H3, LO4-074H3, and LO4-062H3 are redundant and are not printed, although they are included in the enclosed CD.

Photos LO4-078H, LO4-073H, LO4-066H, and LO4-054H are printed as complete frames.

Latitude Range	Photo Number									
56 N–90 N	104		092		080	068					191H1
27 N–56 N	110	103	098	091	086	079	074	067	062	055	191H2
0–27 N	109	102	097	090	085	078	073	066	061	054	191H3
0–27 S	108	101	096	089	084	077	072	065	060	053	046
Longitude at Equator	3 W	4 E	10 E	16 E	24 E	30 E	38 E	43 E	49 E	57 E	63 E

Table 9.1. The cells shown in white represent the high-resolution photos of the Serenitatis Basin Region (LO4-XXX H1, -H2, and -H3, where XXX is the Photo Number). The Imbrium Basin Region is to the west, the Nectaris Basin Region is to the south, the Eastern Basins Region is to the west, and the North Polar Region is to the north.

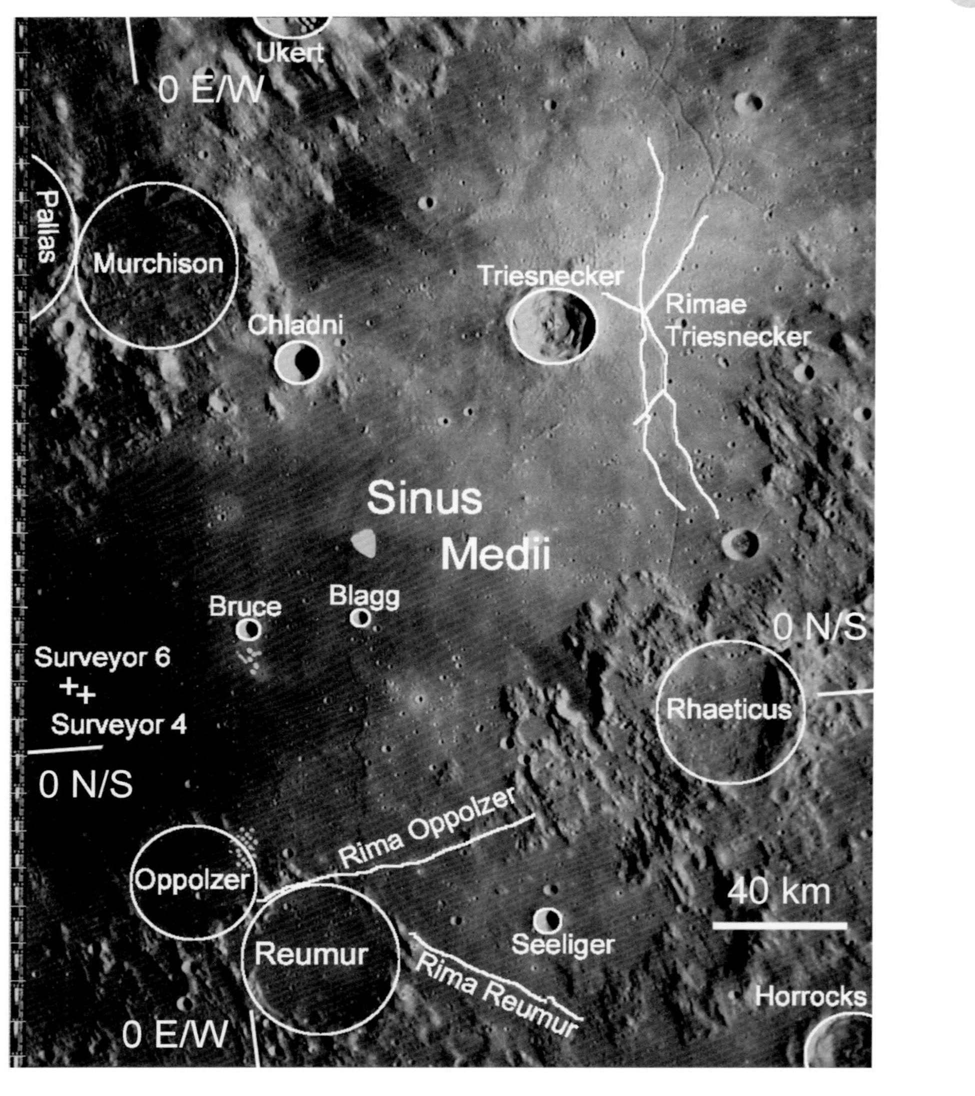

LO4-102H1 | Sun Elevation: 21.6° | Altitude: 2699.13 km

Surveyors 4 and 6 are discussed in the note for LO4-108H3, in the Humorum Basin Region. The complex pattern of Rimae Triesnecker reflects stretching of the surface of Sinus Medii away from its border as lava cooled. The last major flow of lava came from the east, as indicated by the scarp that runs north from Reaumur and continues past Blagg. A large positive gravity anomaly in the Clementine data, centered about 50 km northeast of the early Copernican crater Triesnecker, may mark the source of the flow. The ridges and valleys are Fra Mauro Formation, ejecta from the Imbrium Basin whose main ring is about a radius away. A detailed examination of Murchison indicates that its floor has been flooded from near its walls.

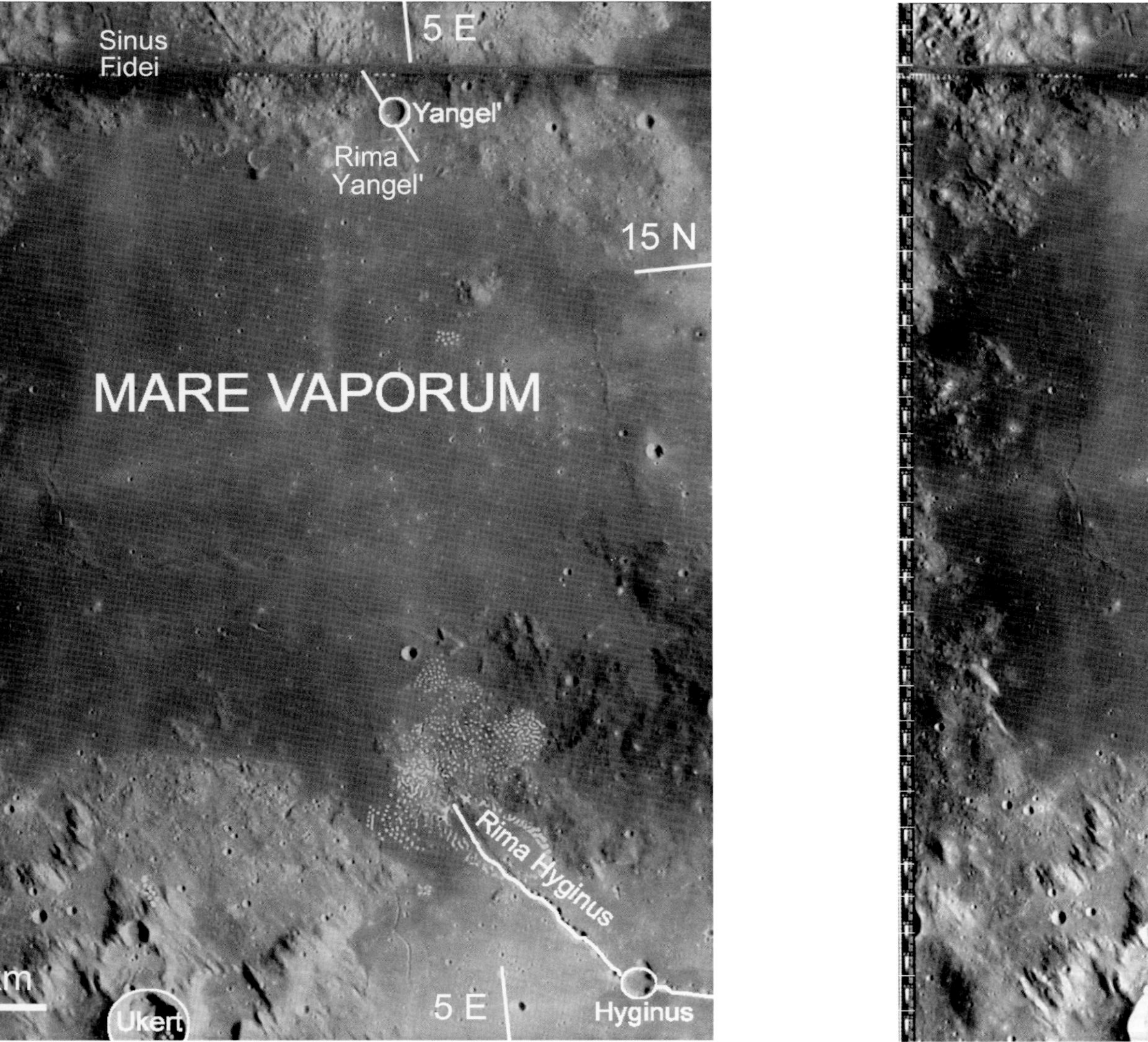

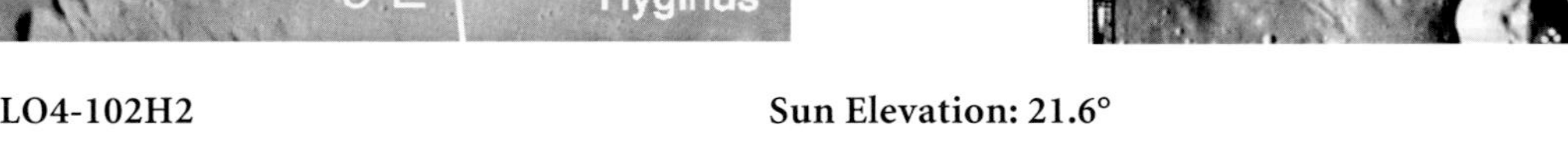

LO4-102H2 | Sun Elevation: 21.6° | Altitude: 2699.13 km

Mare Vaporum, south of the foothills of the Montes Apenninus, has filled the trough outside the main ring of the Imbrium Basin. Its southern boundary is the lower hills of the next outer ring of the Imbrium Basin, 1700 km in diameter. The nearest gravity anomaly in Clementine data that could mark the source of the lava is near Triesnecker (LO4-102H1) to the south. Rima Hyginus may be a collapsed lava tube that ran to the northwest to help form Mare Vaporum. The craters along Rima Hyginus, notable for their lack of rims, may have been individual collapse events. Hyginus itself, a rimless crater with a dark halo, may have been an endogenic source for some of the lava carried by the tube.

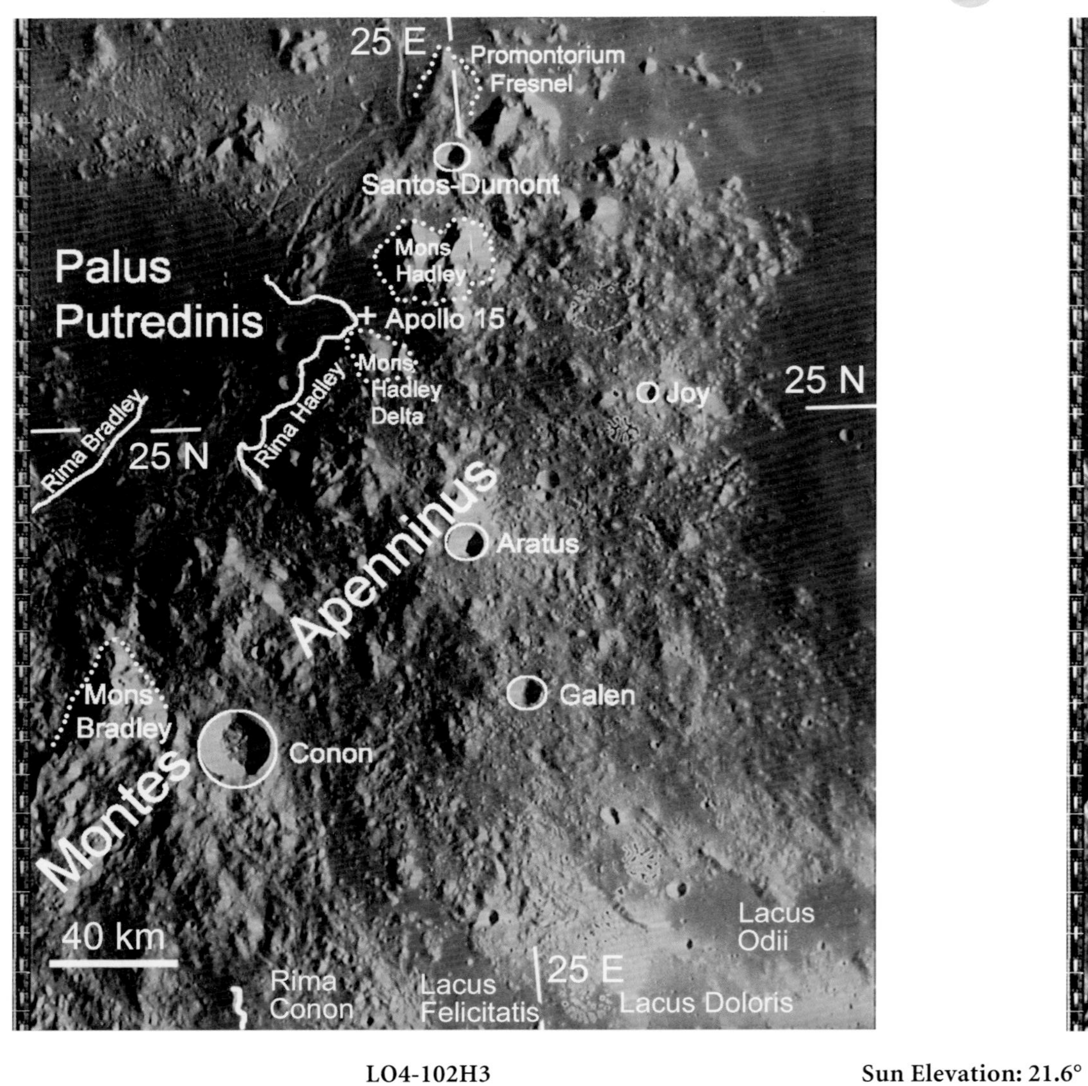

LO4-102H3 Sun Elevation: 21.6° Altitude: 2699.13 km

The landing site for Apollo 15, Palus Putredinis ("Swamp of Decay") near the edge of Mare Imbrium, was chosen for the opportunity to explore both highlands formed by the main ring of the Imbrium Basin and the deep cut through mare material of Rima Hadley. As in other basin rims, rising lava seems to be blocked by a massive rim range, flowing out through a vent and down the side of the range. Once it reaches the flatter terrain of Palus Putredinus, Rima Hadley (like Vallis Schroteri) meanders like terrestrial rivers in nearly flat land. The astronauts saw layers in the walls of Rima Hadley, indicating a series of lava flows. One of the rocks sampled from the slope of Mons Hadley delta is the "Genesis rock," containing white plagioclase. This low-density mineral is typical of the original crust formed 4.5 billion years ago. However, it appears to have been shocked and recrystallized by an impact, resetting its "atomic clock" so that its measured age is 4.15 billion years.

LO4-103H1 Sun Elevation: 22.3° Altitude: 2926.55 km

See LO4-102H3 for notes on the Apollo 15 landing site. The bubbly development artifact lies along a "strait" of mare material that connects Mare Imbrium (left) with Mare Serenitatis (right). In this area, the Imbrium main ring, marked by Montes Apenninus and Montes Caucasus, crosses inside of the main ring of Serenitatis. A melt sheet there may have inhibited formation of mountains in the strait. Although the Serenitatis Basin was formed before the Imbrium Basin, the last major mare flow seems to have been from Serenitatis, as shown by the scarp that runs north from Promontorium Fresnel.

LO4-103H2 Sun Elevation: 22.3° Altitude: 2926.55 km

The hummocky (bumpy) area in the upper left quadrant of this photo is the Alpes Formation of the Imbrium Basin, similar to the Montes Rook Formation of the Orientale Basin. This area is between the main ring of the Imbrium Basin (Montes Caucasus) and an inner ring. Eudoxus is fresh looking and shows no sign of being affected by ejecta from the large crater Aristoteles just to the north (LO4-103H3), so it is probably Copernican. Alexander formed before the Imbrium Basin; it is overlain with Imbrium striations. Cassini has occurred after the Imbrium Basin was formed, or it would have been obliterated. Subsequently flooded by one of the first lava flows in Mare Imbrium, it is assigned to the Early Imbrian Period.

LO4-103H3 Sun Elevation: 22.3° Altitude: 2926.55 km

The Eratosthenian crater Aristoteles (87 km) provides an excellent example of the ejecta pattern of a crater of this size. A heavy ejecta blanket with radial ridges extends beyond the rim to a distance of about one rim radius. Beyond that, the thinning ejecta becomes an array of secondary impactors, which often form chains of craters. Galle has apparently had a gap in its northern rim invaded by mare lava, which cascaded to its floor in one or possibly two episodes. The lava flow(s) must have been the last flows in the area, or the crater would have been completely inundated. The mare area is part of Mare Frigoris. In this area, it has formed in an intersection of the first outer trough of the Imbrium Basin and the first outer trough of the Northern Serenitatis Basin.

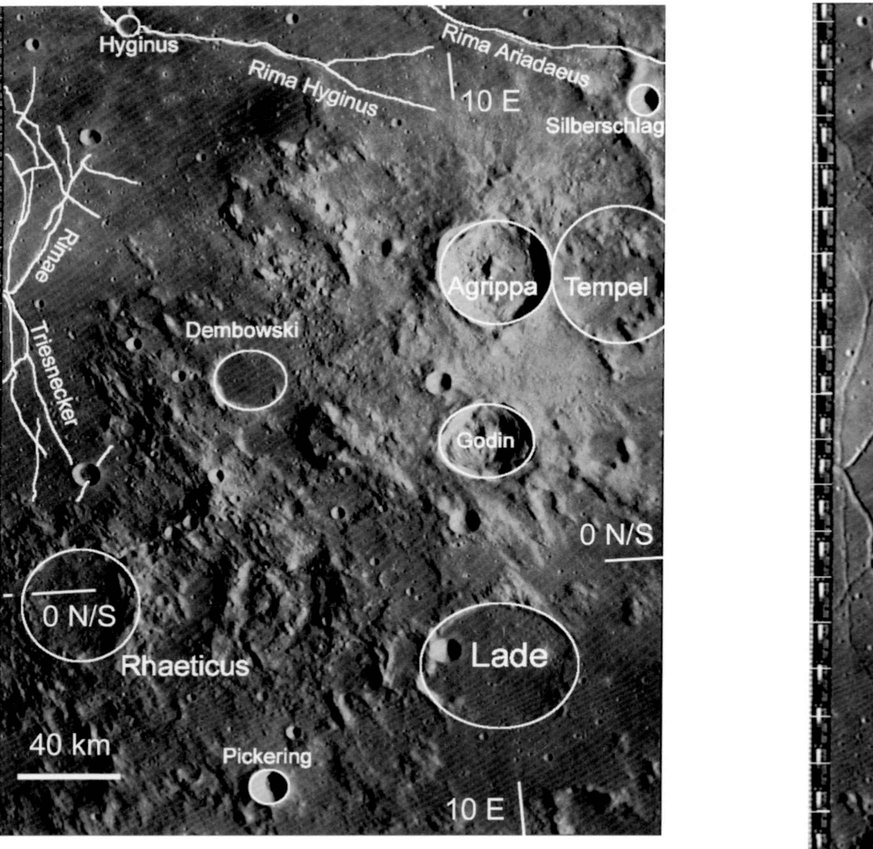

LO4-097H1 Sun Elevation: 21.6° Altitude: 2705.28 km

The striated ridges of the Fra Mauro Formation from Imbrium have impacted rugged highlands about one radius away from the main ring of Imbrium. Rimae Triesnecker are interlocking radial and circumferential stress faults at the edge of Mare Vaporum. The Copernican crater Godin has sent its ejecta blanket over that of the Eratosthenian crater Agrippa, which in turn, along with Imbrium, impacted the ancient crater Tempel. The floor of Lade is covered with hummocky material that may have been ejected from Imbrium and been decelerated by glancing off the rim of Lade. Some of this material may be from the rim itself. Northeast of Rhaeticus, there is a chain of fresh-looking secondary craters that is radial to Orientale.

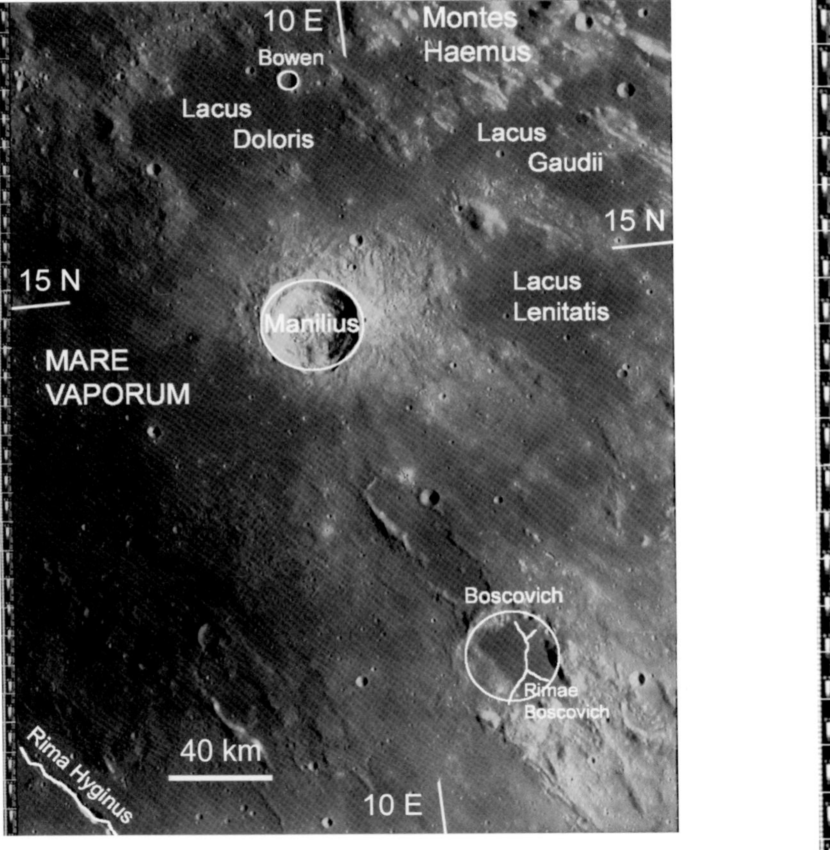

LO4-097H2 Sun Elevation: 21.6° Altitude: 2705.28 km

Montes Haemus marks the main ring of the Serenitatis Basin. This area of Mare Vaporum formed in the first outer trough of the Serenitatis Basin, along with Lacus Gaudii and Lacus Lenitatis. Once thought to be volcanic, the elongated formation near Boscovich is now considered to be a chain of secondary craters from an early stage of the Imbrium impact, modified by an ejecta flow from the same impact. These two stages would have been separated in time by only a few minutes. Flooding by mare would have been millions of years later. Opening of the stress cracks under Rimae Boscovich would have been still later. Manilius is considered to be of the Late Imbrian Epoch because its ejecta overlies the mare.

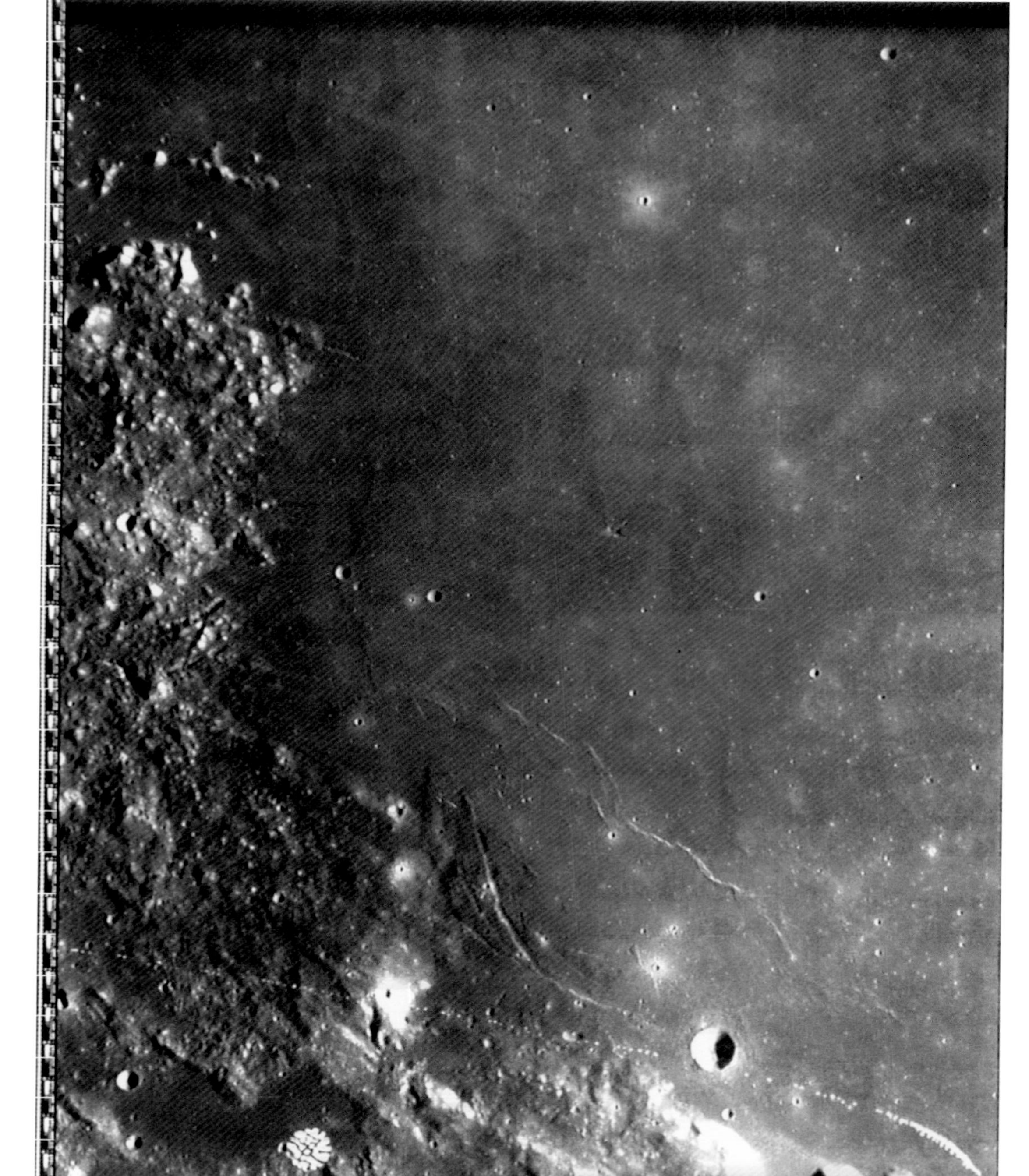

LO4-097H3 Sun Elevation: 21.6° Altitude: 2705.28 km

At the border of Mare Serenitatis, between Joy and Sulpicius Gallus, is a deposit that darkens the mare, the bumpy terrain between Montes Haemus and the mare (Alpes Formation; see the note for LO4-103H2), and Lacus Odii. This deposit is characterized as dark mantling material, a type that is found on the border of other maria. It may be due to "fire-fountain" volcanic eruptions from depth or possibly result from volatiles released where a thin layer of lava flows over crustal material. Circumferential ridges such as Dorsum Buckland and Dorsum Gast and fractures such as Rimae Sulpicius Gallus are common at the edges of maria. Linne is a Copernican crater that has impacted Mare Serenitatis. A chain of such craters marches across the edge of Mare Serenitatis into Montes Haemus.

LO4-098H1 Sun Elevation: 22.3° Altitude: 2926.55 km

The projection in the lower left of the photo marks the intersection of the main ring of the Serenitatis Basin with that of the Northern Serenitatis Basin. The section of the Montes Caucasus in the upper left corner of this photo is the main ring of the Northern Serenitatis Basin. Mare Serenitatis shows few indications of drowned craters, except for some faint circular ridges, indicating that the mare is deep. There is a strong mascon associated with Mare Serenitatis, indicating that a large plume of lava from the mantle fed the flooding of the overlapping pair of basins.

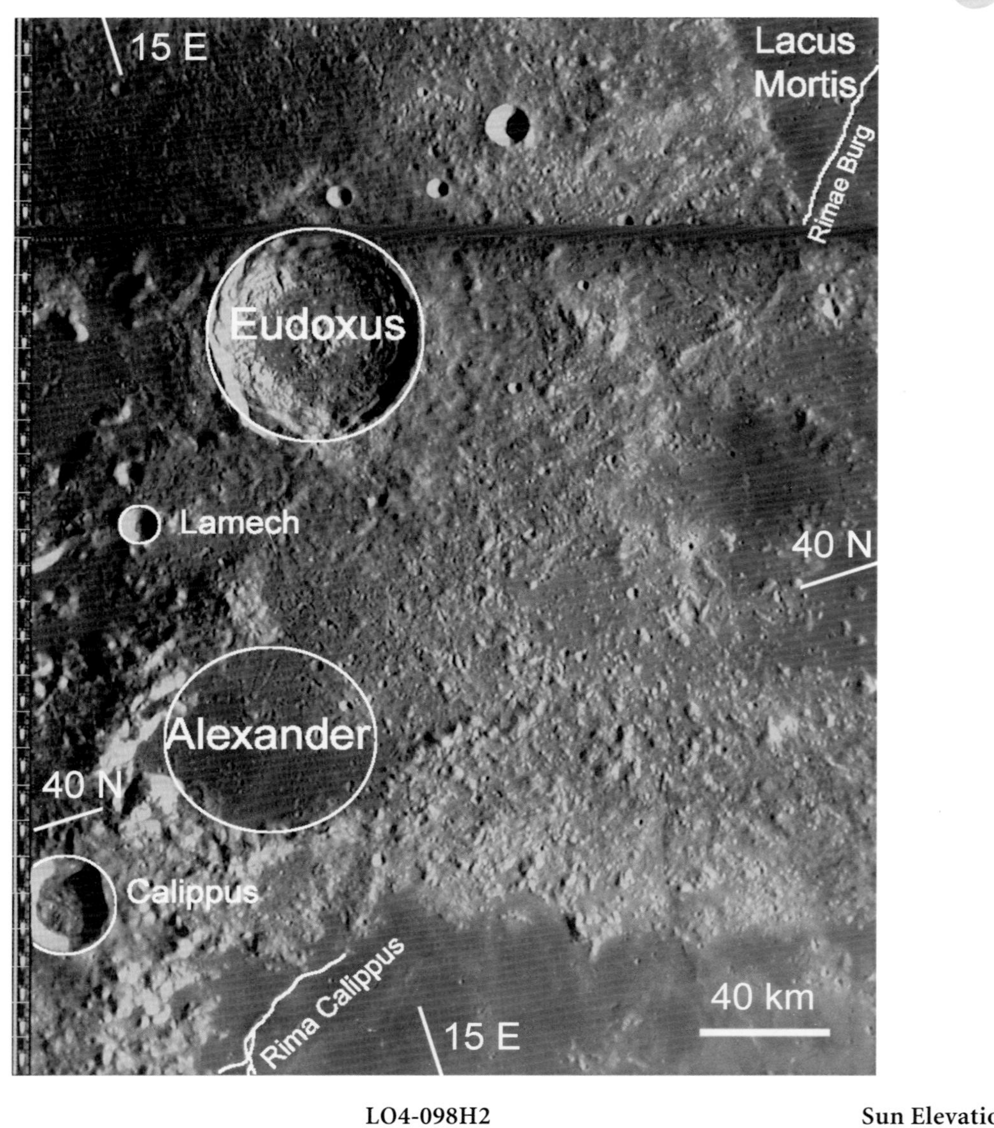

LO4-098H2 | Sun Elevation: 22.3° | Altitude: 2926.55 km

In this area north of Mare Serenitatis there are overlapping deposits ejected from the Serenitatis, Northern Serenitatis, and Imbrium Basins (with the Imbrium ejecta uppermost). Because of its hummocky character, it is mapped as an extension of the Alpes Formation of the Imbrium Basin. However, it may have a different forming mechanism than the area inside of the Imbrium rim.

The portion south of 40° north latitude may be primarily Northern Serenitatis material within its rim, and the northern portion may not really be "hummocky"; it may be simply crosshatched with striations from Imbrium and Serenitatis.

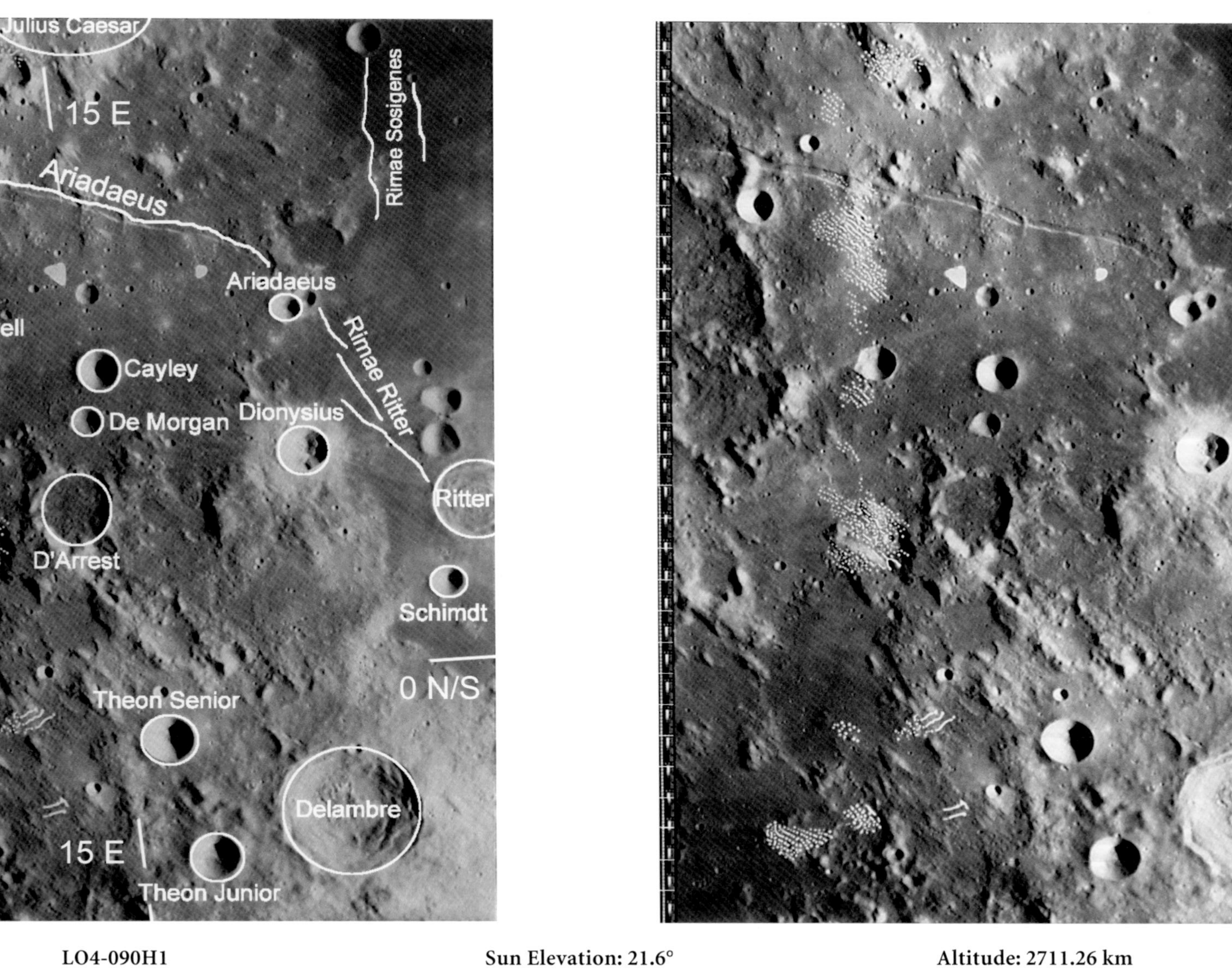

LO4-090H1 **Sun Elevation: 21.6°** **Altitude: 2711.26 km**

In many parts of the Moon, there are small patches of smooth plains in otherwise rugged highland regions similar to this area surrounding Cayley, the archetype of the Cayley Formation. These regions are associated with the outer ejecta blanket from major basins such as Imbrium, Orientale, and Nectaris. These highland plains may result from relatively uniform blankets of finely pulverized ejecta that are shaken down by local tectonic forces. To quote Ken Mattingly, the Command Module Pilot for Apollo 16, "The Cayley represents a pool of unconsolidated material which has been 'shaken' until the surface is relatively flat." Rima Ariadaeus, a large fault with an offset, may indicate strong tectonic action in this area around Cayley.

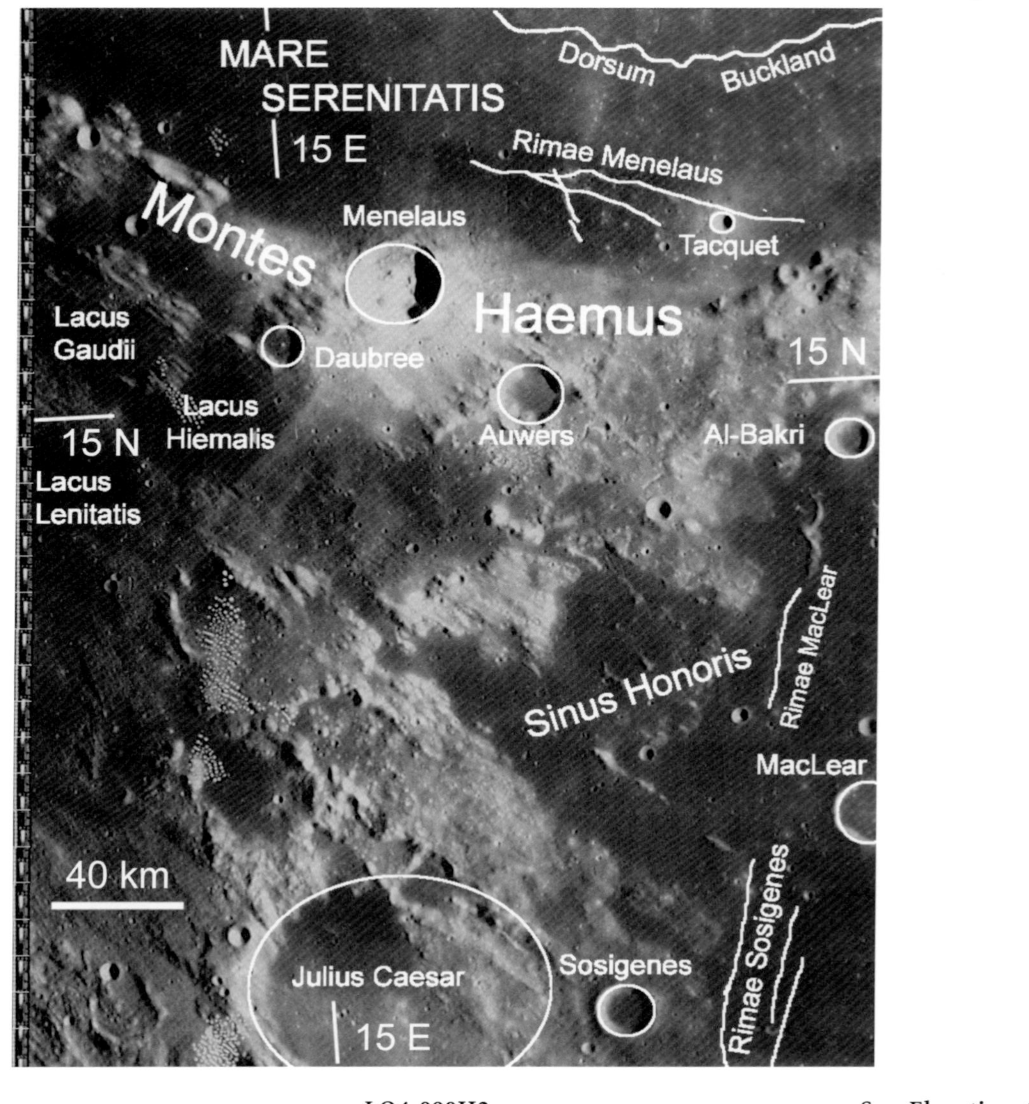

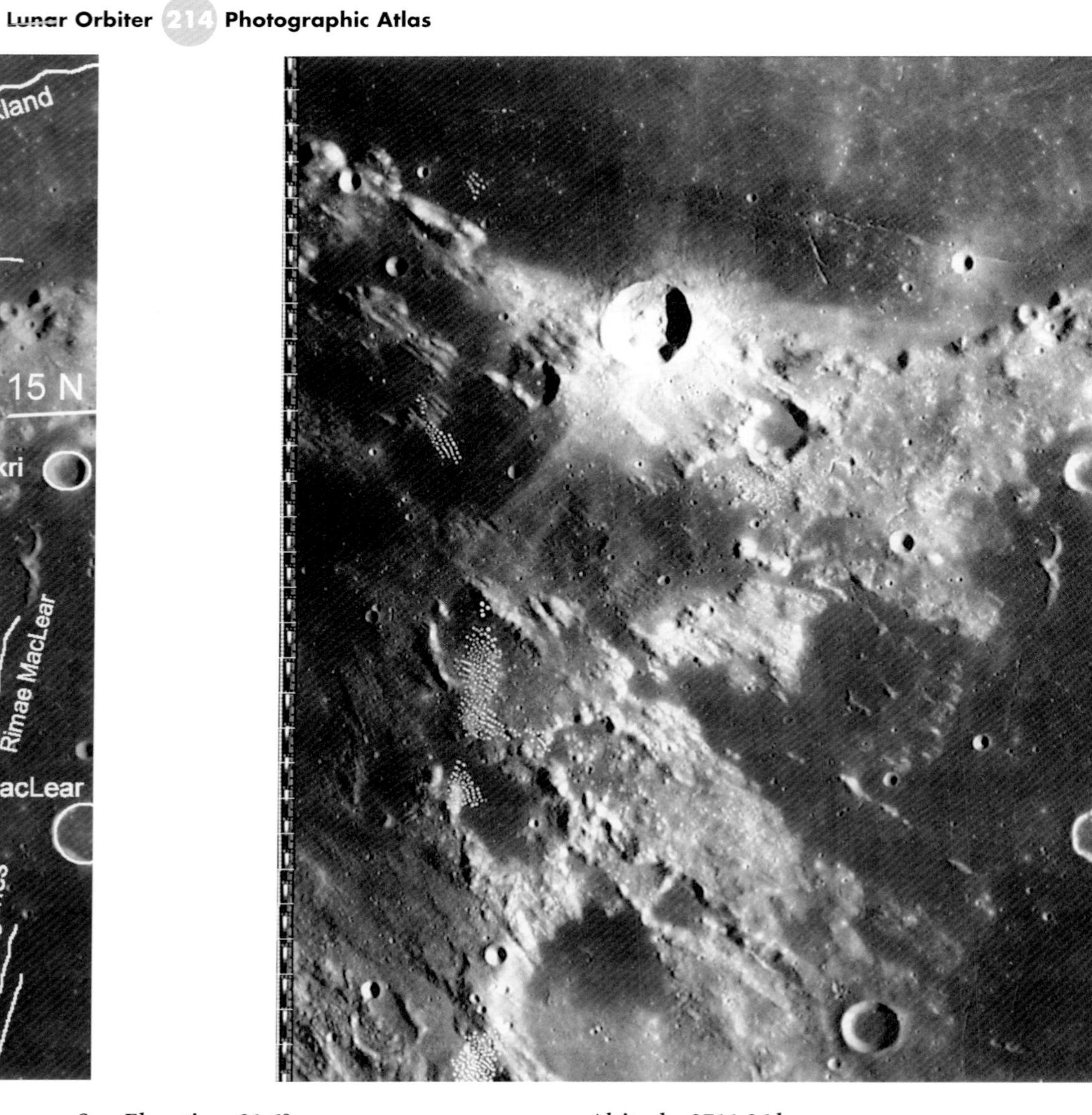

LO4-090H2 Sun Elevation: 21.6° Altitude: 2711.26 km

This interesting region south of Mare Serenitatis shows Montes Haemus, the southern main ring of the Serenitatis Basin, impacted by striations from the Imbrium Basin. The outer trough of the Serenitatis Basin has been flooded by Lacus Gaudii, Lacus Hiemalis, Lacus Lenitatis, and Sinus Honoris. Mare Tranquillitatis is to the southeast of Sinus Honoris; the much-degraded main ring of the Tranquillitatis Basin traverses Julius Caesar. Rimae Maclear and Rimae Sosigenes are circumferential faults at the edge of Mare Tranquillitatis. Menelaus has impacted precisely on the inner rim of the main ring of the Serenitatis Basin, projecting rays in all directions.

LO4-090H3 Sun Elevation: 21.6° Altitude: 2711.26 km

The flooded floor of the Serenitatis Basin shows only subtle variation. Dorsum Azara may represent a flow boundary or may be part of a pattern marking a submerged crater. The bright streaks radiate not from Bessel, but from Menelaus to the south (LO4-090H2).

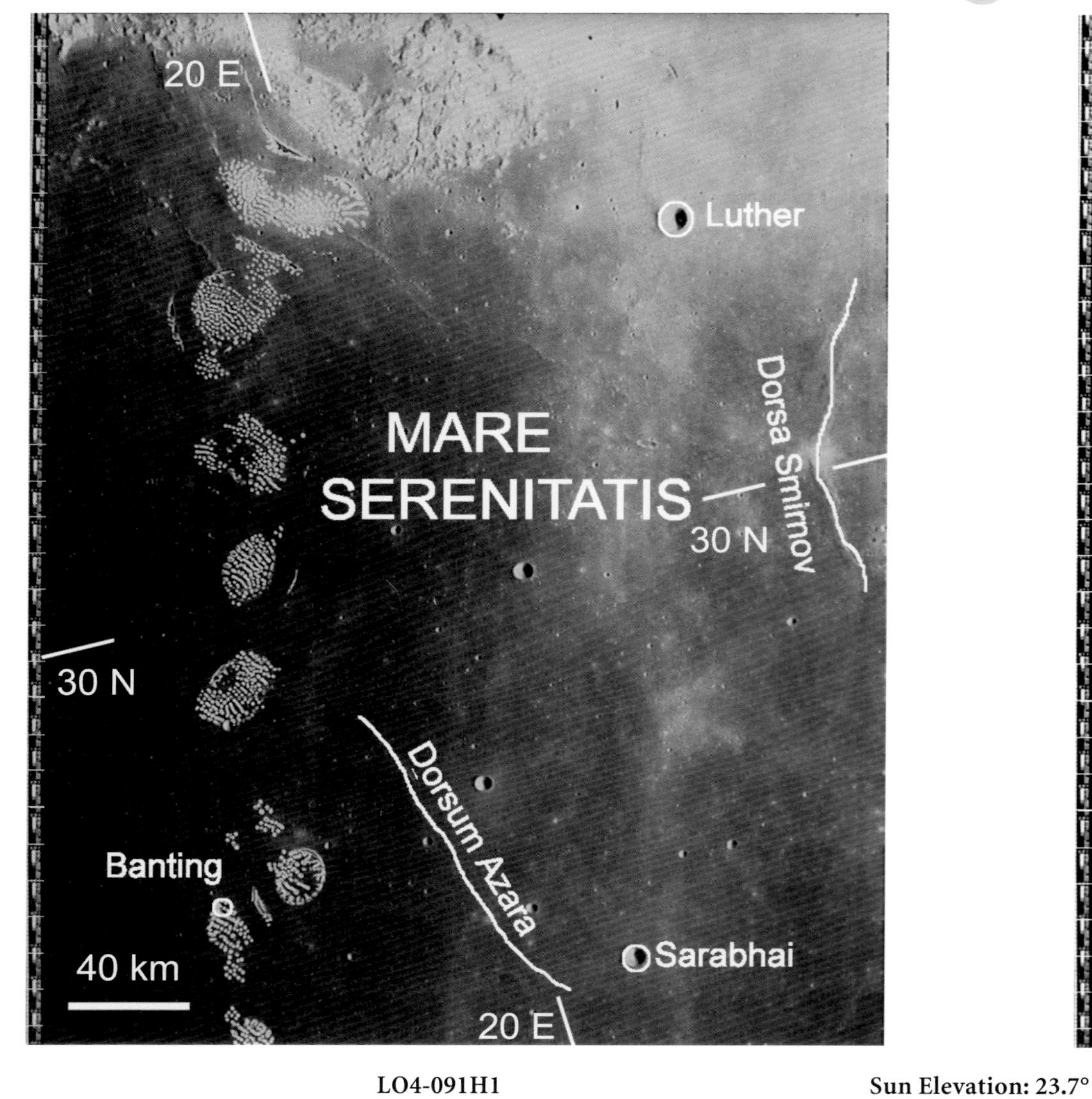

LO4-091H1

Sun Elevation: 23.7°

Altitude: 2947.25 km

In this northeastern part of Mare Serenitatis, an increased frequency of ridges indicate a transition from the deep lava of Mare Serenitatis to the shallower Lacus Somniorum (LO4-091H2), northeast of Luther. The brightness of the mare surface comes from rays from Copernican craters Burg (to the north) and Thales (far to the northeast).

LO4-091H2 | Sun Elevation: 23.7° | Altitude: 2947.25 km

Lacus Somniorum has been formed by the flooding of a complex depression that shows signs of multiple underlying craters. A single large crater seems to underlie Lacus Mortis. Burg is a Copernican crater with a well-developed ray pattern. The floor of Lacus Mortis shows extensive fracturing, including a scarp that is about 800 m in height, judging by its shadow at 23.7° sun elevation. There is some evidence of dark deposits in the vicinity of faults (rimae) in Lacus Mortis, suggesting the release of volatiles after the deposit of rays from the Burg impact.

LO4-091H3 | Sun Elevation: 23.7° | Altitude: 2947.25 km

This eastern part of Mare Frigoris is older than the western part, according to crater counts. There are signs of a possible basin (360 km, 55° N, 30° E) under this part of Mare Frigoris. The ridge running through Baily and around to just southeast of Gartner, could be the rim of a basin underlying western Mare Frigoris (LO4-098H3). This area shows a number of mare ridges (dorsa) that are influenced by the underlying topography. Baily seems to have been formed by a pair of impactors arriving side by side into a highland area at a low angle from the east-northeast. The compound crater was subsequently flooded by lava and then its floor was fractured, forming the rimae in the crater floor.

LO4-085H1 Sun Elevation: 22.5° Altitude: 2716.89 km

Mare Tranquillitatis was a prime choice for the first Apollo landing site because of its large smooth area and equatorial position. Further, its eastern location made it a first choice within a launch window. Ranger 8 was the first to return detailed pictures from this site in February 1965, confirming the topographic suitability for a manned landing but leaving open the question of soil strength. Surveyor 5 (September 1967) confirmed the firmness of lunar soil here. An alpha-scattering experiment found that the chemical composition of the mare soil was characteristic of basalt. As can be seen from this Lunar Orbiter photo, the landing site for Apollo 11 was chosen to be as flat and featureless as possible, nearly free of large craters and ridges that could jeopardize a landing. On July 20, 1969, Eagle landed at Tranquility Base. The mission returned 22 kg of samples collected from the mare by Armstrong and Aldrin, including rocks 3.6 billion years old.

LO4-085H2 Sun Elevation: 22.5° Altitude: 2716.89 km

Near Plinius, the main ring of the Serenitatis Basin intersects the main ring of the preexisting Tranquillitatis Basin. Serenitatis doubtless demolished and dispersed the main ring of Tranquillitatis. The melt sheet of Tranquillitatis may have suppressed the sector of the main ring of Serenitatis beyond Promontorium Archerusia. An extensive fracture zone, marked by Rimae Plinius, passes where the main ring of Serenitatis would be expected. The last layer of lava to flow between the two basins has come from Tranquillitatis; its boundary can be seen northwest of Promontorium Archerusia. Ridges in Mare Tranquillitatis show evidence of the underlying topography, indicating that the mare may be shallow here.

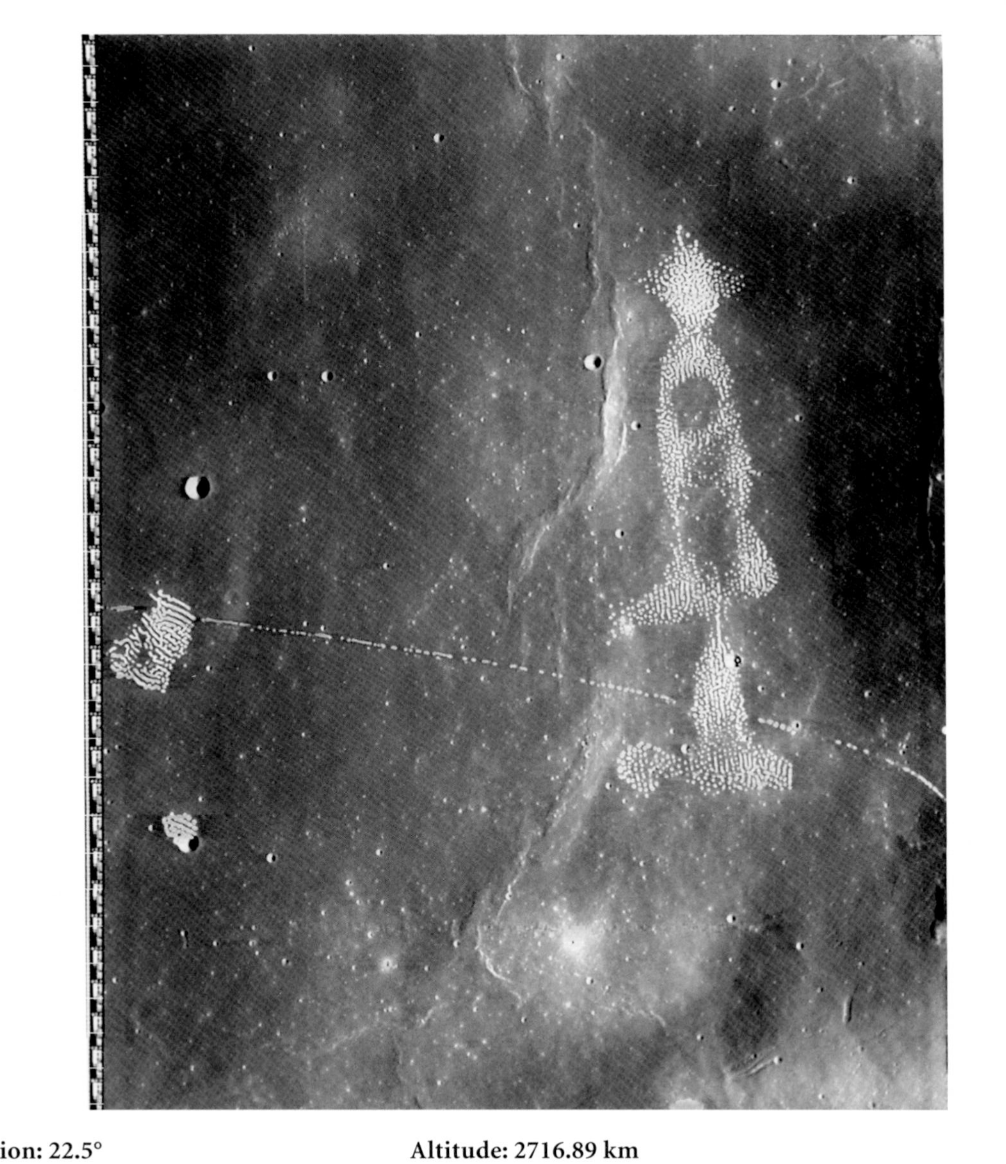

LO4-085H3 Sun Elevation: 22.5° Altitude: 2716.89 km

The area to the right of this photo, toward the eastern edge of Mare Serenitatis, is not only darker but of a different spectral class than the central mare surface. It may be covered with dark mantling material. See also LO4-078H3, whose view continues to the east.

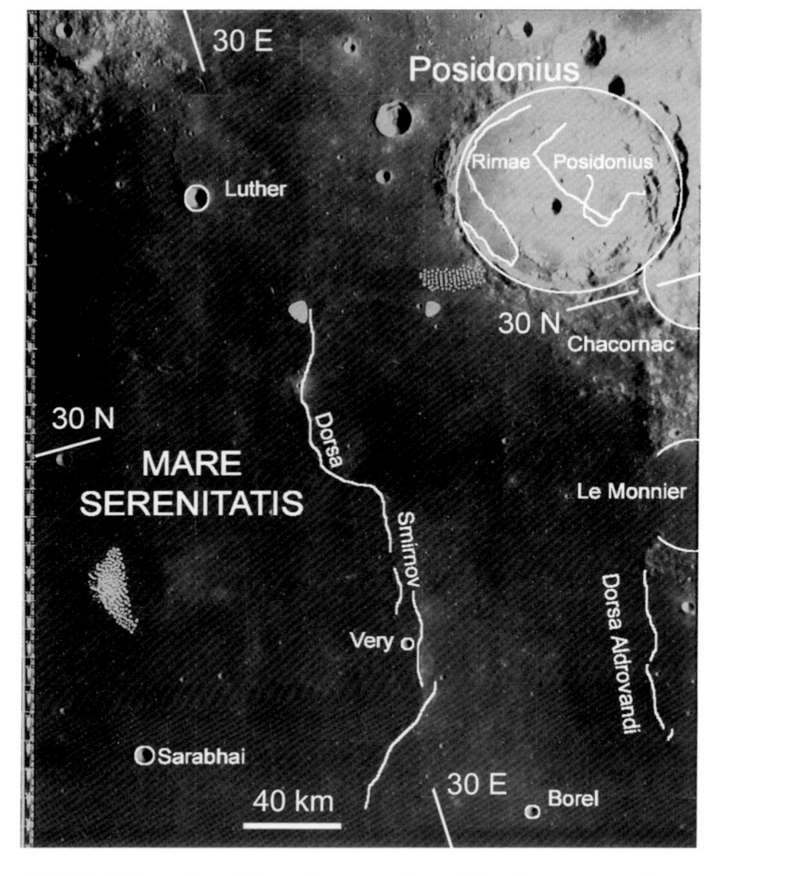

LO4-086H1 Sun Elevation: 23.7° Altitude: 2956.08 km

The central and southern sections of Dorsa Smirnov may mark an inner ring of the Serenitatis Basin. The extension of Mare Serenitatis to the north may represent a flow of lava north from Mare Serenitatis through Lacus Somniorum and Lacus Mortis toward Mare Frigoris. This photo was exposed for the mare, not for the brighter highlands. There is better coverage of Posidonius in LO4-079H1.

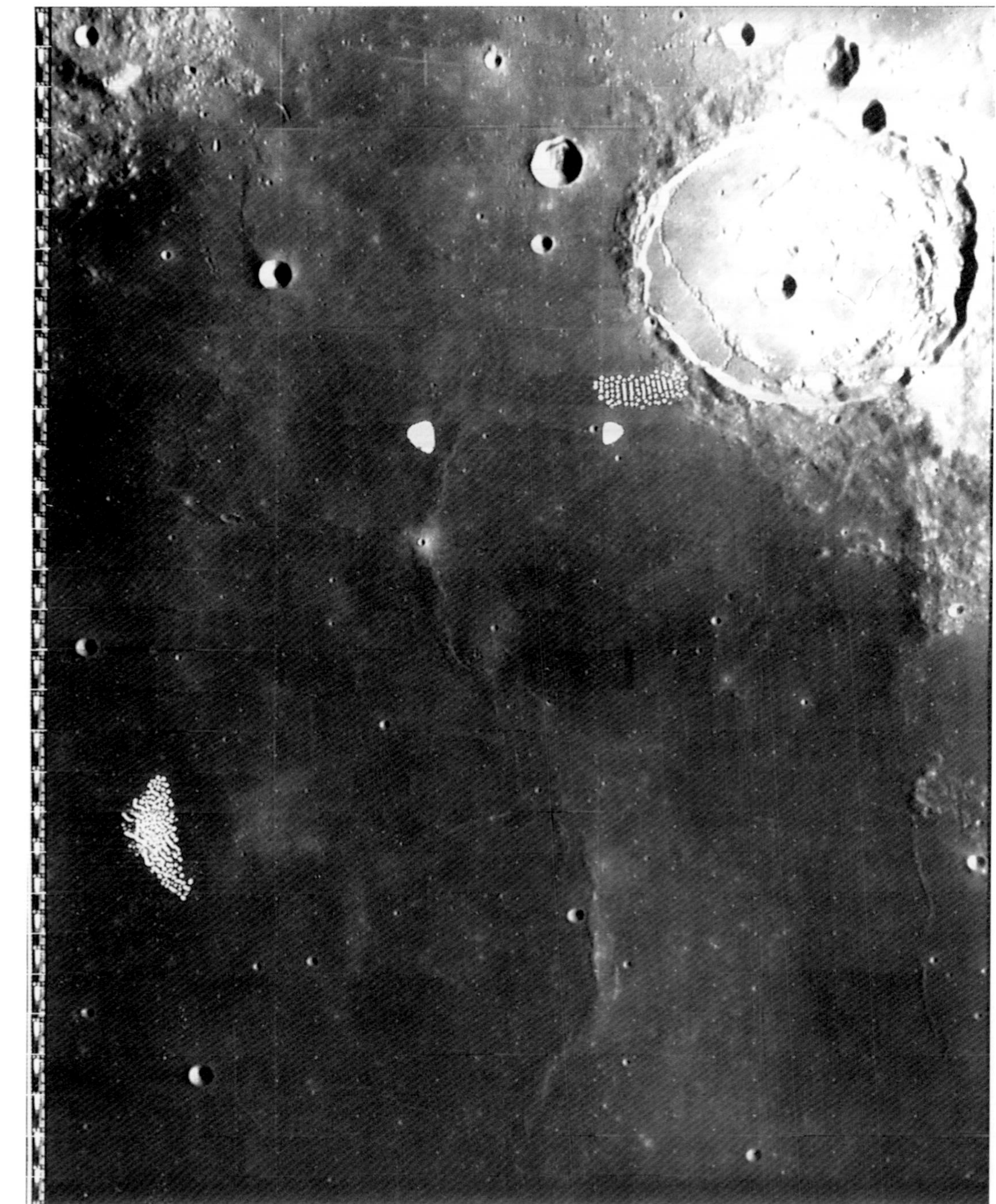

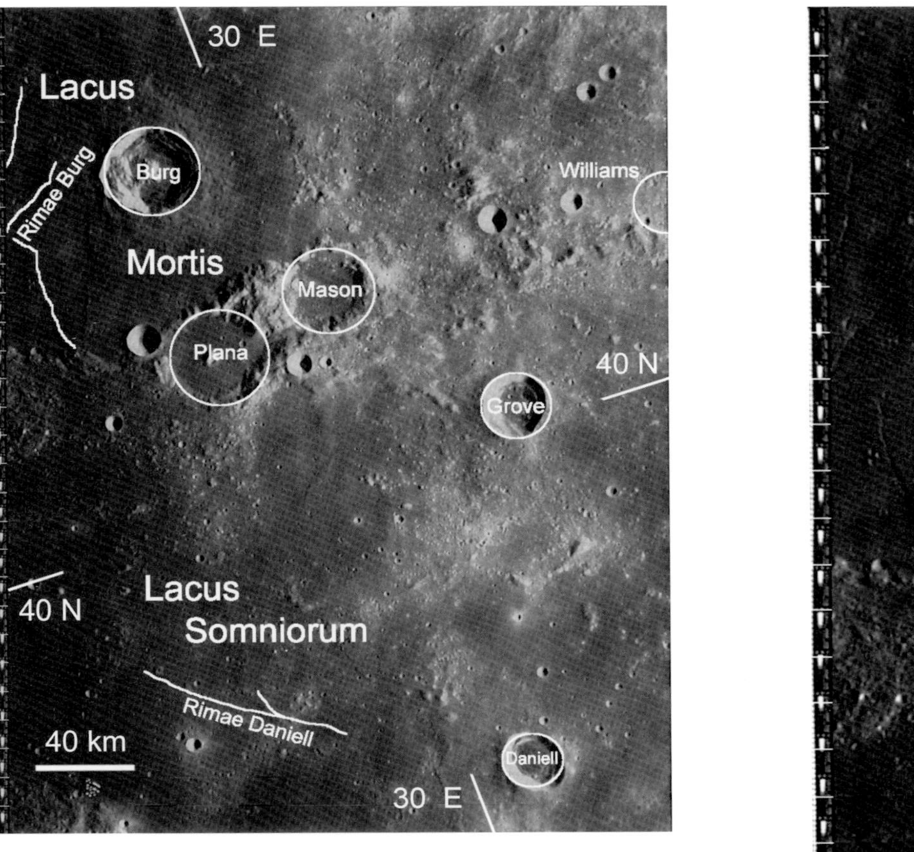

LO4-086H2 Sun Elevation: 23.7° Altitude: 2956.08 km

Craters Plana and Mason have impacted the main ring of the large crater beneath Lacus Mortis (see LO4-091H2), causing the strange pattern of redistributed material surrounding them. In particular, crater Mason has thrown heavy lobes of debris to the southeast, modifying ejecta from the crater whose rim it has impacted.

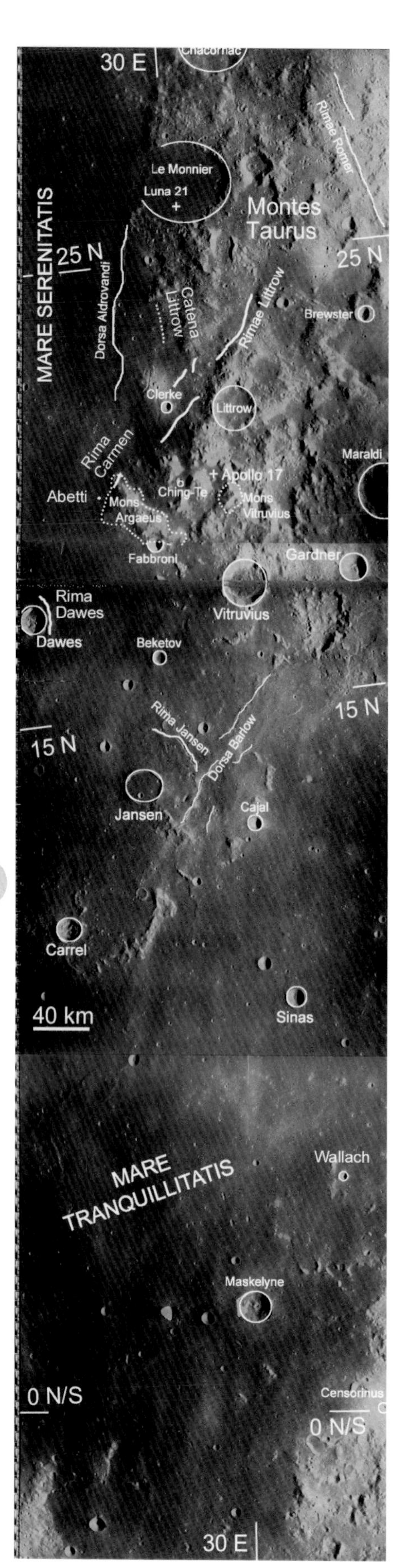

LO4-078H
Sun Elevation: 23.6°
Altitude: 2722.07 km

One objective of the Apollo 17 Mission (December 1972) in the Taurus-Littrow valley was to sample the dark mantling material in this area. Although the material was mostly covered, it was excavated and exposed by the impact of local craters. Sampled by Schmitt and Cernan, it is composed of beads of a titanium-rich composition that is dark orange if a glass and black if crystallized. Aged at 3.4 billion years old, it is generated by volcanic fountains driven by an unknown volatile material.

Lava may have risen through a hardening surface at Dorsa Barlow, running through Rima Jansen to the northwest and filling a crater about 50 km north of Jansen. The main rings of Tranquillitatis and Serenitatis each pass between Dawes and Fabbroni, but there are no mountains there. Perhaps Serenitatis destroyed this sector of the Tranquillitatis ring and the melt sheet of Tranquillitatis suppressed the Serenitatis ring.

The highlands near Copernican crater Censorinus in the bottom right corner of this photo mark the remains of the southwest sector of the much-degraded main ring of the Tranquillitatis Basin (see Photos LO4-077H3 and LO4-072H3 in the Nectaris Basin Region).

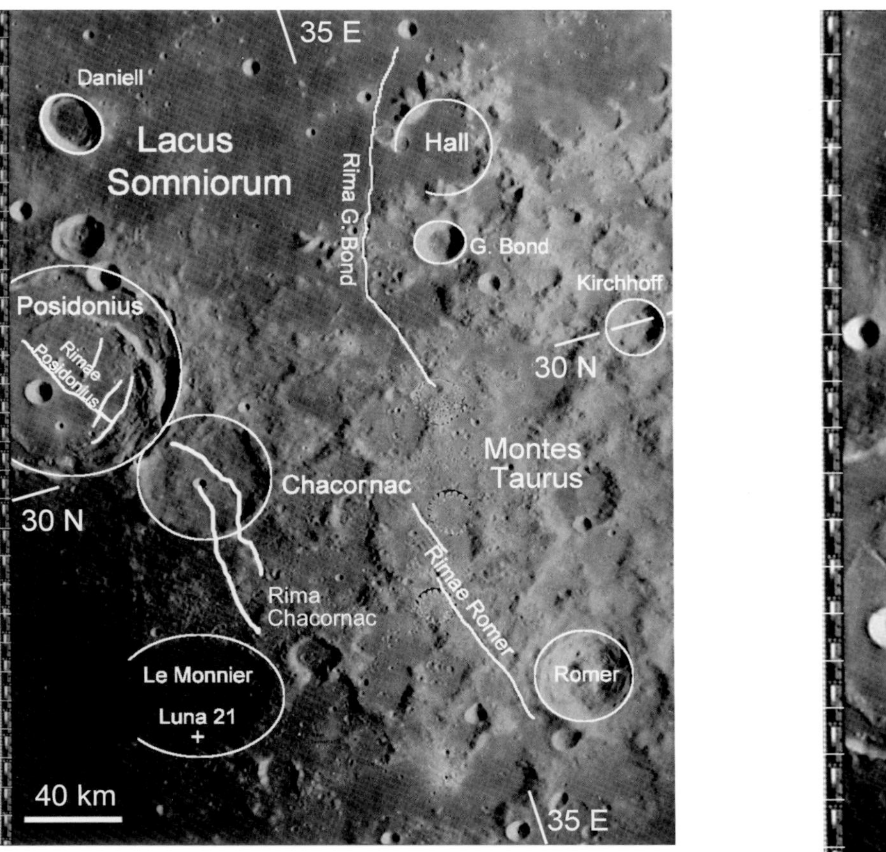

LO4-079H1 **Sun Elevation: 24.5°** **Altitude: 2963.78 km**

Luna 21, the first of three post-Apollo unmanned successes of the USSR lunar program, landed in January 1973. It deployed the second Lunokhod rover. On the floor of Le Monnier, this rover encountered slippery going (80% wheel slip) and sank to its axles at one point inside a crater. This is consistent with Le Monnier being heavily covered with uncompacted dark mantling material. The rover moved from the landing site into the foothills of Montes Taurus, finding a decrease in iron content from 9% to 4%. It also explored an open crack in the basalt more than 300 m wide. Uplift from rising lava has raised the greater part of the floor of Late Imbrian Posidonius nearly to the top of its rim (LO4-079H3).

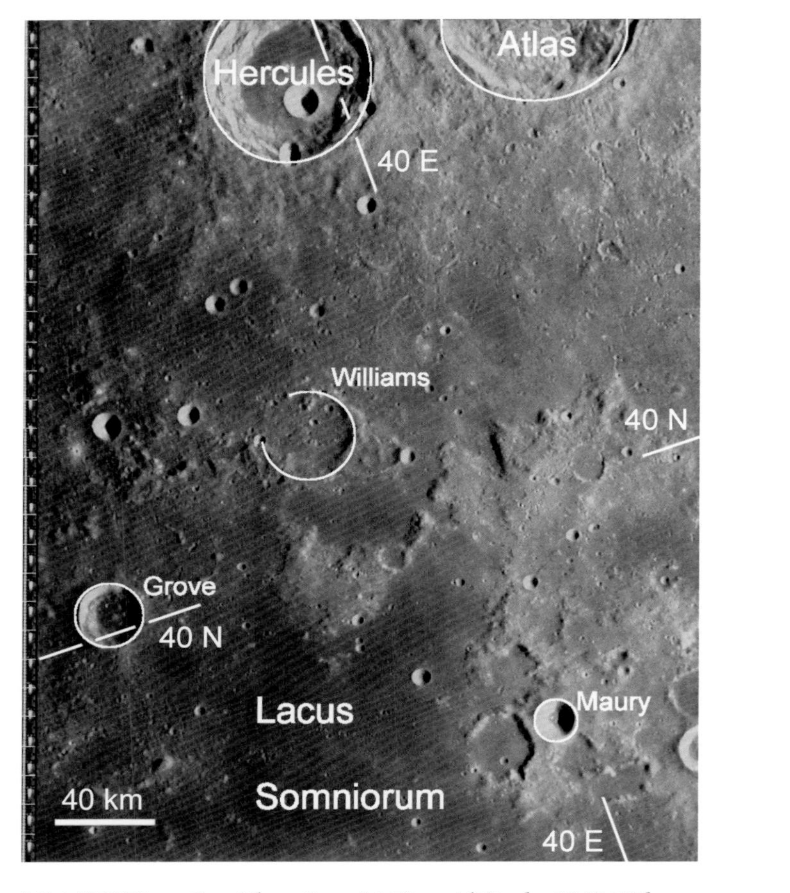

LO4-079H2 **Sun Elevation: 24.5°** **Altitude: 2963.78 km**

Atlas is believed to be Late Imbrian in age and Hercules to be Eratosthenian (the next younger period). It is interesting to compare the degree of degradation and interplay of ejecta between the two craters of comparable size, impacting similar terrain. More of these craters can be seen in LO4-079H3. Lacus Somniorum is quite shallow, as indicated by the many partially flooded craters and islands above the mare surface.

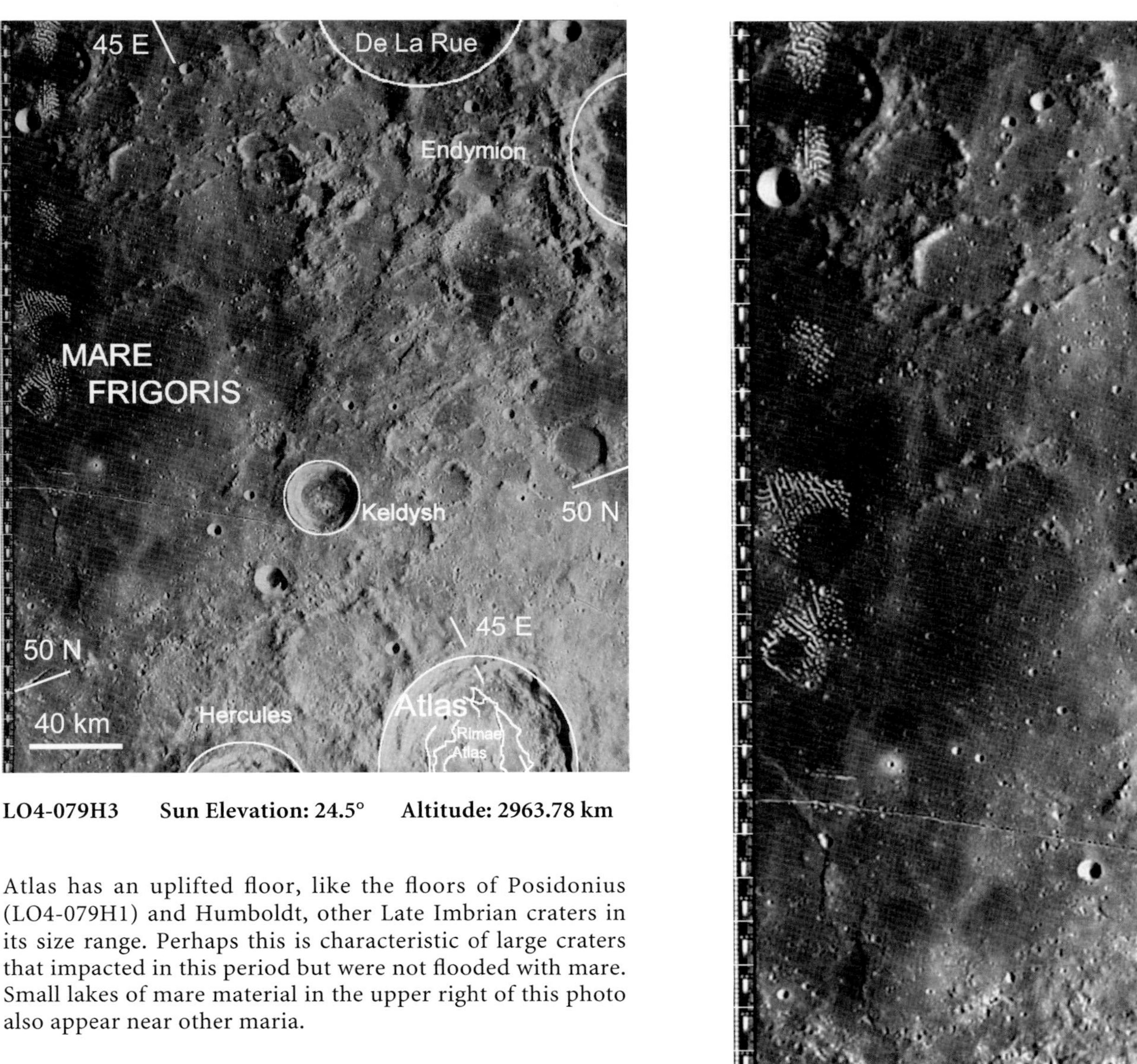

LO4-079H3 **Sun Elevation: 24.5°** **Altitude: 2963.78 km**

Atlas has an uplifted floor, like the floors of Posidonius (LO4-079H1) and Humboldt, other Late Imbrian craters in its size range. Perhaps this is characteristic of large craters that impacted in this period but were not flooded with mare. Small lakes of mare material in the upper right of this photo also appear near other maria.

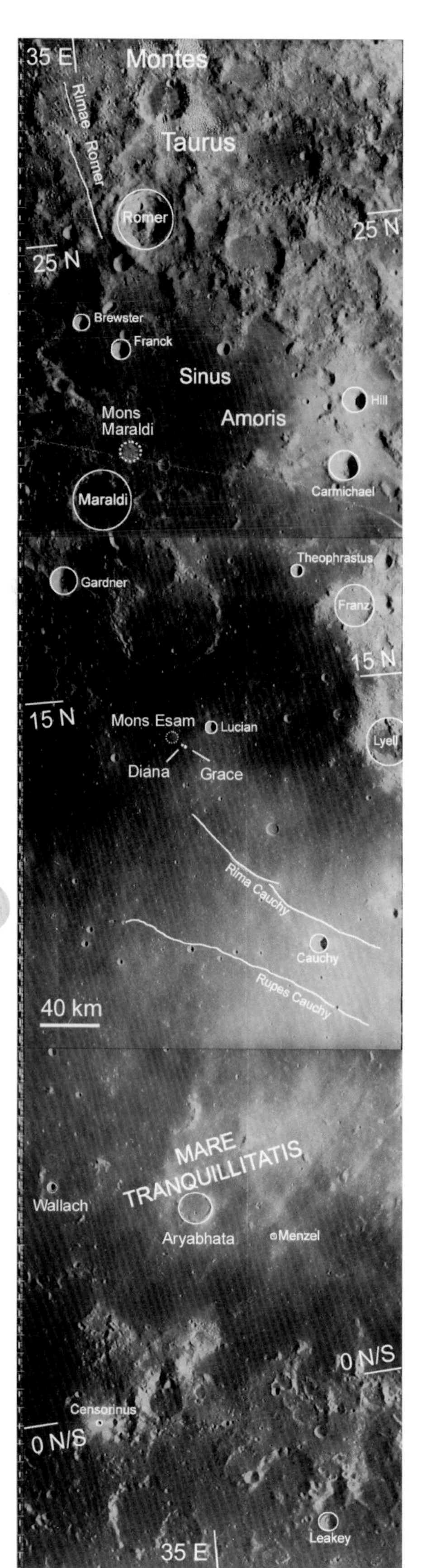

LO4-073H
Sun Elevation: 24.9°
Altitude: 2756.52 km

Montes Taurus separates the Serenitatis and Crisium Basins; it is within a radius from the rim of each, and no doubt received ejecta from both, overlaid with Imbrium ejecta. An 85-km crater may underlie Sinus Amoris. Romer, Hill, and Carmichael are fresh Copernican craters. Chains and clusters of Imbrium secondary craters are east of Romer.

The IAU named Diana and Grace in 1979. Were the IAU representatives thinking of two commoners who married royalty? Mons Esam is a mound surmounted by a crater, in the middle of a mare. It looks like an ash cone or fumarole, a possible source of dark mantling material. Rima Cauchy (a rille) and Rupes Cauchy (a scarp) probably are both surface expressions of faults in the mare.

The Censorinus Highlands are located at a triple intersection of the main rings of the Tranquillitatis and Fecunditatis Basins and the first outer ring of the Nectaris Basin. Shaking may have leveled Imbrium ejecta in the terra plains unit between Censorinus and Leakey. There are deposits of dark mantling along the shoreline of Mare Tranquillitatis and on the Censorinus highlands.

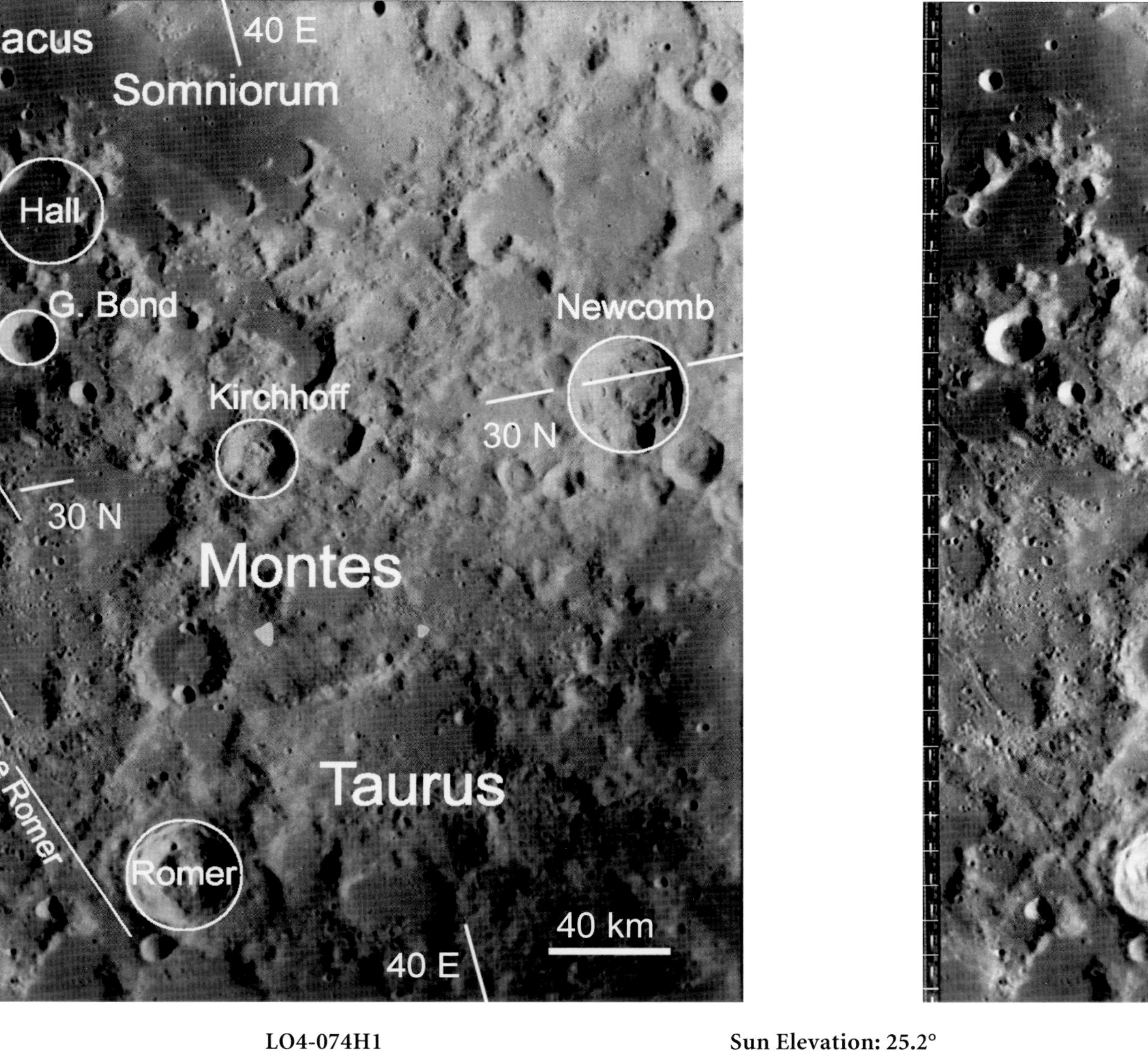

LO4-074H1 Sun Elevation: 25.2° Altitude: 2969.81 km

Ejecta with striations radial to the Crisium Basin can be seen in the lower right corner of this photo, south of Newcomb. Newcomb and Romer seem free of such ejecta, but Kirchoff shows probable secondaries from Crisium. Newcomb is a very complex crater. Extensive slumping has left deep scallops on its rim. The impact of the 20-km crater just southwest of Newcomb may have caused the extensive collapse of the nearby wall.

LO4-074H2 Sun Elevation: 25.2° Altitude: 2969.81 km

The structure north of Maury is very unusual. It looks like a shield volcano that has erupted in about four episodes, each with lava that is much more viscous than other flows on the Moon. More likely, however, it is molten ejecta or a salvo of lava bombs from Crisium to the south. Primary craters in this area are assigned to diverse age ranges. Franklin is of the Early Imbrian Epoch, Atlas is of the Late Imbrian Epoch, Hercules is of the Eratosthenian Period, and Cepheus is Copernican.

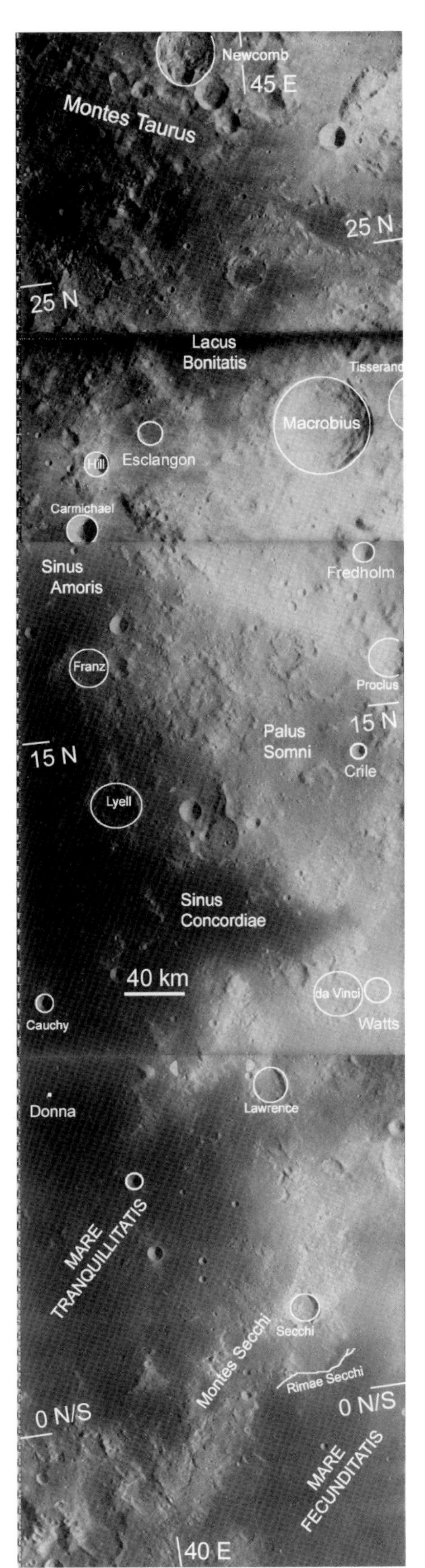

LO4-066H
Sun Elevation: 24.2°
Altitude: 2730.73 km

This part of Montes Taurus is a highlands area between the Serenitatis, Tranquillitatis, and Crisium Basins. The ridges near Lacus Bonitatis are radial to the Crisium Basin, whose main ring runs through Macrobius. Lacus Bonitatis, running at right angles to the ridges, is in an outer trough of Crisium.

The complex region east of Mare Tranquillitatis has been influenced by many basins. Sinus Concordiae has flooded part of an outer trough of the Fecunditatis Basin, beyond the ring that passes north of da Vinci. Palus Somni has received deposits from Tranquillitatis, Fecunditatis, Crisium, Serenitatis, and Imbrium (in that order). These deposits cover the eastern sector of the main ring of Tranquillitatis as it passes east of Lyell. The highland peaks between Cauchy and da Vinci are at an intersection of rings from the Tranquillitatis and Fecunditatis Basins.

An outer ring of the Fecunditatis Basin passes near Zahringer, and the main ring of Tranquillitatis passes by Montes Secchi. The intersection of these two rings, just north of Lawrence, has raised a plateau (LO4-066H3). The Censorinus highlands are in the bottom left. Dark mantling material has slid off steep slopes of these mountains, leaving bright areas.

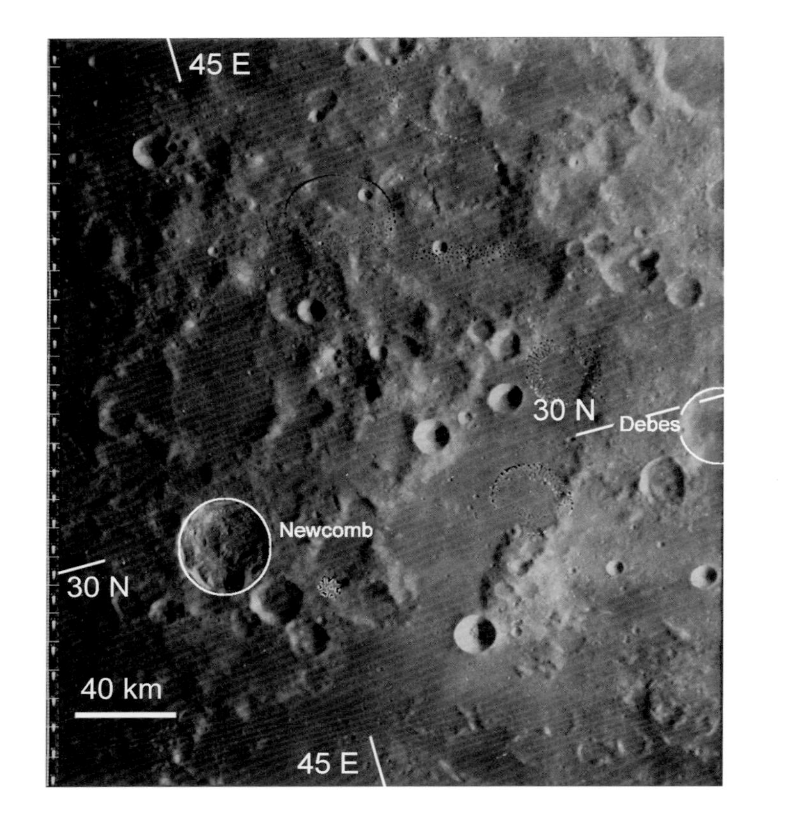

LO4-067H1 **Sun Elevation: 24.7°** **Altitude: 2975.55 km**

The large trough between Newcomb and Debes may have been initially formed by a chain of craters from the Tranquillitatis impact and then been reinforced as an outer trough of the Crisium Basin. Shaking from the Crisium impact may have served to level deposits in the area from the Fecunditatis Basin as well.

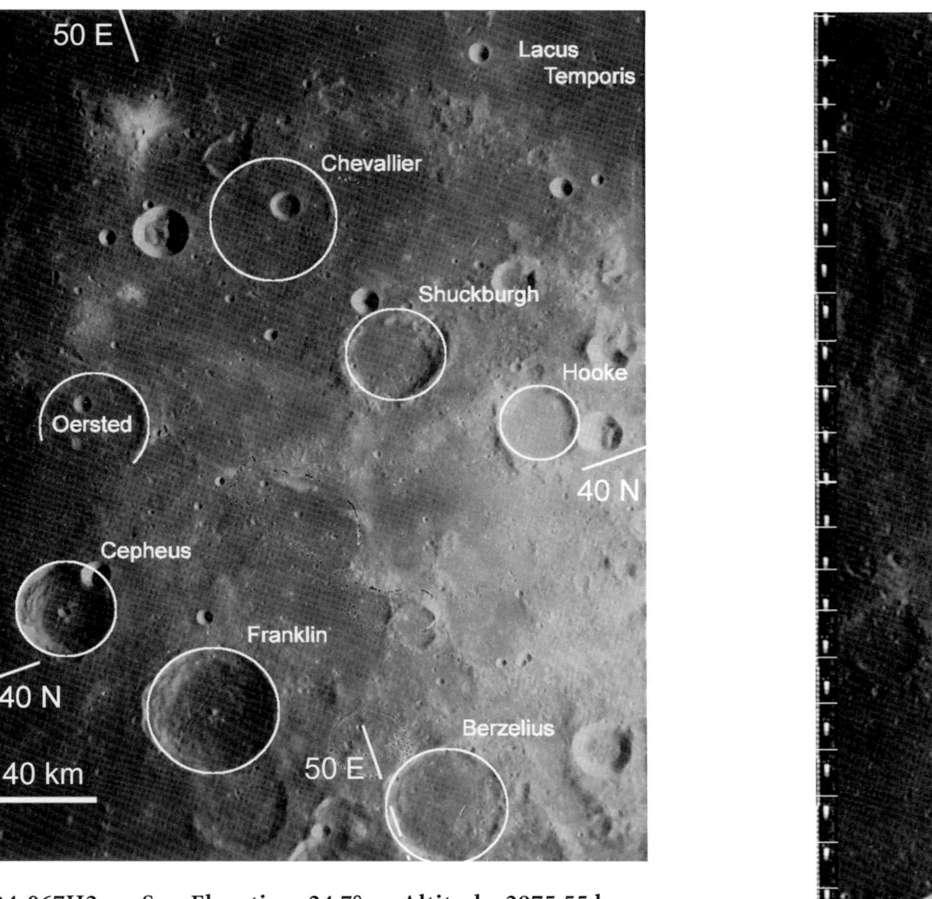

LO4-067H2 Sun Elevation: 24.7° Altitude: 2975.55 km

The area between Franklin and Lacus Temporis is too flat to be crater-saturated highlands, yet it is not completely flooded with mare material even though craters such as Chevallier, Shuckburgh, and Oersted have been partly submerged. Clementine elevation data shows a roughly circular depression centered at 53.5° E and 42.0° N (near Shuckburgh), about 400 km in diameter (including both Franklin and Lacus Temporis). This depression may mark a basin that has leveled this region.

LO4-067H3 **Sun Elevation: 24.7°** **Altitude: 2975.55 km**

The Nectarian crater Endymion has been scarred by ejecta from the Humboldtianum Basin, of the same period. Endymion is located in an outer trough of the Humboldtianum Basin; the main ring of the Humboldtianum Basin can be seen crossing the upper right corner of this photo. The shock of the Humboldtianum impact may have been the cause of the unusually strong slumping of the wall of Endymion.

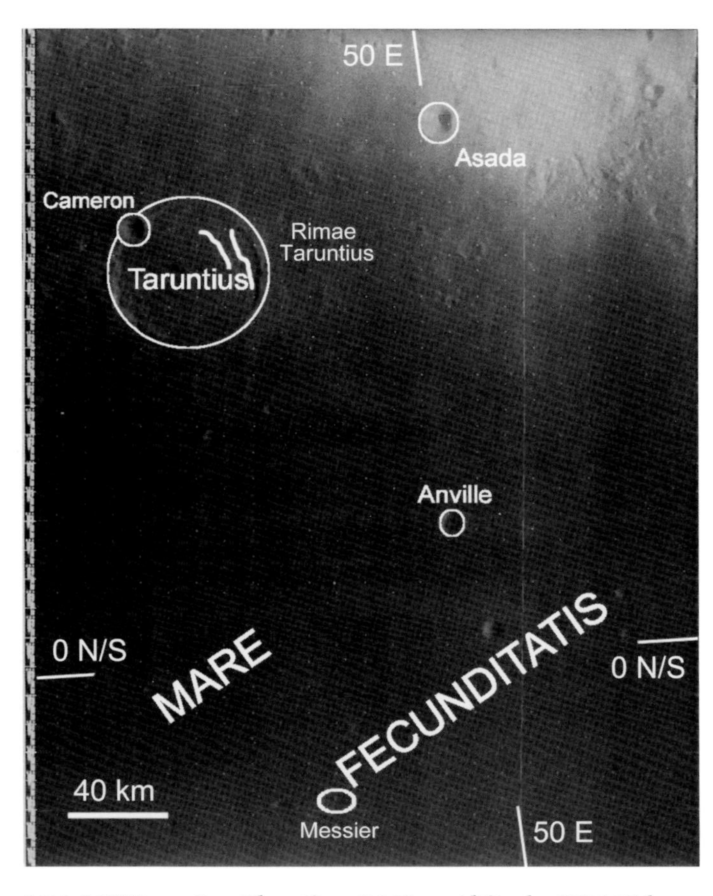

LO4-061H1 Sun Elevation: 24.6° Altitude: 2734.00 km

Failure of Lunar Orbiter's thermal door (lens cap) caused glare that strongly affected this photo. This area includes the northwestern portion of the floor of Mare Fecunditatis (the southern portion is in the Nectaris Basin Region). The brighter area in the upper right-hand corner is a highland area close to Mare Crisium (LO4-061H2). The Copernican crater Taruntius impacted after the mare flooding. Rising lava has uplifted a melt sheet on its floor; the resulting fractures underlie Rimae Taruntius. Catena Taruntius lies between Taruntius and Anville but cannot be seen in this photo.

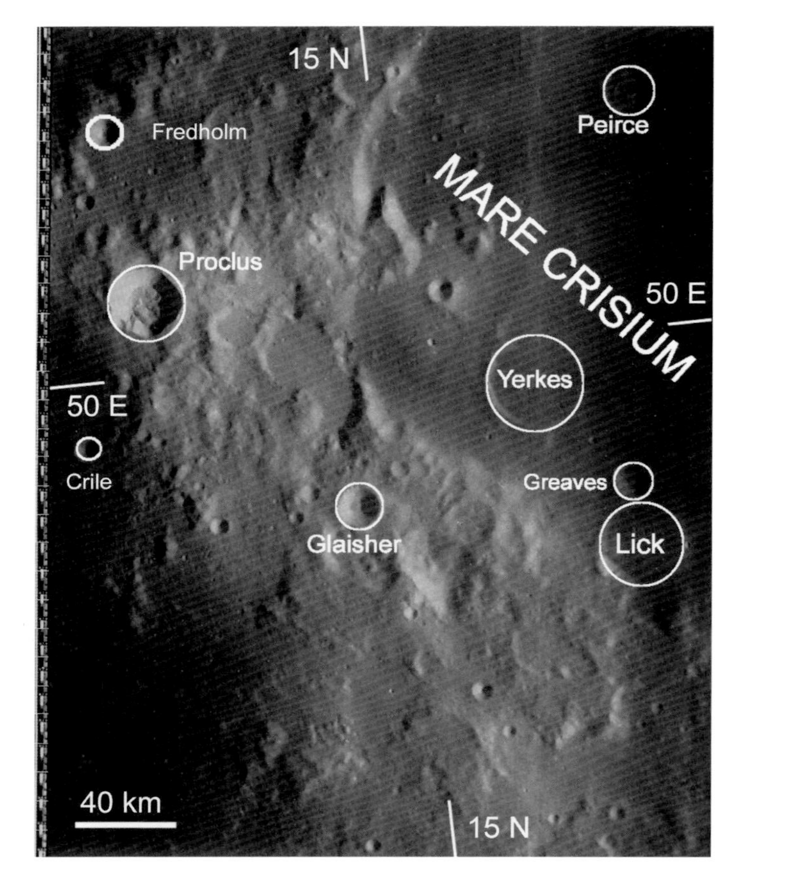

LO4-061H2 Sun Elevation: 24.6° Altitude: 2734.00 km

This view of the western shore of Mare Crisium shows striations of the limited ejecta pattern in this direction. Other coarser striations (and remote sensing data from Clementine) indicate that this Crisium ejecta has been subsequently covered with ejecta from the Imbrium and/or Serenitatis Basin to the northwest (both in the same direction and both rims being two basin radii away). The smooth sloping walls of Proclus indicate that it has impacted relatively unlayered, uncompacted material. Such is typical of the ejecta blanket just outside a basin rim, in this case the Crisium Basin. Proclus is a young Copernican crater with a spectacular ray pattern. Part of the pattern can be seen on the floor of Mare Crisium.

50 E
Debes
Tralles
Cleomedes
25 N
25 N
Tisserand
40 km
MARE CRISIUM
Macrobius
50 E
Swift

LO4-061H3 Sun Elevation: 24.6° Altitude: 2734.00 km

This area to the northwest of Mare Crisium probably shows ejecta from the Tranquillitatis Basin to the southwest, only about one basin radius from the main ring, as well as ridges and valleys from the direction of the Imbrium and Serenitatis Basins. The outer trough of the Crisium Basin is flooded with mare, which continues into the floor of Cleomedes. The broad elongated depression in the upper left corner is radial to Tranquillitatis (LO4-067H1). A valley halfway between Tisserand and Cleomedes is radial to Crisium.

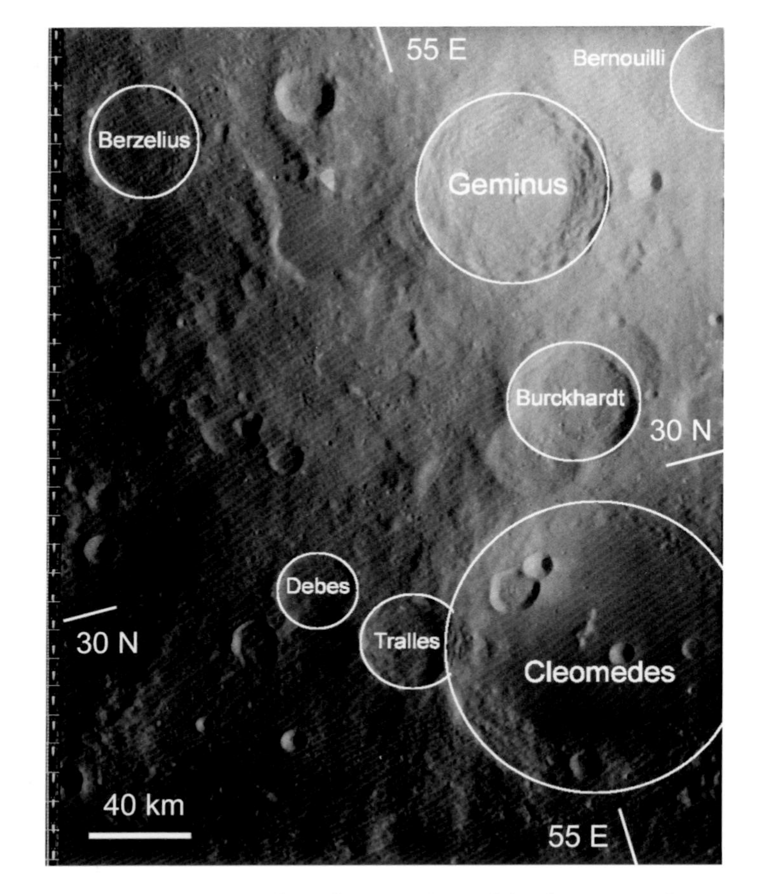

LO4-062H1 Sun Elevation: 25.4° Altitude: 2979.32 km

This area is about two basin radii away from the Imbrium, Serenitatis, Tranquillitatis, and Humboldtianum Basins and is about one basin radius away from the Crisium Basin. It is likely that rugged highlands are covered with a thick blanket of pulverized ejecta that has somewhat leveled the terrain. A very smooth area in a depression that may have been formed by a chain of craters from the Crisium Basin lies between Geminus and Berzelius. Perhaps it is ejecta that has been further leveled by local tectonic forces. Geminus has a complex, terraced wall, suggesting that it impacted layered material. Cleomedes, somewhat degraded, but with its own ejecta pattern intact, is thought to be Nectarian.

LO4-062H2 Sun Elevation: 25.4° Altitude: 2979.32 km

The western parts of Messala have received heavy deposits from Crisium to the south. Additional but finer deposits from Crisium can be seen between Carrington and Schumacher. Lacus Temportis, Shuckburgh, and Hooke may be located in the floor of an ancient crater or basin (see LO4-067H2). Messala and the 30-km crater in its western floor are both identified as of the Pre-Nectarian Period.

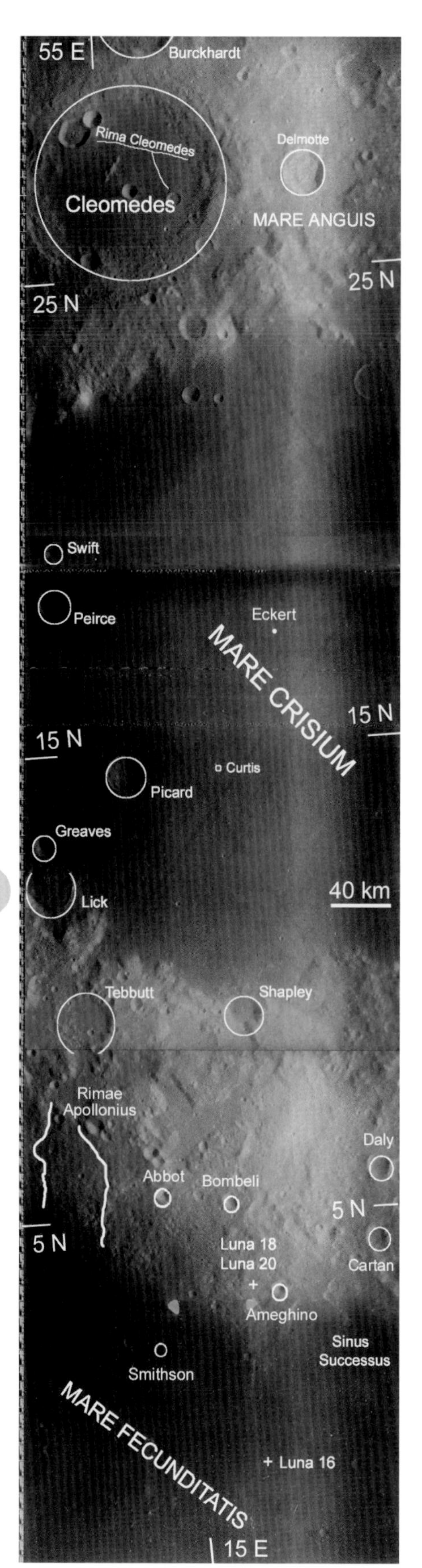

LO4-054H
Sun Elevation: 25.5°
Altitude: 2736.65 km

The ring that bounds Mare Crisium has a flat-topped structure like a plateau. In comparison with Orientale rings, this ring is more like one of the Montes Rook rings (inner rings) than like Montes Cordillera (the main ring). There is sporadic lava flooding beyond this inner ring, just as in the Orientale Basin.

The Crisium Basin has its own mascon, revealing an internal source of lava from beneath the crust. An "isthmus" of about 200 km separates Mare Crisium and Mare Fecunditatis. This narrow strip of highlands is deeply scored by ejecta from the younger Crisium.

Luna 16 (September 1970) landed on Mare Fecunditatis, drilled a 35-cm core, and returned a 101-gram (101-g) sample to earth. It showed a mineral composition similar to that of other maria samples, with some quantitative differences. Luna 18 (September 1971) failed upon landing, possibly due to the roughness of its landing site. Luna 20 (February 1972) returned 50 g of samples, probably Crisium ejecta. The samples are distinctly different from mare material; lighter in color and higher in density, the samples are similar to those gathered by Apollo from other basin rims.

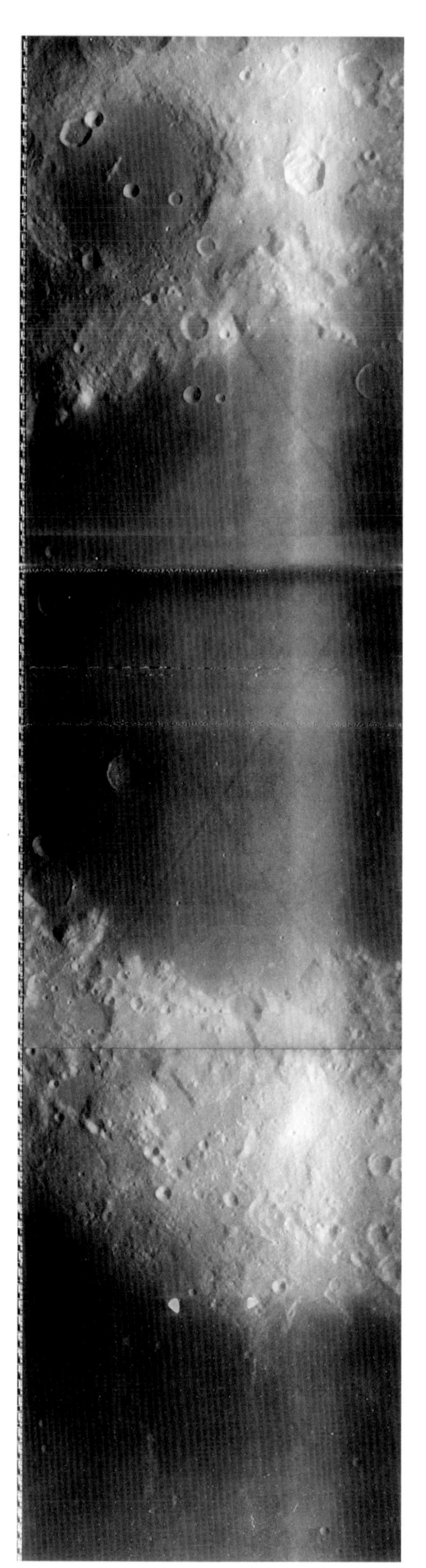

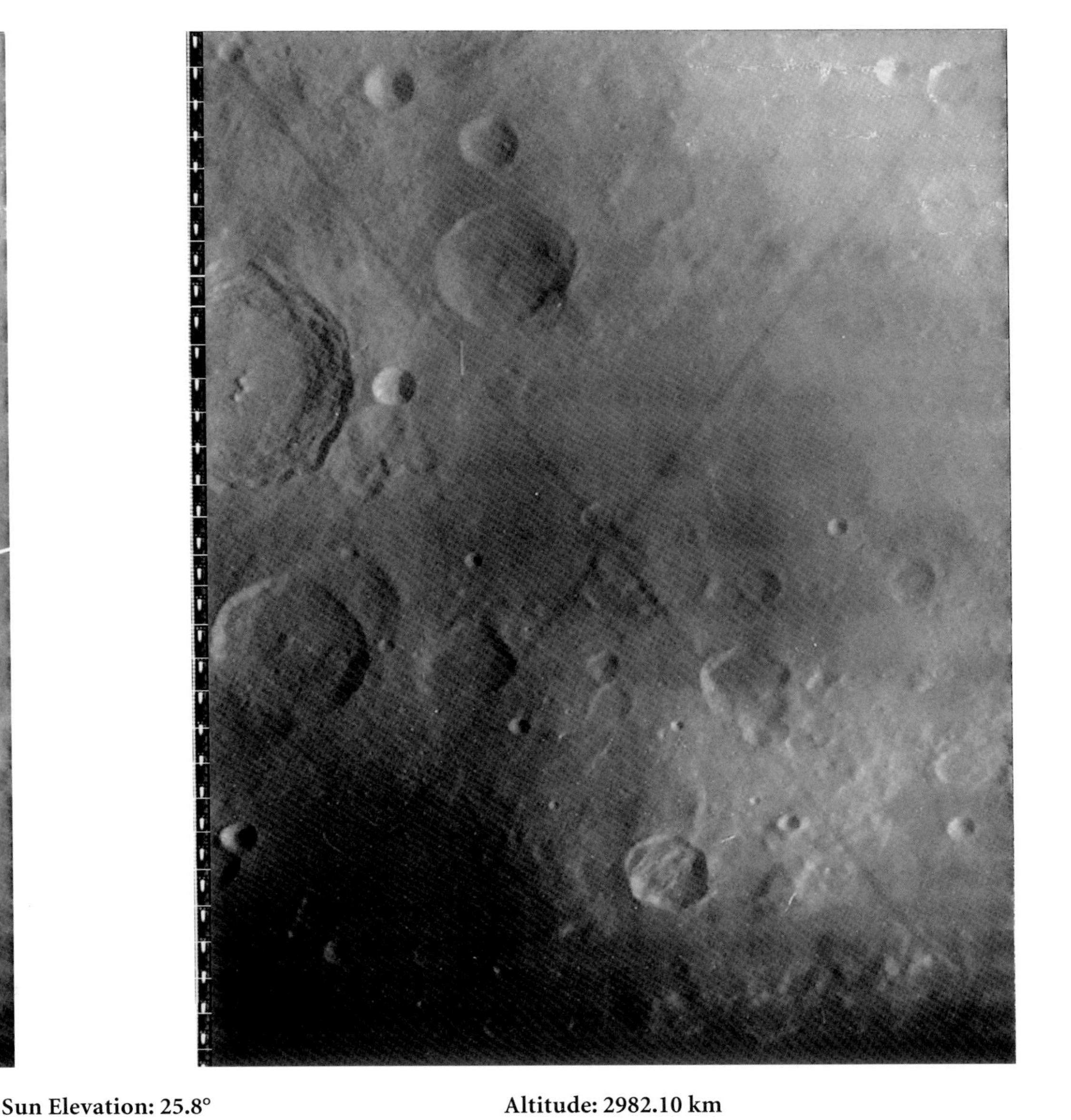

LO4-055H1 Sun Elevation: 25.8° Altitude: 2982.10 km

The large flat area east of Burckhardt is in an outer trough of the Crisium Basin. In Clementine albedo data, it has a dark, mottled look as if it is incompletely flooded with mare material. The flattening may have been partly caused by the formation of the trough of Crisium, partly by tectonic shaking of unconsolidated highlands, and partly by mare flooding.

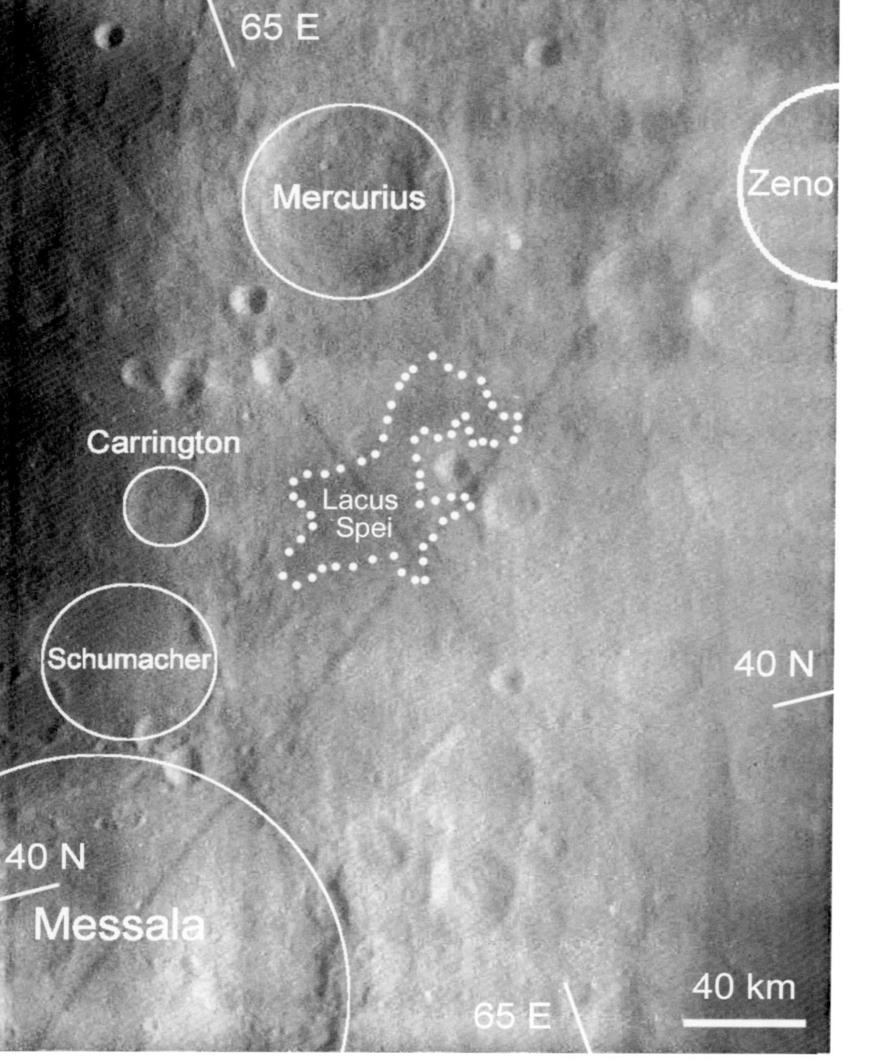

LO4-055H2

Sun Elevation: 25.8°

Altitude: 2982.10 km

Chains of secondary craters in this area radiate from the Humboldtianum Basin to the northeast. In Clementine albedo data, Lacus Spei is a dark and well-defined area of mare flooding. Schumacher is also flooded with mare material. For a clearer photograph of this area see LO4-165H2 in the Eastern Basins Region, Chapter 10.

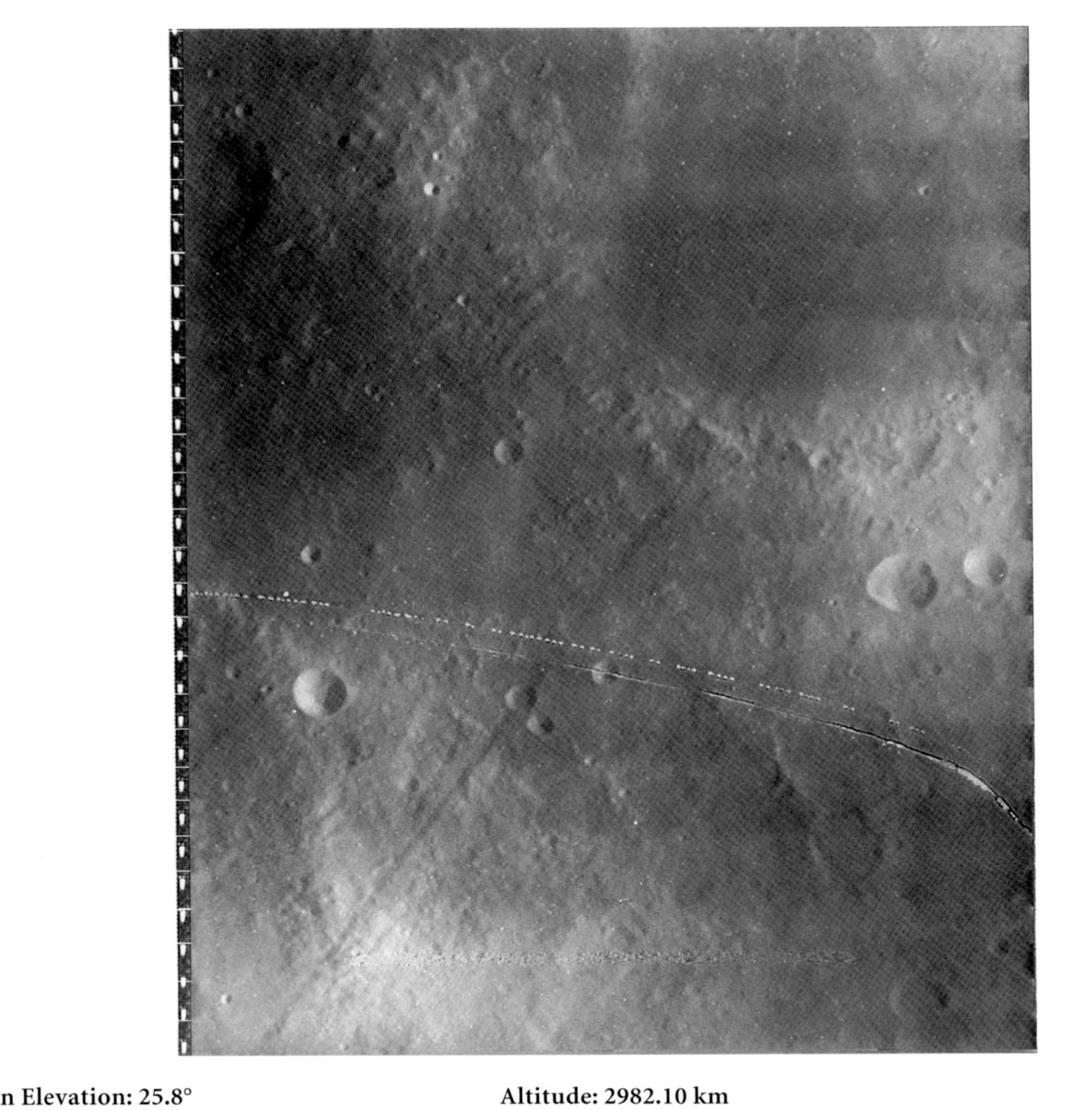

LO4-055H3 Sun Elevation: 25.8° Altitude: 2982.10 km

Rays from Hayn to the northeast appear vertical in this picture because of its orientation. The main ring of the Humboldtianum Basin passes across the lower left corner of this photo; an inner ring bounds Mare Humboldtianum. Between the two rings is an area that has been flattened by the forces that created the trough.

Chapter 10

Eastern Basins Region

10.1. Overview

The Eastern Basins Region covers roughly the area between 60° east longitude and the eastern limb (90° east longitude) and between 60° south latitude and 60° north latitude.

The region contains part of several definite ringed basins, Crisium, Australe, Smythii, and Humboldtianum (southern part) in downward order of size, and the probable basin, Balmer-Kapteyn. Each of these basins has some mare material within it, but (with the exception of Mare Crisium) the mare is irregular, probably because this region is midway between the thin crust of the western and central near side and the thick crust of the far side.

Although the basins in this region are older than Imbrium, they are relatively free of its influence becausee the region is mostly beyond a basin radius away from the main ring of Imbrium.

The Crisium Basin

The Crisium Basin has a topographic rim 740 km in diameter. It is one of the younger basins and is particularly interesting. It represents the transition between the older basins to the west, strongly dominated by ejecta from the giant, relatively young Imbrium Basin, and basins near the eastern limb and on the far side that are relatively free of the influence of Imbrium.

In a first impression, the Crisium Basin appears to be elongated in form, which might be taken as evidence of a glancing impact. On closer inspection, the shape resolves into two circular basins, as shown in Figure 10.1.

Figure 10.1. Crisium Basin (part of LO4-060M). Mare Crisium looks somewhat like a flounder. Mare Fecunditatis and Mare Tranquillitatis are in the lower left corner of the figure. Mare Marginis is far to the right. Mare Undarum, the dark area southeast of Crisium, is in an intersection of outer troughs of the Crisium and Fecunditatis Basins. Mare Crisium has flooded depressions from a complex series of impacts. The circular shape of the largest depression has been modified by a large crater or basin to the west-northwest of the main impact and a smaller crater to the east (the "tail of the fish").

The western, northern, and southern shores of the Crisium Basin are nearly covered with ridges and valleys formed by ejecta from the Imbrium Basin to the northwest. To the east, the topography is dominated by ejecta from Crisium itself. Clementine multispectral data show a strong difference between the mineral signatures to the west, which is upper crust ejected from the Imbrium Basin, and the signatures to the east, which are believed to contain portions of lower crust below the Crisium Basin.

The Australe Basin

The Australe Basin is partly on the near side and partly on the far side; its center is at 52° south latitude and 95° east longitude. Since it was formed in the Pre-Nectarian Period, it has been severely degraded. No rings outside of the main ring have been identified, but at least one inner ring has been traced.

The central part of this basin has been only partially flooded with mare lava. As seen in Figure 10.2, lava has erupted only where impactors have formed large craters in the basin floor. Perhaps the central Australe Basin had an extensive melt sheet that inhibited the rise of lava except where it has been penetrated.

The Smythii Basin

The Smythii Basin, as large as Crisium, is at 2° south latitude and 87° east longitude and straddles the eastern limb. It is best seen in Lunar Orbiter 2 photograph 196M, shown in Figure 10.3. There is a mascon associated with Mare Smythii.

10.2. High-Resolution Images

Lunar Orbiter 4 was planned to start its photographic mission at the eastern limb with a series of photos like the

Figure 10.2. Australe Basin (LO4-009M). The center of the basin is slightly left and below the center of the photo. The main ring, most pronounced in the eastern sector, spans 30° of the lunar circumference. The dark circular feature near the top is Mare Smythii, with radial ridges and valleys of the Smythii Basin ejecta surrounding it. The Balmer-Kapteyn Basin straddles the terminator southwest of Mare Smythii. Near the bottom of the picture is Schrodinger, a small far side basin that nonetheless has strong radial valleys extending from its main ring. The dark far side feature to the right is the flooded crater Tsiolkovskiy. Half of the ancient South Pole–Aiken Basin of the far side is in this picture. Its main ring intersects that of Australe. Schrodinger, but not Tsiolkovskiy, is within the South Pole–Aiken Basin.

Figure 10.3. The Smythii Basin (LO2-196M). The dark compact mare to the upper left is Mare Smythii, with Mare Marginis above it and a small area of mare below it. The southeastern sector of the main topographic ring of the Smythii Basin can be seen curving from the top center and out the lower left side of this photo.

rest of the mission, systematically taking four to six exposures on each orbit from near perilune, at an altitude of about 3000 km.

Unfortunately, the thermal door (a lens cap) locked open, leaving the optics open to the cold of space. As a result moisture condensed on the inside of the lenses (much as dew might settle on the outside of a lens), fogging the photographs. After analysis, the spacecraft was kept pointing at the lunar surface (warmer than deep space) and the lens cleared. Meanwhile, many exposures were lost or badly degraded. Of the planned high-resolution photos in this region, only LO4-009H and LO4-018H are printed in this chapter.

The mission operators compensated for the loss by scheduling a series of photos near the end of the mission, when the apolune (the other end of the orbit, at about 5500 km) came over the eastern limb. These are the photos used to cover the Eastern Basins Region. The resolution is of course lower and the coverage larger. As a result, two subframes from each of these exposures cover parts of this region. The remaining subframes cover parts of the North and South Polar Regions. Table 10.1 shows the high-resolution images of the Eastern Basins Region in schematic form.

The following pages show the high-resolution subframes from south to north and west to east. The apolune photos are in the order LO4-191H3, LO4-191H2, LO4-184H2, LO4-184H1, LO4-178H2, LO4-178H1, LO4-177H3, LO4-177H2, LO4-165H3, and LO4-165H2. This seems like a confusing sequence but it results from the spacecraft flying from north to south, in reverse of the usual direction. For consistency, we follow the usual south to north convention, despite having to present the pictures in the reverse order of their numbers. The perilune photos are presented in the order of LO4-018H1 ... LO4-018H3, LO4-009H1 ... LO4-009H3 (the usual order).

Because of the loss of photography early in Lunar Orbiter Mission 4, high-resolution exposures in the southern part of the Eastern Basins Region (with the exception of LO4-009) have been lost. To identify the features of this region, a medium-resolution photo from Mission 2 (LO2-196M) is provided at the end of this chapter.

Lattitude Range	Photo Number						
56 N–90 N		191H1		177H1	165H1	191H1	
27 N–56 N	062	191H2		177H2	165H2		
0–27 N	061	191H3		177H3	165H3	018	
0–27 S	060		184H1	178H1			
27 S–56 S	059		184H2	178H2			009
56 S–90 S			184H3	178H3			
Longitude at Equator	57 E	54 E	70 E	75 E	81 E	90 E	96 E

Table 10.1. The cells shown in white represent the high-resolution photos of the Eastern Basins Region (LO4-XXX H1, -H2, and -H3, where XXX is the Photo Number). The North Polar Region is to the north, the Serenitatis Basin Region is to the northwest, the Nectaris Basin Region is to the southwest, and the South Polar Region is to the south. The far side is to the east. Exposures 165 to 191 were taken from apolune in afternoon sun, and exposures 009 and 018 were taken at perilune in morning sun. Apolune photos were taken at about twice the altitude of perilune photos, so they have about half the resolution and four times the area of coverage.

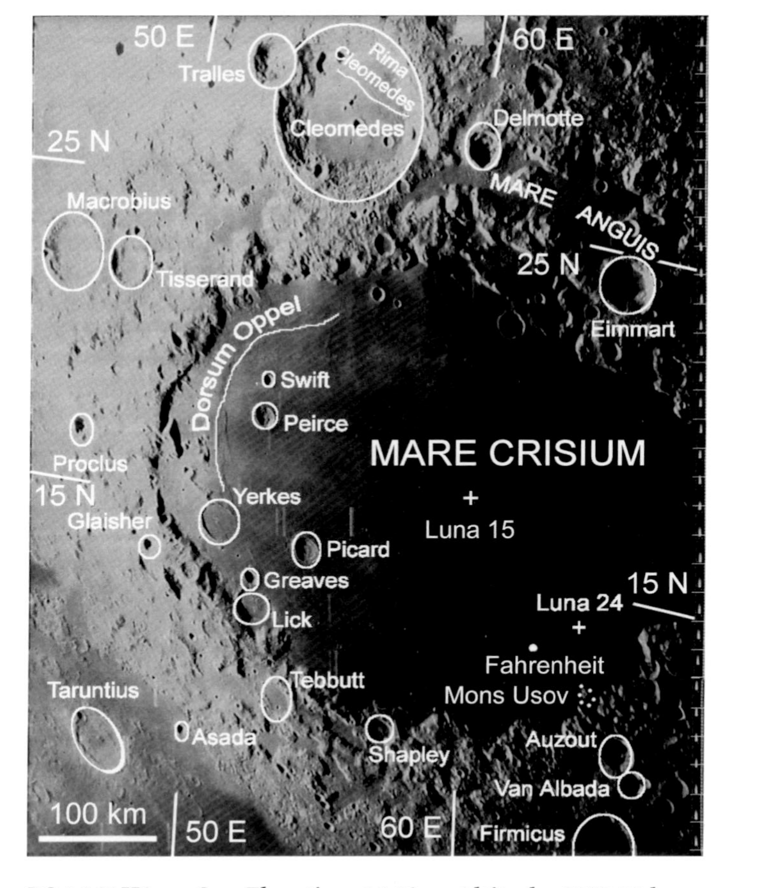

LO4-191H3 Sun Elevation: 16.3° Altitude: 5503.39 km

Mare Anguis has flooded an outer trough of the Crisium Basin. The edge of Mare Fecunditatis can be seen in the lower left corner of this photo. Dorsum Oppel, together with Dorsa Tetyaev (LO4-177H3), may outline an inner ring, along with Dorsum Termier and Dorsa Harker (not visible here). Days before Apollo 11 landed (July 1969), the retro-rocket of Luna 15 failed and it crashed. Luna 24 successfully landed (August 1976) and returned a core sample from a mare unvisited by Apollo. This was the last sample returned to Earth to this day.

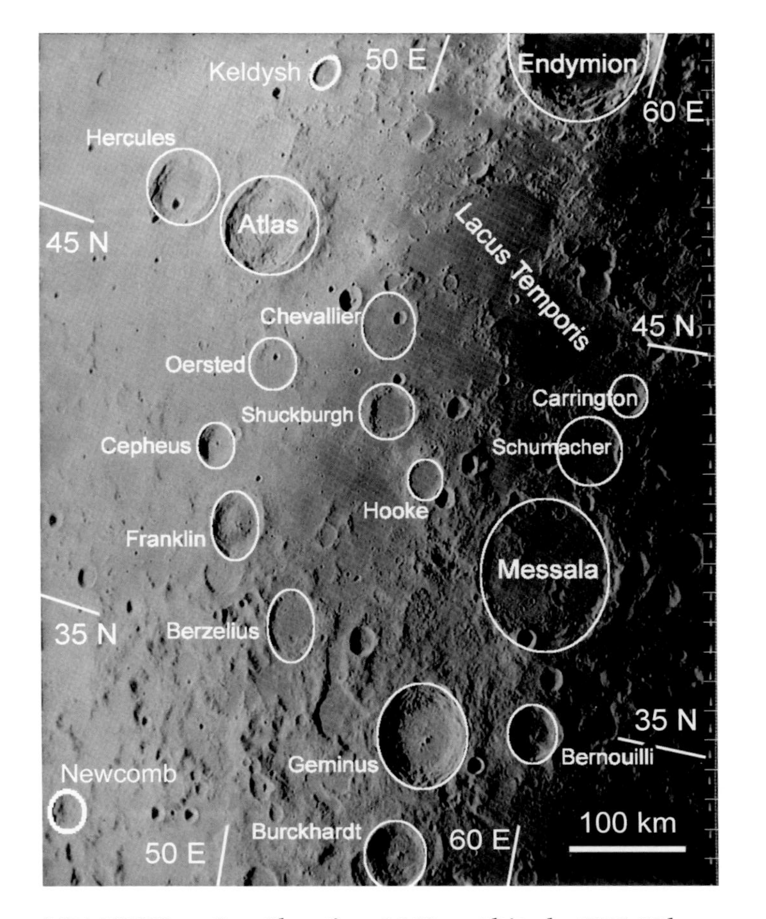

LO4-191H2 Sun Elevation: 16.3° Altitude: 5503.39 km

The eastern part of this area is in the Eastern Basins Region, but the western part is in the Serenitatis Basin Region. Because of the overlap, there is an opportunity to compare the afternoon sun illumination of this photo (coming from the east) with the morning sun of the photos in the Serenitatis chapter (Chapter 9). Striations from the south are radial to Crisium and those from the northeast are from Humboldtianum.

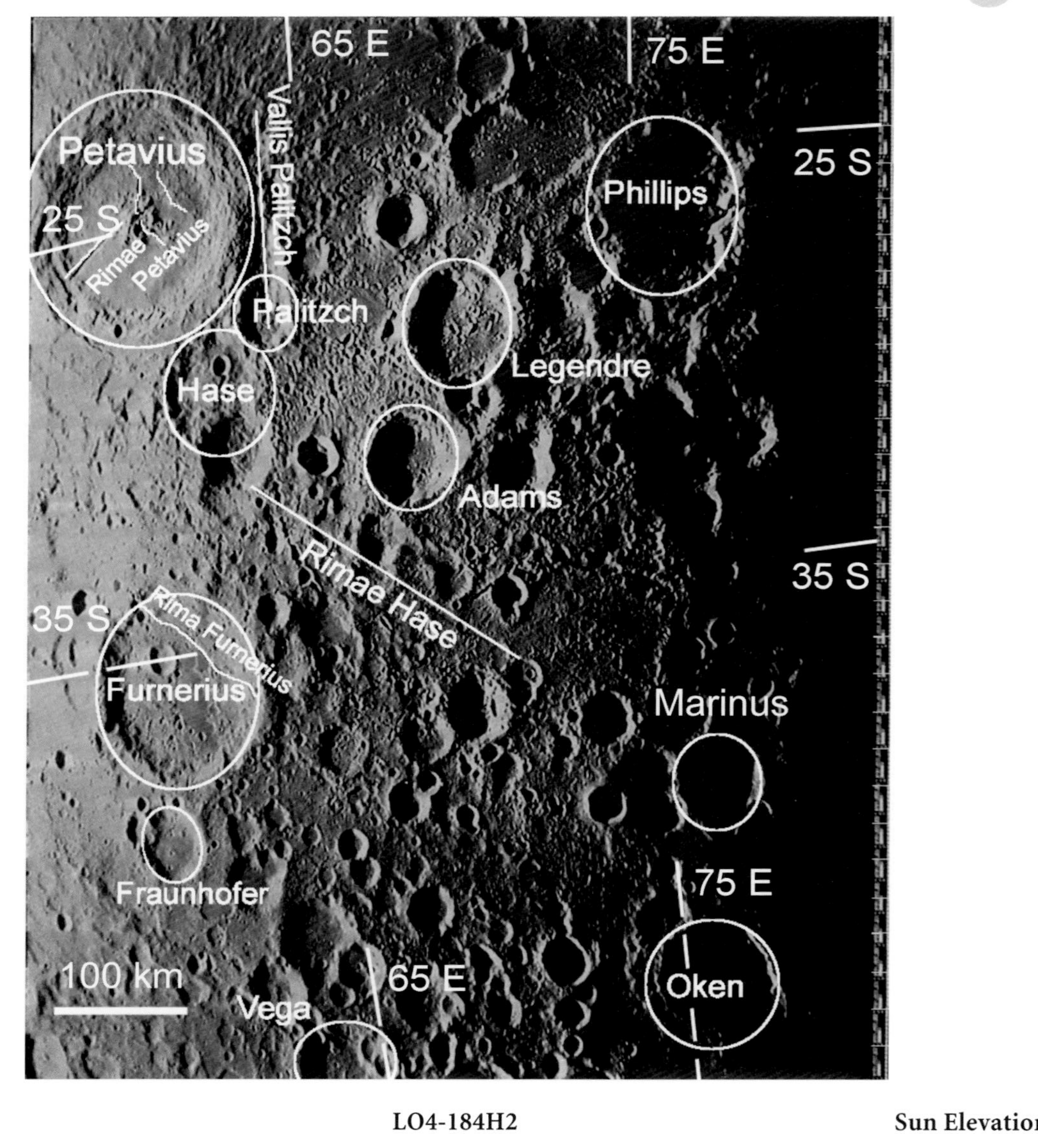

LO4-184H2 Sun Elevation: 6.9° Altitude: 5790.16 km

This highland area southeast of the Fecunditatis Basin has extensive chains of secondary craters, some radial to Petavius and some from the Nectaris Basin. Rimae Hase are stress fractures radial to the Nectaris Basin. The main ring of the much-degraded Australe Basin passes through Marinus. The flat area between these craters and Furnerius (Late Imbrian) may be in an outer trough of this large, degraded, Pre-Nectarian Basin. See LO4-053H1 and LO4-052H3 in the Nectaris chapter (Chapter 8)for a discussion of the fractured-floor craters Petavius and Furnerius.

LO4-184H1 Sun Elevation: 6.9° Altitude: 5790.16 km

This somewhat oblique photo shows an overview of the equatorial region in the vicinity of 70° east longitude. The Pre-Nectarian Balmer-Kapteyn Basin can clearly be seen. Balmer has impacted the main ring of this basin. Striations crossing this basin could be from Langrenus or the Fecunditatis Basin. Note the large difference in albedo between Pre-Nectarian Vendelinus and the Eratosthenian crater Langrenus, whose terraced wall is relatively sharp. Lame has pushed material from the wall of Vendelinus onto its floor. The chain of large craters between Kapteyn and MacLaurin and the smaller chain crossing Lame radiate from Mare Crisium. See LO4-054H1 (the Serenitatis chapter, Chapter 9) for Luna 16, 18, and 20.

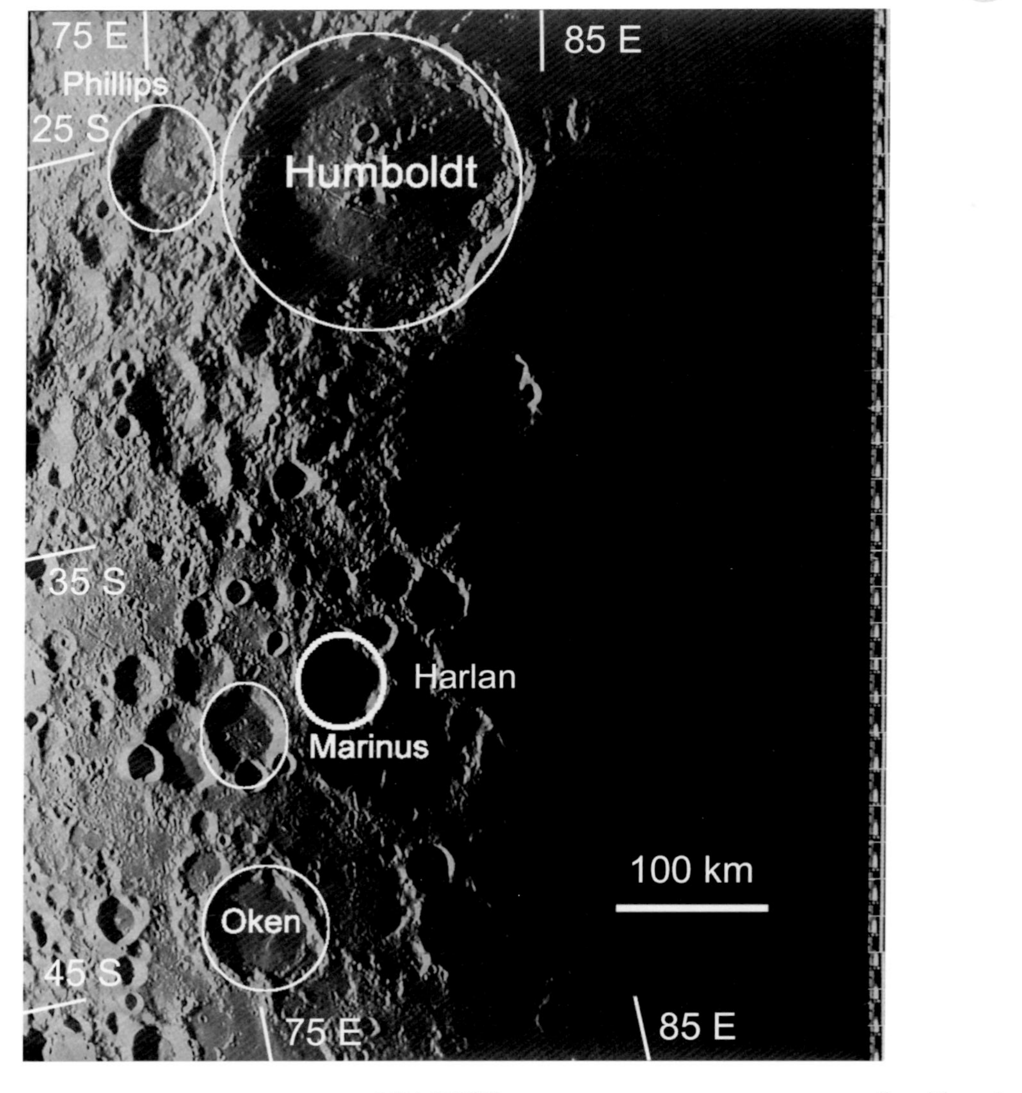

LO4-178H2 Sun Elevation: 1.5° Altitude: 5796.24 km

Humboldt is a crater with an extensive deposit that extends into the Australe Basin, whose main ring passes near Oken and Harlan. Humboldt has a floor with a fracture pattern that reveals signs of uplift (see the inset from LO4-027H1, a perilune photo with morning light). Dark mantling material can be seen on the floor of this crater, near the wall. Striations in this photo radiate from the Nectaris Basin.

LO4-178H1 Sun Elevation: 1.5° Altitude: 5796.24 km

This shows the western sector of the main ring of the Smythii Basin, along with an outer trough (underlying Kastner) and outer ring. Mare Undarum fills an outer trough of the Crisium Basin, to the northwest of this photo. Hecataeus and the smaller crater to its north show signs of a simultaneous impact. Specifically, the eastern rims of the two craters interlock; neither can be said to be superimposed on the other.

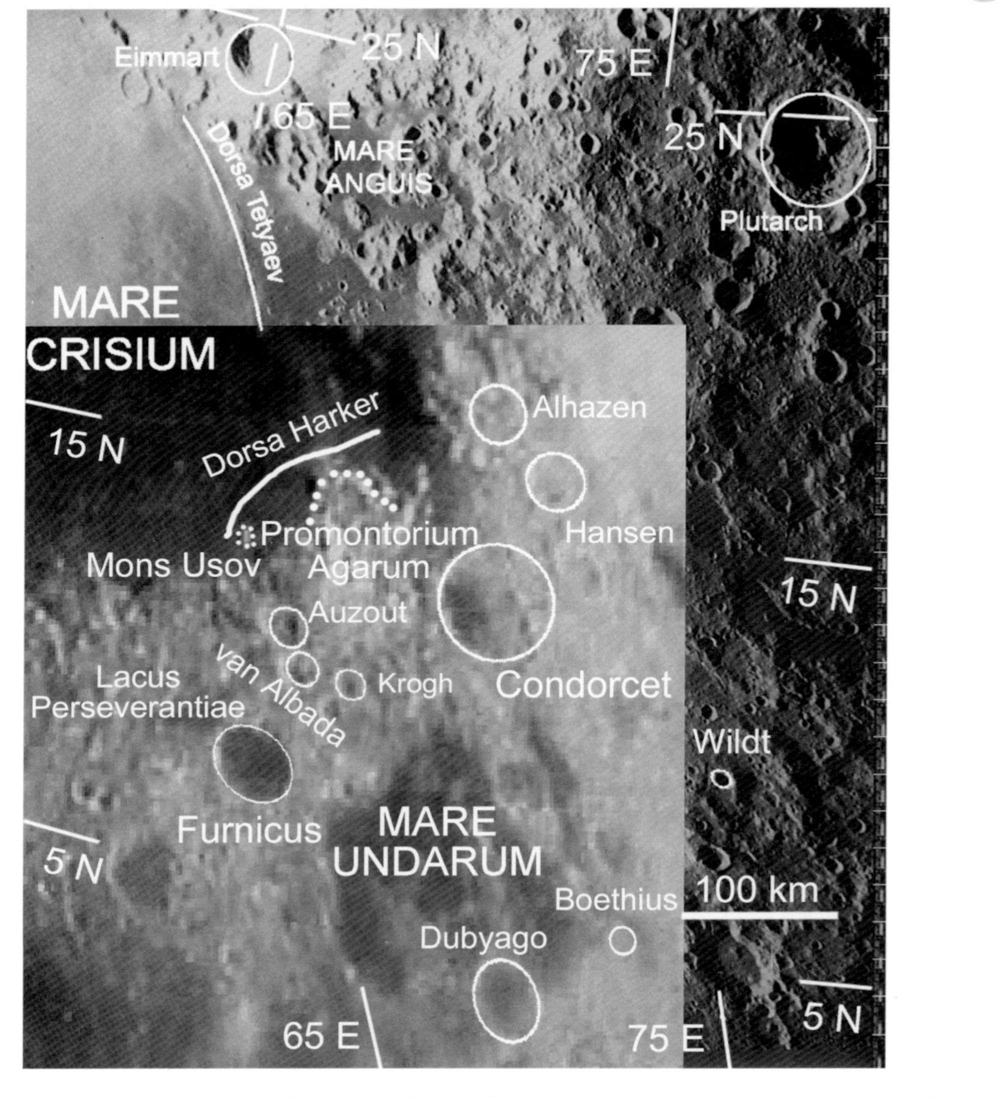

LO4-177H3 and LO4-060M

Sun Elevation: 14.8°

Altitude: 5491.85 km

LO4-177H3 was taken near the end of the mission (sun from the west) to replace coverage lost at the beginning but glare has destroyed most of the photo. The lost area has been replaced here with part of medium-resolution photo LO4-060M (sun from the east). Dorsa Tetyaev is part of a circle of mare ridges that may overlie an internal ring of the Crisium Basin. The ridges and valleys near Wildt are Crisium ejecta, probably overlying deposits from the Smythii Basin. This is the only area where Crisium ejecta is exposed: ejecta from Imbrium and Serenitatis overlies Crisium ejecta to the west. Mare Undarum lies in an outer trough of the Crisium Basin.

LO4-177H2 Sun Elevation: 14.8° Altitude: 5491.85 km

Note the dark areas throughout this region, which may indicate dark mantling material. The strong ridges and valleys in the center of this photo are part of the ejecta blanket from the Humboldtianum Basin directly to the north, in the North Polar Region. Humboldtianum is classified as Nectarian, like Crisium, but is believed to be older. Zeno, Gauss, and ejecta from Gauss are superimposed on the Humboldtianum ejecta and thus must be younger, but they are still classified as Nectarian features. Note the difference in texture south of Berosus, as Humboldtianum ejecta is covered with that of Crisium.

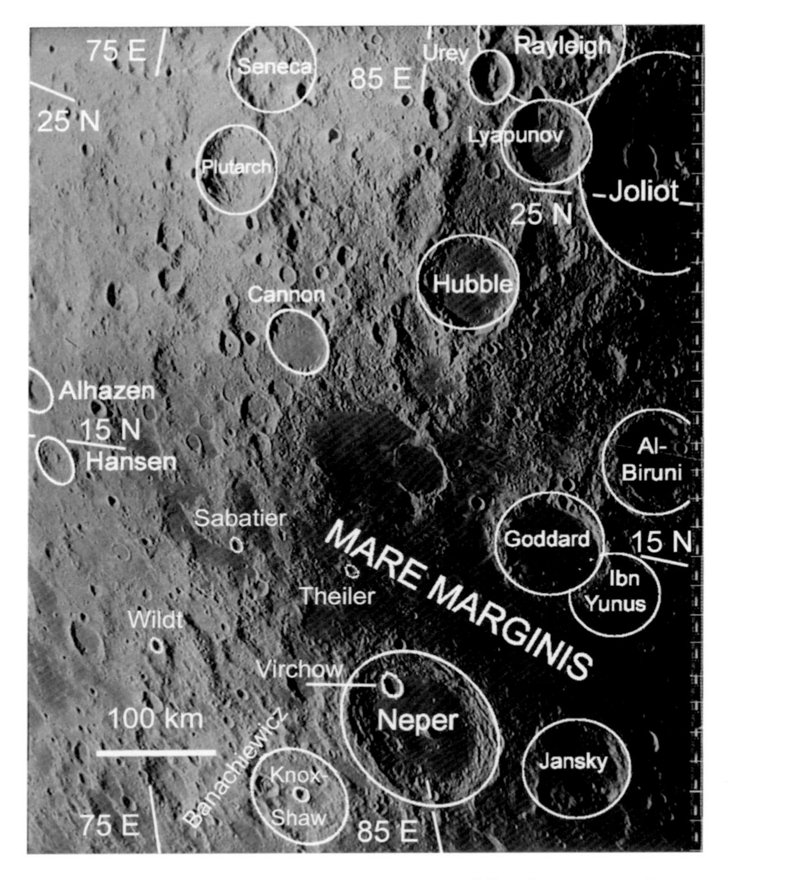

LO4-165H3 **Sun Elevation: 13.9°** **Altitude: 5486.51 km**

(Jean-) Frederic Joliot married Irene Curie, the daughter of Pierre and Marie Curie (LO4-165H3). Irene and Frederic Joliot-Curie worked together and shared the Nobel Prize for synthesizing new radioactive elements by bombarding targets with alpha particles. Mare Marginis is not deep, and Clementine elevation data show a pattern that is more like a collection of craters such as Neper than a basin floor. The pitted area south of Cannon and Hubble is at the antipodes of the Orientale Basin (see also LO4-18H2 and -H3). As in other such cases, rocks ejected in all directions from Orientale have converged here after traveling halfway around the Moon, leading to a local high density of small secondaries.

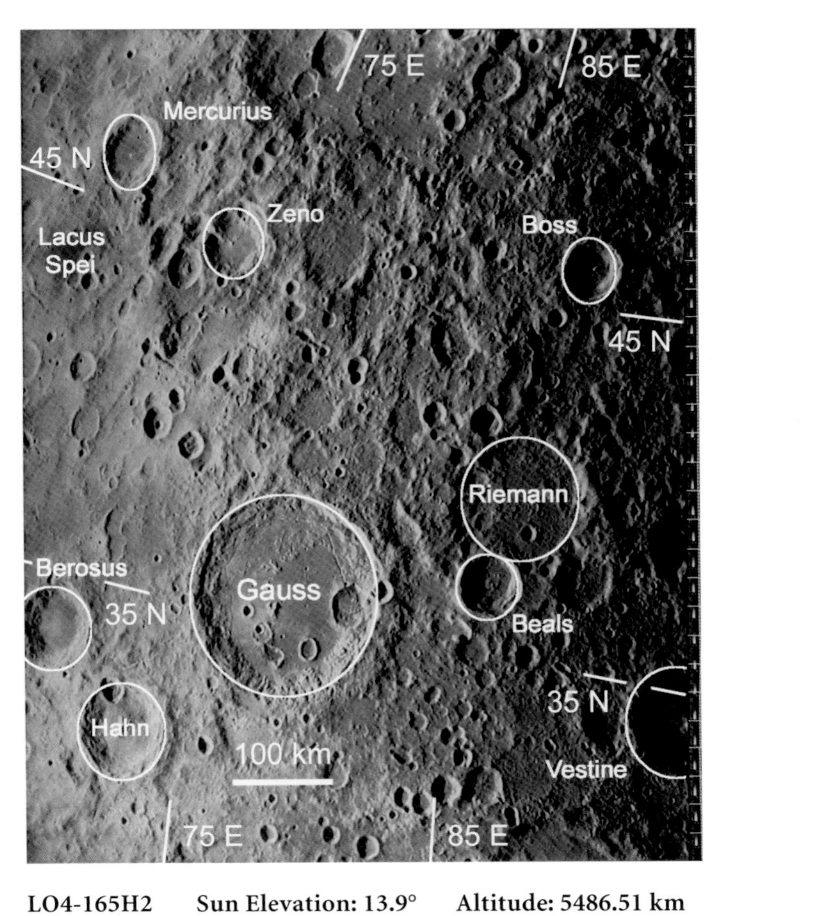

LO4-165H2 **Sun Elevation: 13.9°** **Altitude: 5486.51 km**

This region is dominated by ridges, valleys and secondary craters of the Humboldtianum Basin to the north, in the North Polar Region. Zeno and Gauss have been superimposed on the Humboldtianum ejecta. Secondaries from Crisium to the south have landed near Zeno. Gauss is unusual, for its size, in having a minimal central peak.

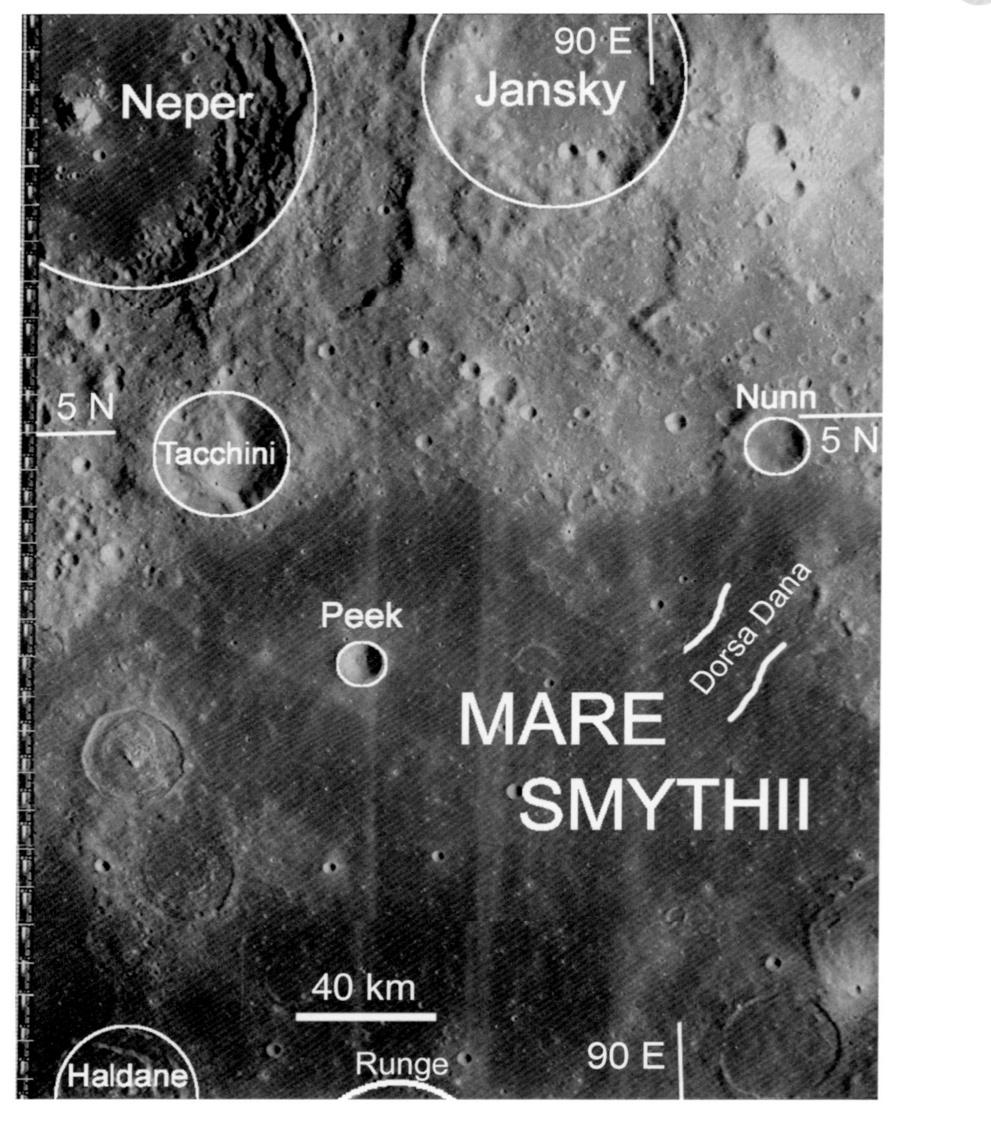

LO4-018H1 Sun Elevation: 28.4° Altitude: 2739.66 km

Mare Smythii is in a well-formed Pre-Nectarian basin. Clementine data show a circular depression (centered near the middle bottom edge of this photo) with the floor flooded by mare material. A mascon indicates a plume of lava rose from the mantle. Nectarian crater Neper, which impacted the main ring of the Smythii Basin, may have been flooded through subsurface lava flows under the channel west of Tacchini. Similar structures to the plateaus and valleys south of Jansky are seen in the northern part of the rim of the Orientale Basin (Montes Cordillera). Other similarities to Orientale may reflect the deeper crust under these basins. Tacchini and Jansky have extensive deposits of ejecta from the Smythii Basin.

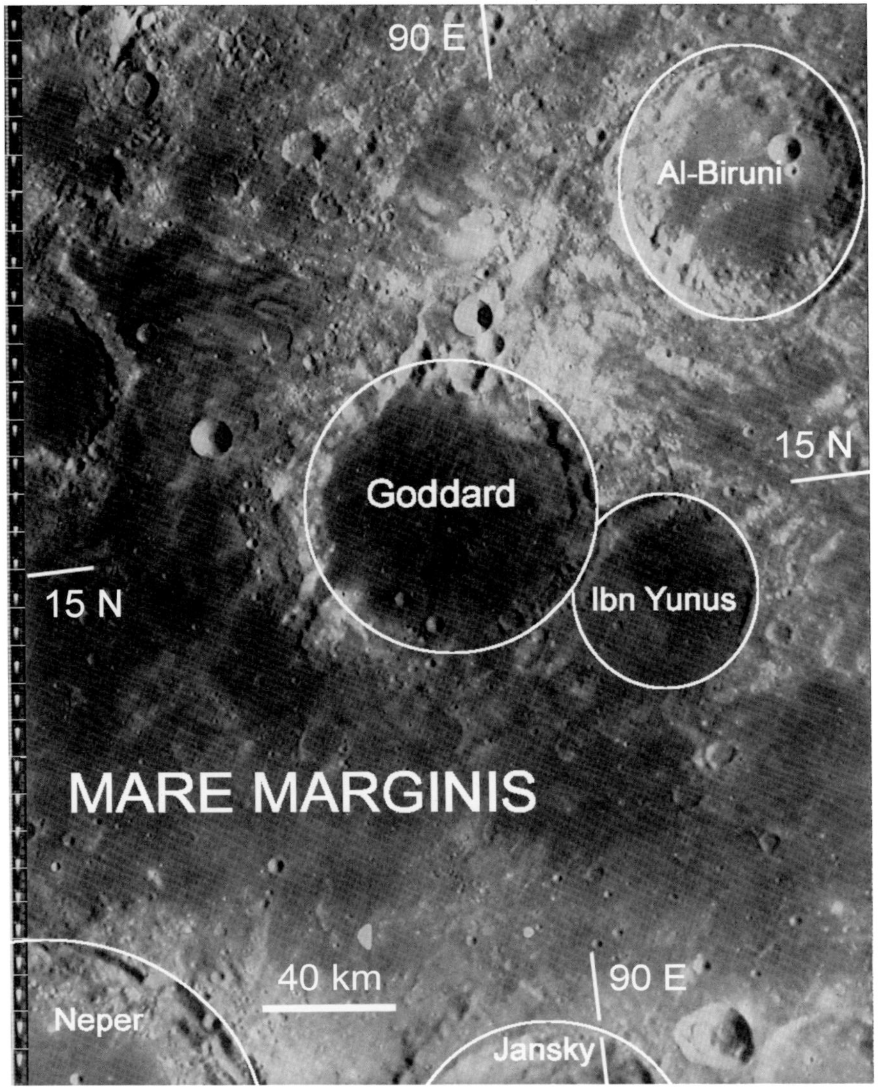

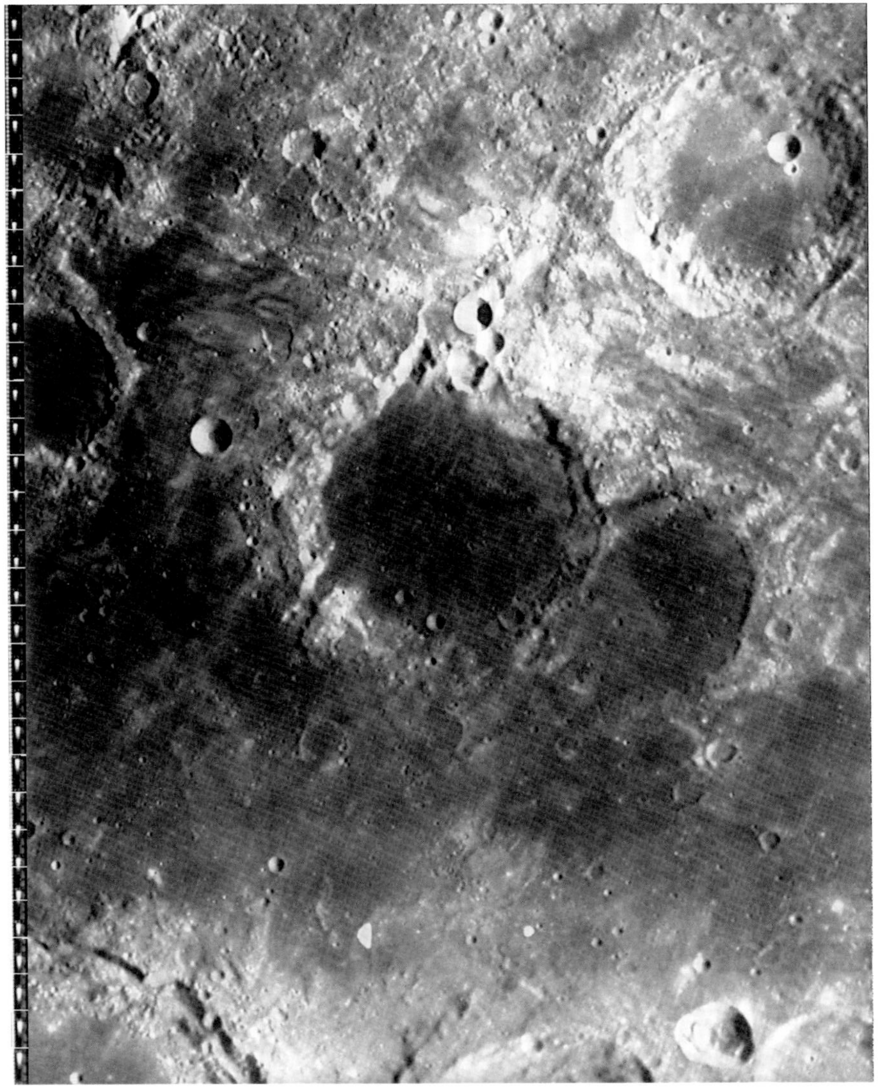

LO4-018H2

Sun Elevation: 28.4°

Altitude: 2739.66 km

The valleys and ridges crossing Neper and Goddard radiate from the Smythii Basin. Copernican crater Goddard A (12 km, just north of Goddard) has spread its ray pattern as far as Mare Marginis. The northwest to southeast striations in the vicinity of Goddard are radial to the Imbrium Basin. The crater walls of Al-Biruni have been overlaid with these striations and with those from the Smythii Basin, so Al-Biruni's crater must have formed before those basins. The mare surface in Al-Biruni has been assigned to the Late Imbrian Period.

LO4-018H3 Sun Elevation: 28.4° Altitude: 2739.66 km

Joliot's central peaks form what may be the beginning of an inner ring. Joliot and Hubble have had their floors flooded with lava, despite being far from any mascon that may mark a plume from the mantle. Striations on the northeastern wall of Joliot are from the far side crater Lyapunov (adjacent and much younger), whose floor is also flooded with lava. Very little mare material is involved because the craters are not fully flooded. The crust is thicker here, along the eastern limb, and that may explain why relatively little lava is available.

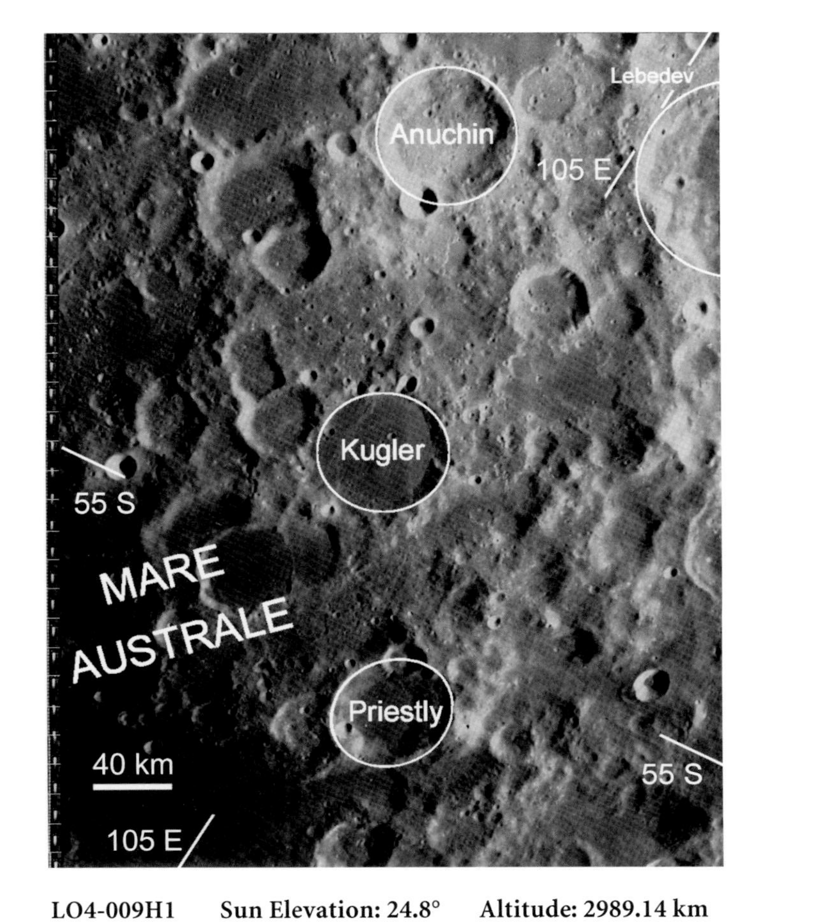

LO4-009H1 **Sun Elevation: 24.8°** **Altitude: 2989.14 km**

The center of the Australe Basin (52° S, 95° E) is near the upper left corner of this photo. The circular segment of a ridge passing through Priestly and Lebedev is an inner ring of this basin. The basin is obviously very old (Pre-Nectarian); many craters of various levels of degradation have impacted its floor. Kugler and nearby craters have been flooded with lava, resurfacing those areas much later (especially in terms of crater counts) than the formation of the floor of the basin. Clementine elevation data show this southern portion of the basin floor to be higher than the northern floor.

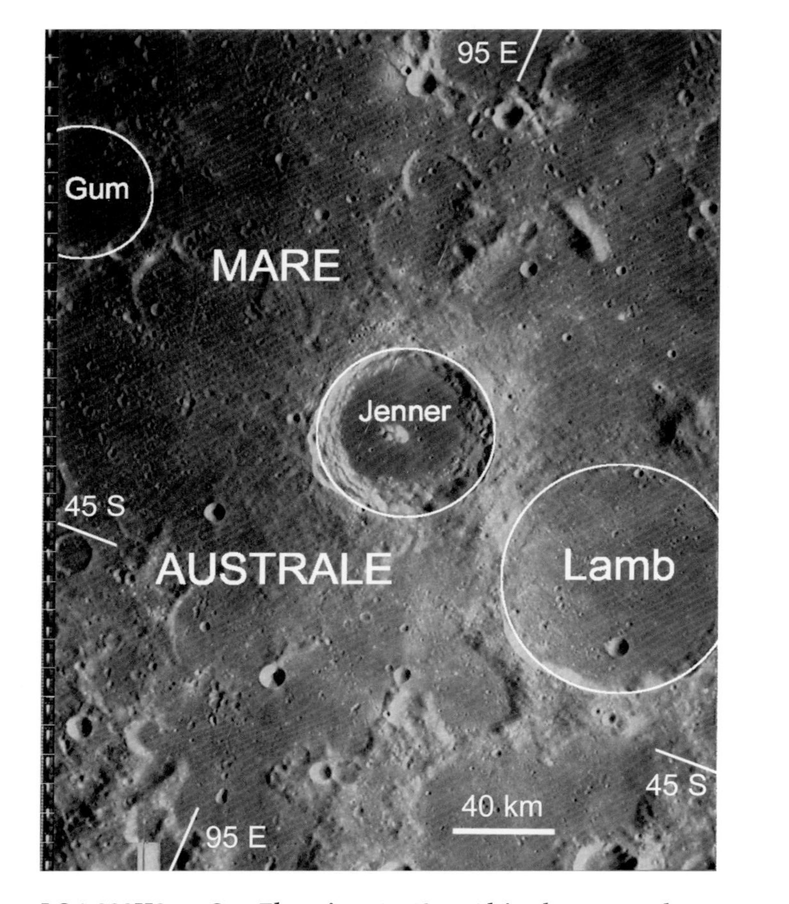

LO4-009H2 Sun Elevation: 24.8° Altitude: 2989.14 km

Ejecta of Late Imbrian crater Jenner overlies the mare material here, but lava flooding the floor of Gum has interrupted the secondary field of Jenner, and Jenner itself is flooded. Apparently Jenner impacted Mare Australe between two or more lava flooding events. Another example of two Australe lava flows straddling a crater event can be seen in LO4-009H3 near Humboldt. Like the South Pole–Aiken Basin of the far side, the Australe Basin has only a very weak positive gravity anomaly, suggesting either a weak, diffuse source of lava or isostatic adjustment of the surface of these very old basins: perhaps the crust was still plastic when they were formed.

LO4-009H3 Sun Elevation: 24.8° Altitude: 2989.14 km

A sector of the main ring of the Australe Basin passes between Abel and Barnard, forming the northern shore of Mare Australe. The secondary field of Late Imbrian crater Humboldt has been covered by a young mare flow. Compare this with that of Jenner (LO4-009H2). Neither Barnard nor its smaller companion to the south dominates the crater wall between them, suggesting a simultaneous impact. Abel has had its wall pushed onto its floor by Barnard's companion. Marie Curie (born Sklodowska) and her husband Pierre Curie received a Nobel Prize for studies of radioactivity. After Pierre's death, Marie won a second Nobel Prize for the discovery of radium and polonium (see LO4-165H3 for more Curie family history).

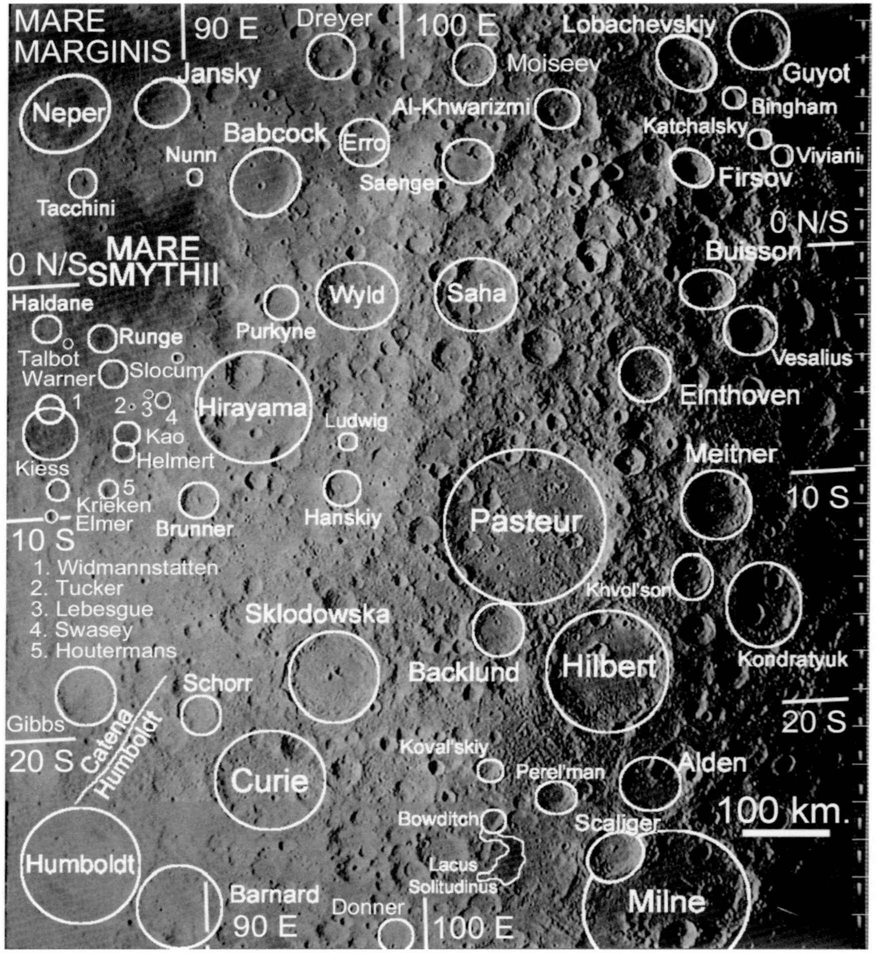

LO2-196M **Sun Elevation: 19.9°** **Altitude: 1519.03 km**

The plan for Lunar Orbiter Mission 4 was to take extensive high-resolution photos of the eastern limb. However, many of these photos were lost, especially in the south, because of the failure of the lens cap. This medium-resolution photo from Lunar Orbiter Mission 2 (also shown in Figure 10.3) is provided to fill in the gap and locate the named features near this part of the limb. The southeast sector of the main ring of the Smythii Basin can be seen entering the photo at the top and curving around toward the left edge. There is a mascon below Mare Smythii (which nearly fills an inner ring of the basin) but not below Mare Marginis. Beyond the main ring lies the transition to the highland region of the far side, with its thicker crust. Lacus Solitudinus may be a part of the spotty extrusions of mare material associated with the Australe Basin to the south.

Chapter 11

North Polar Region

11.1. Overview

The North Polar Region covers the area between the North Pole and about 55° south latitude and between the western and eastern limbs.

Mare Frigoris extends along the southern boundary of the region. The area near the pole is heavily cratered highlands. Mare Humboldtianum is near the eastern limb of this region at the center of the Humboldtianum Basin.

Mare Frigoris

Mare Frigoris (Figure 11.1) extends about 1500 km in the east-west direction but only about 300 km in the north-south direction. Its western section is clearly circumferential to the Imbrium Basin, filling the outer trough of that basin. It connects with Sinus Roris, a large bay at the northern extremity of Oceanus Procellarum. The eastern section of Mare Frigoris is northeast of the Serenitatis Basin. The lava surface of Mare Frigoris connects to that of Mare Serenitatis through Lacus Somniorum.

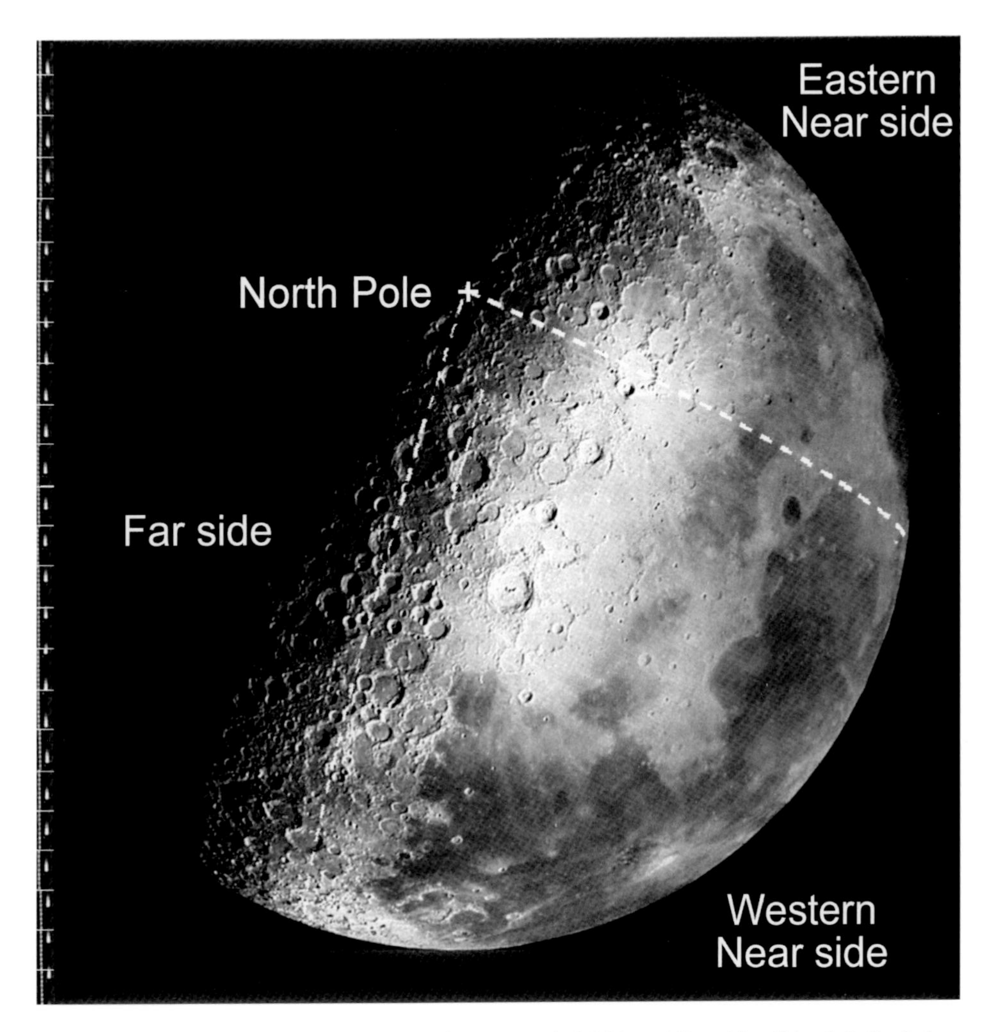

Figure 11.1. Mare Frigoris (LO4-190M). The white dashed lines are the 90° W and 0° E/W meridians. Mare Frigoris is the dark area stretching around the North Pole, a little more than halfway to the edge of the Moon. Mare Imbrium is near the bottom of the photo and Mare Serenitatis is to its right.

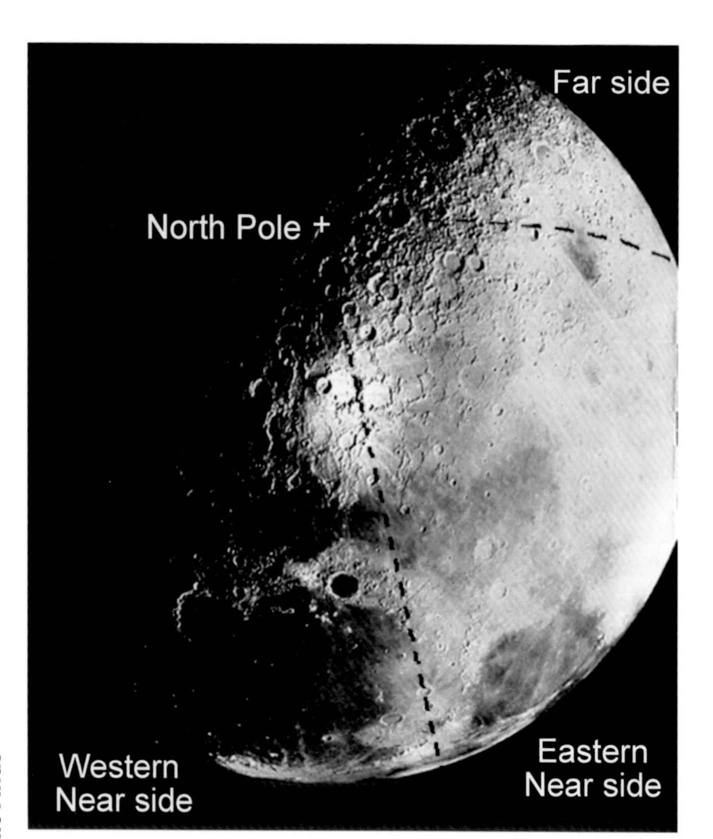

Figure 11.2. Humboldtianum Basin (LO4-128M). The black dashed lines mark the 0° E/W and 90° E meridians. The small dark area in the upper right of this picture (near the 90° E meridian) is Mare Humboldtianum in the floor of the Humboldtianum Basin. The basin structure is obscured in this photo by the bright ejecta and rays of the nearby Copernican crater Belkovitch. The large dark area near the bottom left of the picture is Mare Imbrium; Mare Serenitatis is to the east. Mare Frigoris is north of the two maria.

The Humboldtianum Basin

The Humboldtianum Basin (Figure 11.2), near the northeastern limb of the Moon, is a ringed basin that, like the Orientale and Nectaris Basins, is sufficiently far from younger basins that its structure is well preserved. The ringed structure can be seen in LO4-065H1 in this chapter.

The North Polar Highlands

The heavily cratered region at the polar cap (see Figures 11.1 and 11.2) extends across the pole into the far side highlands. The crust is thicker here, relative to the mare-filled areas to the south. Even though there are large craters (not quite basin sized), there is no sign of lava flooding beyond 70° north latitude on either the near side or far side.

The North Pole

The best parts of the photos taken near the North Pole on each of the orbits of Clementine have been carefully assembled and are shown in Figure 11.3.

The highest points in this area, such as on the rims of large craters, are in sunlight all or nearly all the time because the axis of the Moon is within 1.5° of a right angle to the plane of its travel around the sun (the ecliptic plane). Such a site is especially interesting for a long-term base on the Moon because continuous or nearly continuous solar cell power would be available. Conceivably, solar panel farms could be established there (or near the South Pole) and the power could be beamed to Earth.

Nearby, on the floors of smaller, deeper craters within Rozhdesvenskiy and Peary, there are very cold areas that never receive light or heat from the sun (see Chapter 12). Analysis of data from the neutron spectrometer instruments on the Lunar Prospector spacecraft has detected deposits of hydrogen in the vicinity of these cold traps near both poles. If the hydrogen is in water molecules, it is in the form of crystals distributed in the lunar soil, not as solid ice. The neutron spectrometer detects the energy of solar wind neutrons that rebound from the nuclei of atoms on the lunar surface. A neutron that rebounds from the single proton in the nucleus of a hydrogen atom loses much more energy than if it rebounds from a nucleus of a heavier element.

See the overview in the chapter on the South Polar Region (Chapter 12) for a discussion of how these hydrogen concentrations can be used.

11.2. High-Resolution Images

The Lunar Orbiter photos of the North Polar Region, like those of the South Polar Region, have a high degree of overlap because they were taken from a near-polar orbit. All the rectangular sets of three subframes in a high-resolution exposure have the pole in or near the top of the northern subframe. The illumination angle in the middle subframes is about 10° but the angle approaches zero toward the pole. Of course, the pole is always within 1.5° of the terminator.

Although the polar photos were not taken on every orbit over the poles, there is still so much overlap in the middle and northern subframes that only every other photo is printed here (all the photos are in the enclosed CD). Table 11.1 shows the high-resolution images of the North Polar Basin Region in schematic form.

The following pages show the high-resolution subframes from south to north and west to east. The photos are in the order LO4-190H1, LO4-190H2, LO4-190H3, LO4-178H1, LO4-164H1, LO4-164H2, ... The last photo is LO4-165H1, an apolune photo.

To fill in an area in the southwestern part of the North Polar region, which was not photographed in high resolution, the first following page shows medium-resolution photo LO4-189M. The planned high-resolution Lunar Orbiter photos of the far eastern part of the North Polar Region were spoiled by the failure of the thermal door. Near the end of the mission, apolune photos were taken of that area. One of these (LO4-165H1) is presented at the end of this chapter to complete coverage of the near side North Polar Region.

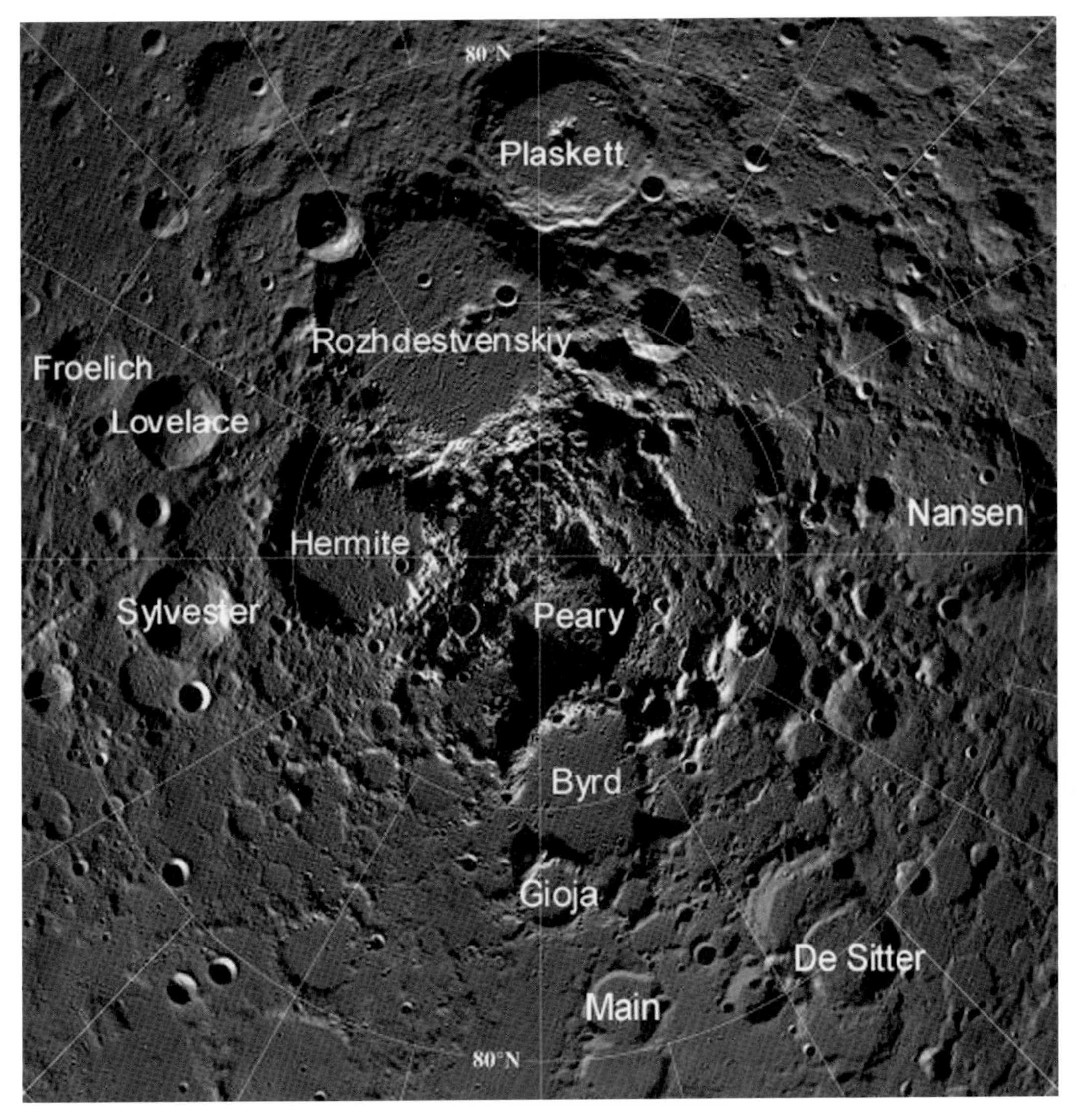

Figure 11.3. North Pole (USGS Astrogeology, PIA00002, NASA). This mosaic covers the North Polar Region within 10° of the pole. The near side is in the lower half of the photo.

Lattitude Range	Photo Number													
75 N–90 N	190H3	176H3	164H3	152H1	140H3	128H1	116H3	104H1	092H3	080H1	068H3	165H1		
60 N–75 N	190H2	176H2	164H2	152H1	140H2	128H1	116H2	104H1	092H2	080H1	068H2			
50 N–60 N	190H1	176H1	164H1	152H1	140H1	128H1	116H1	104H1	092H1	080H1	068H1			
27 N–56 N	189, 183	175, 170	163, 158	151, 145	175, 170	139, 134	127, 122	115, 110	103, 098	091, 086	079, 074	067, 062	055	165H2

Longitude at 50 N	75 W	63 W	51 W	38 W	22 W	12 W	0 E/W	7 E	15 E	28 E	42 E	56 E	66 E	85 E

Table 11.1. The cells shown in white represent the high-resolution Lunar Orbiter photos of the North Polar Region (LO4-XXX H1, -H2, and -H3, where XXX is the Photo Number). The shaded cells within the North Polar Region represent photos that are not printed in this chapter because of redundancy; they are included in the CD. South of this region are the Imbrium Basin Region, the Serenitatis Basin Region, and the Eastern Basins Region. Exposure 165H1 was taken from apolune in afternoon sun. The apolune photo was taken at about twice the altitude of perilune photos, so it has about half the resolution and four times the area of coverage.

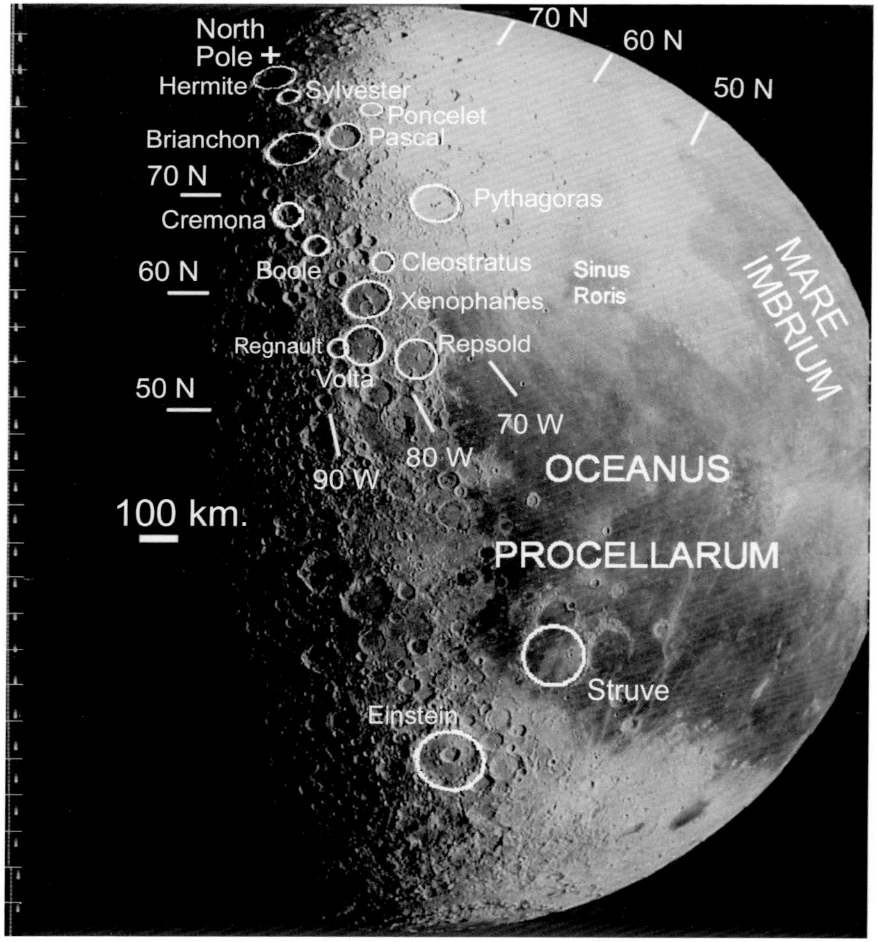

LO4-189 M

Sun Elevation: 18.4°

Altitude: 2877.87 km

Because of the design of Lunar Orbiter Mission 4, the spacecraft orbit was tilted slightly toward the east of the North Pole. As a result, high-resolution photos of the western part of the North Polar Region are not available. The medium-resolution photo above is provided to fill in this area and show the location of the named features within it. The mare surface of northern Oceanus Procellarum is dominated by the ray pattern of a small crater south of Pythagoras (LO4-190H1). The subdued nature of the topography near the western shore of Oceanus Procellarum may be due to an outer trough of the proposed Lavoisier-Mairan Basin.

LO4-190H1 **Sun Elevation: 13.5°** **Altitude: 3378.38 km**

The mare in the lower part of this photo floods two large craters, one between Oenopides and Repsold and another to the southeast of Repsold. Pythagoras is an example of a young Eratosthenian crater; there is detailed, sharply delineated structure in its crater wall and ejecta blanket, but the ray structure has been suppressed by a combination of gardening by meteor impacts and darkening by the solar wind. The rays in this area, extending into northern Oceanus Procellarum, are due to a small crater on the north rim of Oenipedes. Note the secondary field of Pythagoras in the vicinity of Oenopides.

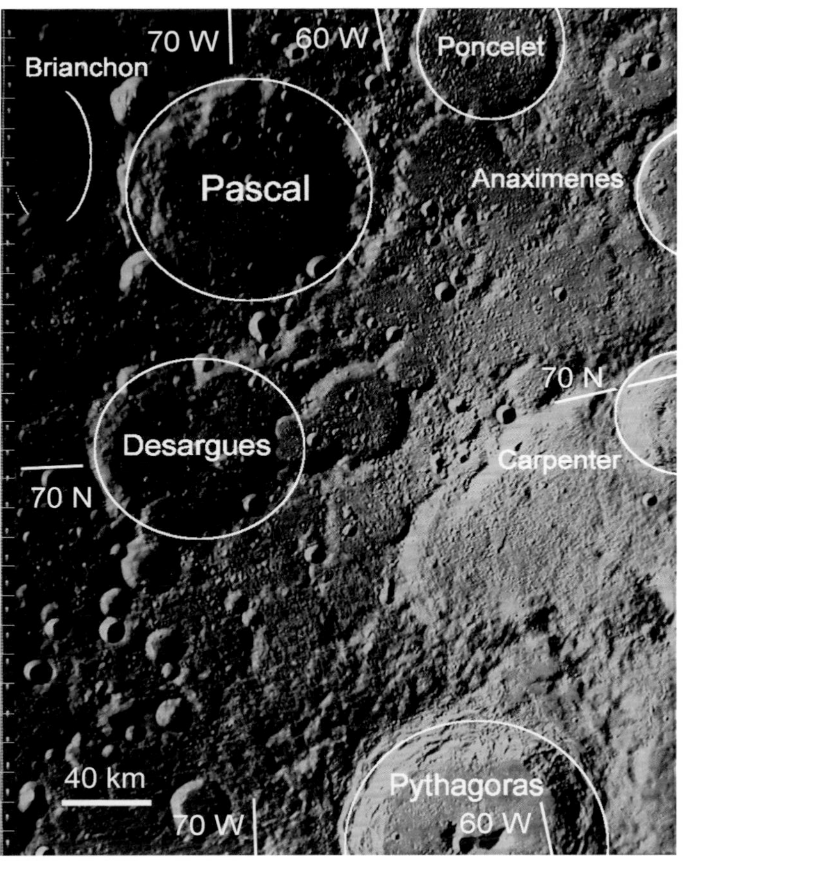

LO4-190H2 **Sun Elevation: 13.5°** **Altitude: 3378.38 km**

Pythagoras has spread its ejecta blanket to the west and north as well as to the south (LO4-190H1). In the northern latitudes, above 70° from the equator, the terrain becomes clearly highland, with craters of every size overlapping.

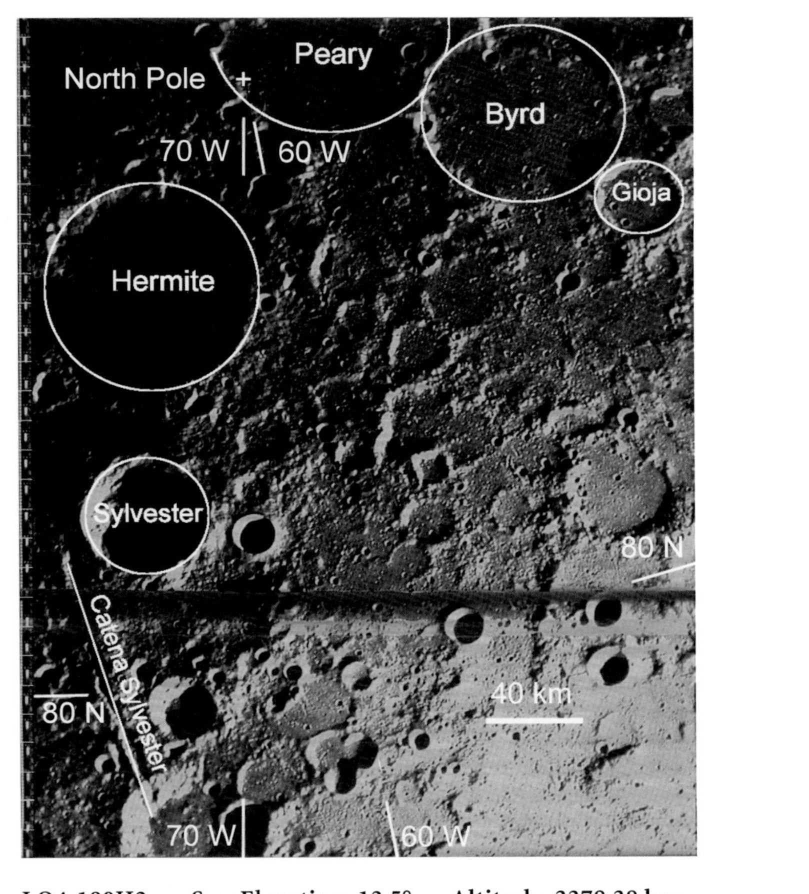

LO4-190H3 Sun Elevation: 13.5° Altitude: 3378.38 km

This is the western approach to the North Pole. Hermite, deeply shadowed in this picture, is about to have its floor illuminated with sunlight. Near the North Pole, the sun does not illuminate parts of crater floors during entire lunar rotations (months). However, the sun illuminates more of the floors over a year's time because of the slight tilt of the Moon's axis relative to the ecliptic plane. Young, sharp, small craters near the pole have more of their floors permanently dark because they are relatively deep and their floors are more shaded by their rims than the floors of old, degraded, large craters.

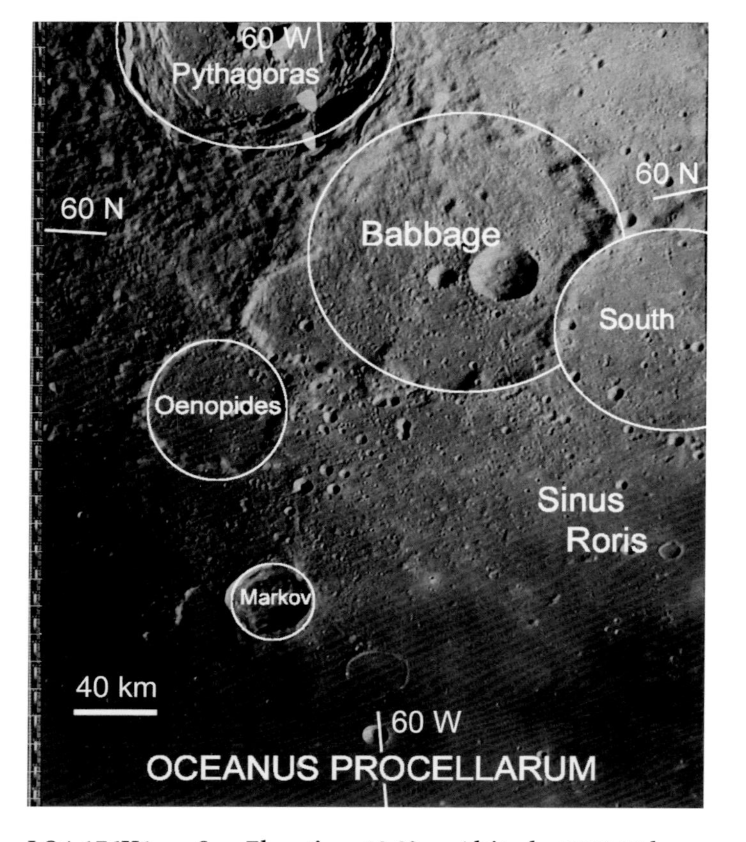

LO4-176H1 **Sun Elevation: 12.8°** **Altitude: 3353.98 km**

The ancient crater Babbage has influenced the formation of the Pythagoras crater in this area. crater South, although relatively degraded in comparison to Babbage, is nevertheless younger, because a sector of its wall traverses Babbage. Sinus Roris has thinly flooded this region, leaving traces of crater walls such as that southeast of Markov. Secondary craters from Pythagoras have fallen on much of this area. A mare unit (south of crater South) has flooded part of the secondary field of Pythagoras, showing that that mare unit formed later. Both Pythagoras and the mare unit are assigned to the Eratosthenian Period.

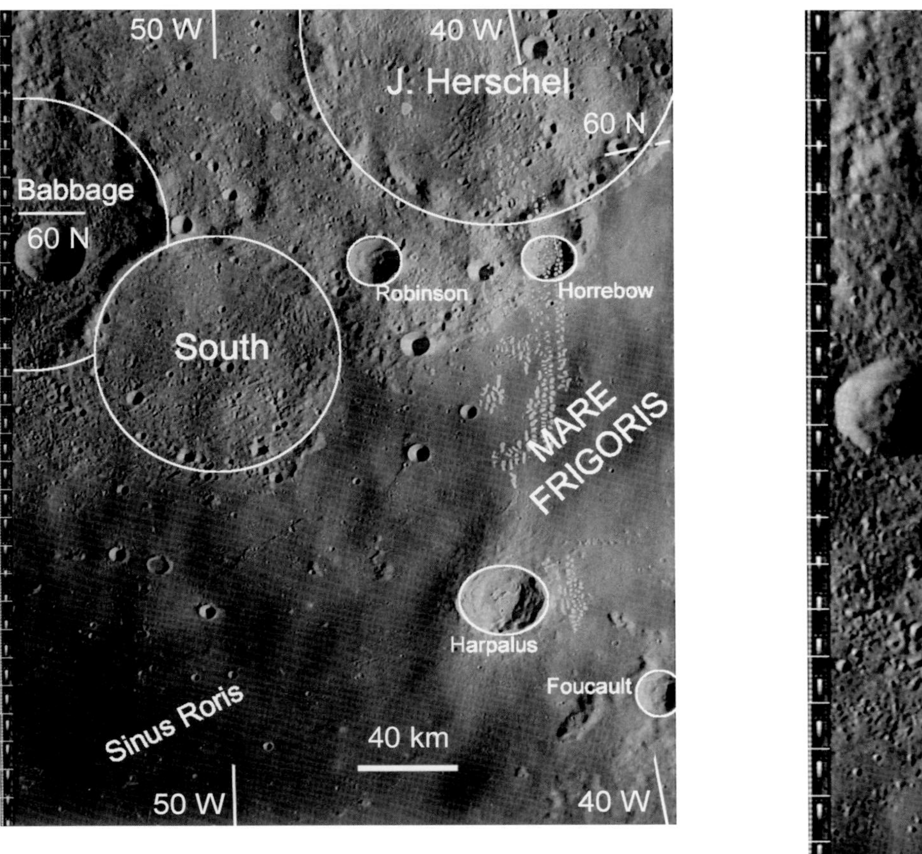

LO4-164H1 **Sun Elevation: 12.9°** **Altitude: 3345.81 km**

The area north of Mare Frigoris, such as the floors of craters South and J. Herschel, is heavily covered with ejecta from the Imbrium Basin to the east-southeast. In particular, ridges and valleys in the crater wall of J. Herschel point to the center of the Imbrium Basin. The crater chain in the south-eastern floor of crater South is likely to be formed by secondary impactors from a crater that has been flooded by Sinus Iridum (to the southeast).

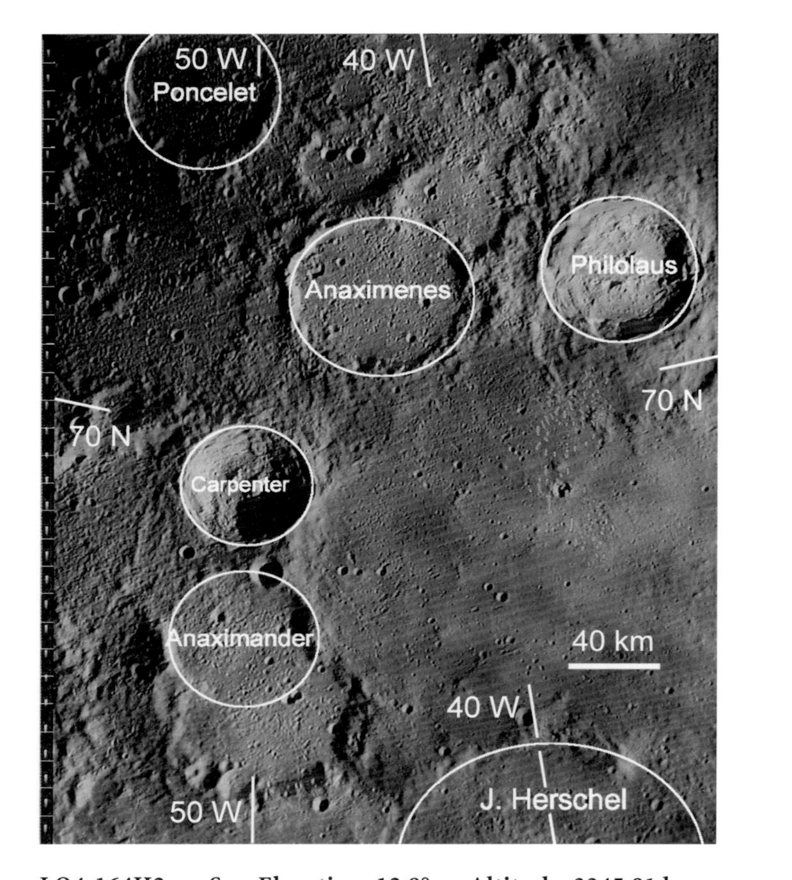

LO4-164H2 Sun Elevation: 12.9° Altitude: 3345.81 km

The interesting highland plains between J. Herschel and Anaximenes have a mottled appearance because they have been overlain with rays from Copernican craters Carpenter and Philolaus. The smooth appearance of this area could be due to unconsolidated ejecta from the Imbrium Basin, shaken down and consolidated by the later impacts of Carpenter and Philolaus, which could have been virtually simultaneous. Or, some major tectonic event could have occurred in this area to shake down the ejecta blanket.

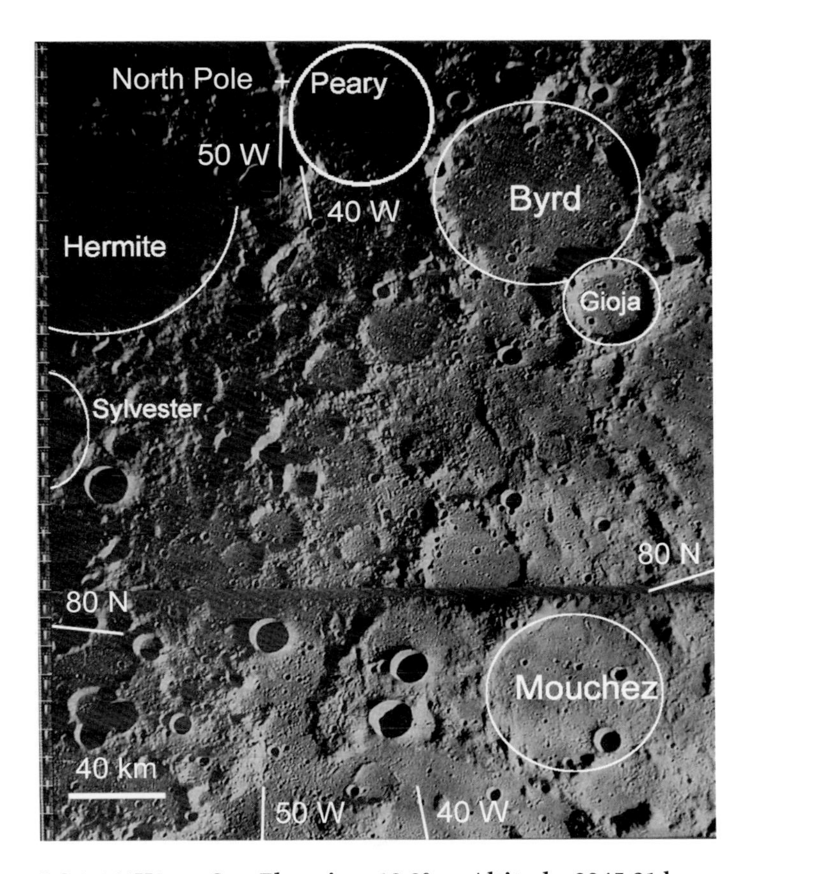

LO4-164H3 **Sun Elevation: 12.9°** **Altitude: 3345.81 km**

The heavy ridges and valleys south of Byrd and Gioja are radial to the Imbrium Basin, part of its outer deposit. The cluster of sharp craters in the 10- to 15-km range of diameters in the southern part of this area could have come from Imbrium. Alternatively, they could have been caused by a salvo of secondaries from the far side crater Rozhdestvenskiy coming over the North Pole.

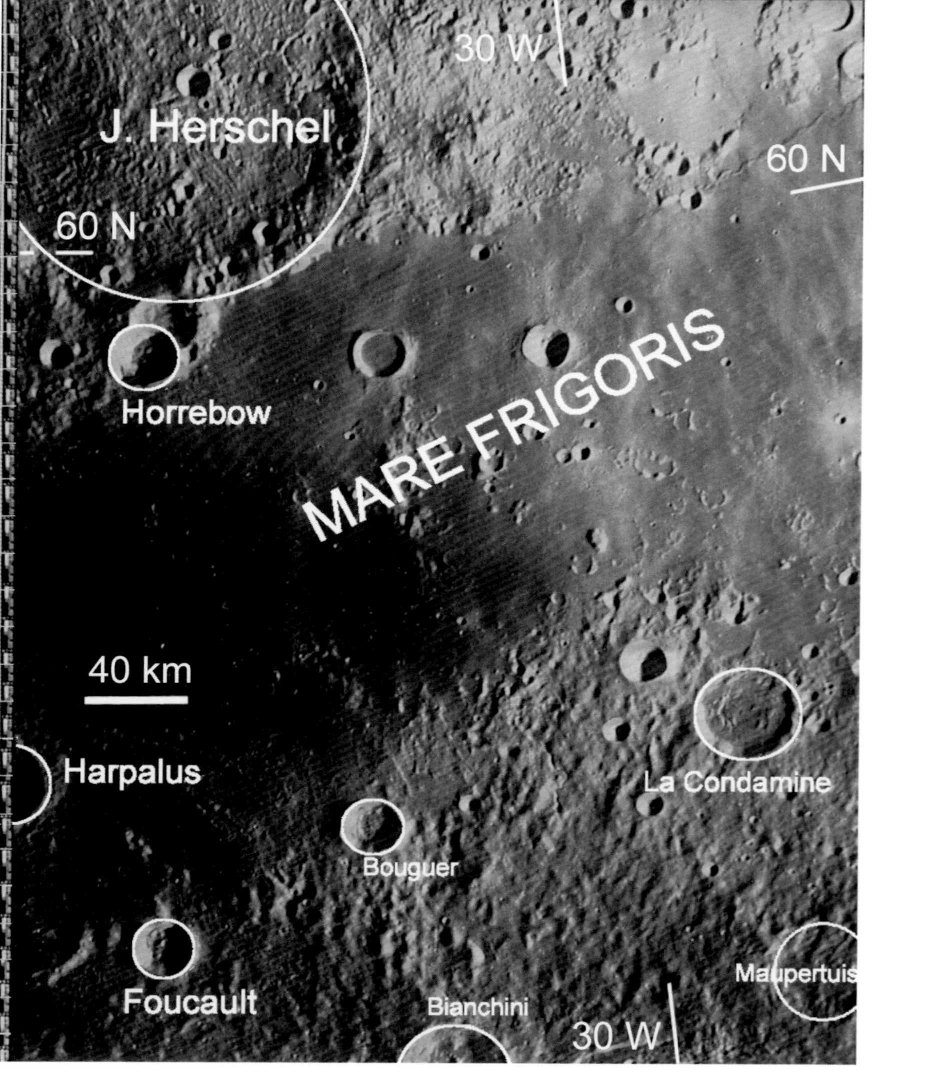

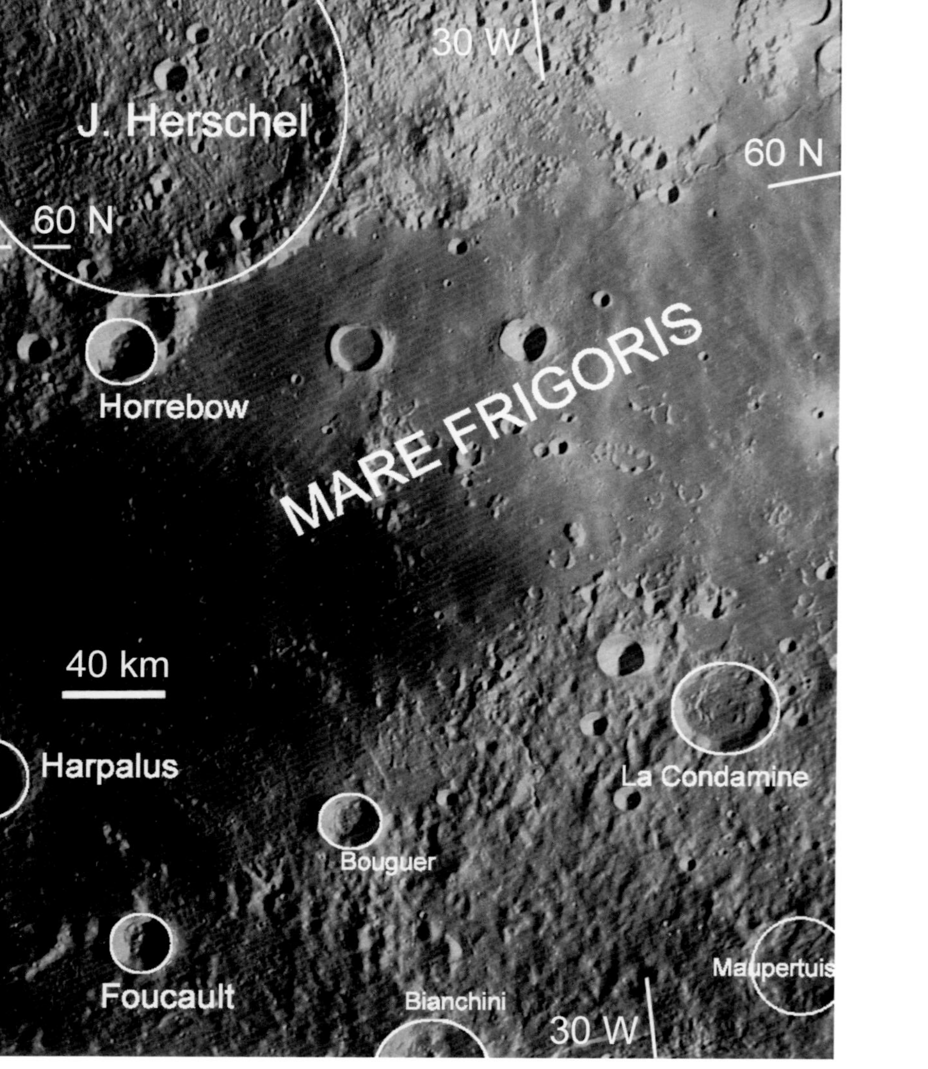

LO4-152H1 Sun Elevation: 13.0° Altitude: 3394.33 km

The ring of the Imbrium Basin that bounds Mare Imbrium passes between Bouguer and Bianchini. Its crosshatched character may be due to fine-textured ejecta from the crater beneath Sinus Roris overlying the coarser ridges and valleys caused by the Imbrium event. The dark area in the eastern part of Mare Frigoris in this photo is typical mare material, which has flooded an outer trough of the Imbrium Basin. In the east, Mare Frigoris appears to have been covered with a layer of lighter rayed material, probably from Copernican crater Anaxagoras (LO4-140H2). Underneath this could be ejecta from Philolaus, also Copernican.

LO4-140H1 **Sun Elevation: 12.5°** **Altitude: 3355.14 km**

The ring of the Imbrium Basin that bounds its mare passes between La Condamine and Maupertuis. Rimae Maupertis are outflow channels or collapsed lava tubes where lava has come up through the ring and flowed north into Mare Frigoris and south into Mare Imbrium. This type of flow suggests that upwelling lava, partially blocked by melt sheets and earlier lava flows (now hardened), can find its way up through the fractured ring material and then downhill.

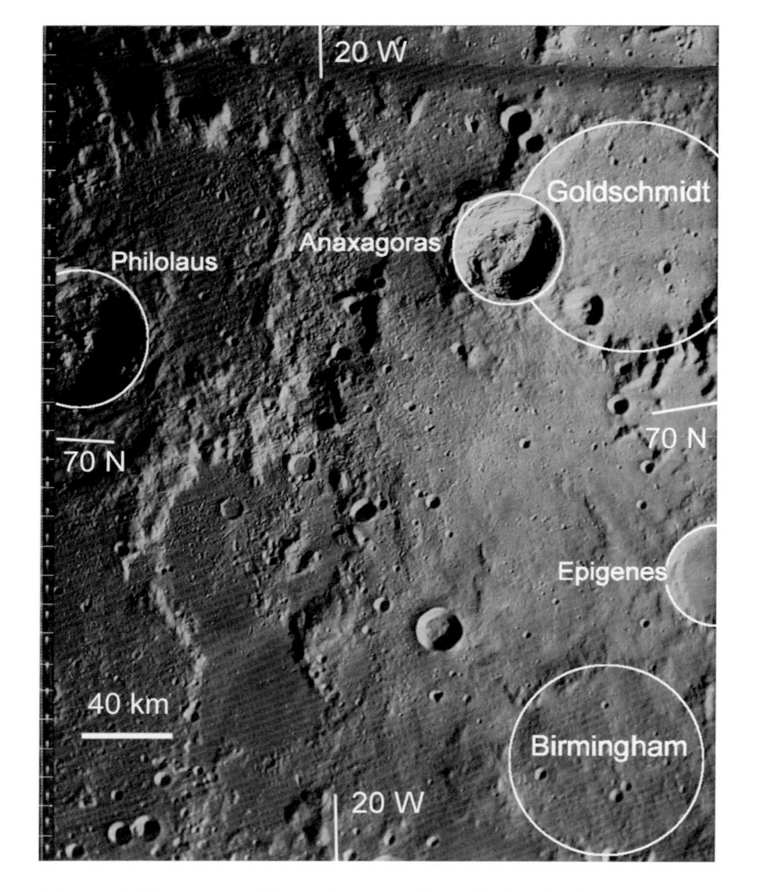

LO4-140H2 **Sun Elevation: 12.5°** **Altitude: 3355.14 km**

Philolaus and Anaxagoras are Copernican craters. Anaxagoras, in particular, has impacted unconsolidated material on the rim of Goldschmidt and ejected particularly massive flows of finely broken material that smooths the floor of Goldschmidt and the area south of Goldschmidt. These rays extend to Mare Frigoris (LO-152H1). Tectonic forces and the shock of the impact of Anaxagoras may also have had a part in the leveling of the plains south of Goldschmidt.

LO4-140H3 Sun Elevation: 12.5° Altitude: 3355.14 km

The ridges that come from the south are radial to the Imbrium Basin. Other striations from the south-southeast, finer in texture, are radial to the Serenitatis Basin. Main and Challis, two craters of approximately the same size and age, could have been caused by nearly simultaneous impacts. Although Main has impinged on the Challis rim, the ejecta blanket of Challis appears to have been better preserved. The area immediately to the west of Main and Challis has been depressed and smoothed, with all signs of ejecta from Main and Challis erased. Tectonic action beneath a thick unconsolidated surface layer could be the cause. Robert Peary led an expedition of 50 Eskimos and reached the North Pole by dogsled in 1909. Richard E. Byrd flew over the North Pole with Floyd Bennett in 1926. He later flew over the South Pole in 1929. He was an active explorer of Antarctica by airplanes and dogsleds from 1928 to 1956.

LO4-128H1 Sun Elevation: 11.7° Altitude: 3369.17 km

This area of Mare Frigoris is rich in ridges and scarps (perhaps the edges of flows). Its surface expression is pinched between the ejecta of Plato and the highland peninsula north of it. Both Plato and the Iridum crater are dated in the Upper Imbrian Epoch. The secondary field of Plato seems to overlie the Iridum ejecta, especially to the northeast, but striations from Iridum seem to overlie Plato ejecta just to the north and south of its rim. Plato's floor has been flooded with mare material, upwelling through the broken material of the main ring of the Imbrium Basin.

LO4-116H1 **Sun Elevation: 11.6°** **Altitude: 3396.94 km**

Just to the east of Plato there is more evidence of lava upwelling through the broken material of the main ring of the Imbrium Basin. Rimae Plato drain pools of dark lava into larger channels such as Vallis Alpes. Vallis Alpes may be a stress crack radial to the Imbrium Basin that has been nearly filled with lava. A lava channel runs through its center. The northern part of Rimae Plato seems to disappear and reappear; it is probably a channel in some sectors and an underground tube in others. Gardening by impactors may have broken through the roof of the tubes in some sectors, leaving channels.

Anaxagoras
Goldschmidt
Meton
Barrow
0 E/W
70 N
70 N
Epigenes
Birmingham
Rima W. Bond
W. Bond
40 km
0 E/W

LO4-116H2 Sun Elevation: 11.6° Altitude: 3396.94 km

The coarse ridges in this photo are radial to the Imbrium Basin. The ejecta blanket of the Imbrium Basin changes character between W. Bond and Meton, shifting from the thick Fra Mauro Formation to thinner outer deposits. Copernican crater Anaxagoras sends its rays across Pre-Nectarian Goldschmidt, W. Bond, Barrow, and Meton. The crater wall of Barrow is more polygonal than circular. It has been strongly influenced by preexisting structures in its target area. For example, the crater wall seems to be further from the center where it penetrates the broken material of the rim of Goldschmidt. Rima W. Bond runs nearly across crater W. Bond, suggesting uplift pressure beneath a melt sheet under its floor.

LO4-116H3 **Sun Elevation: 11.6°** **Altitude: 3396.94 km**

The low sun elevation of this photo near the terminator at the pole shows that the floor of Byrd is very flat.

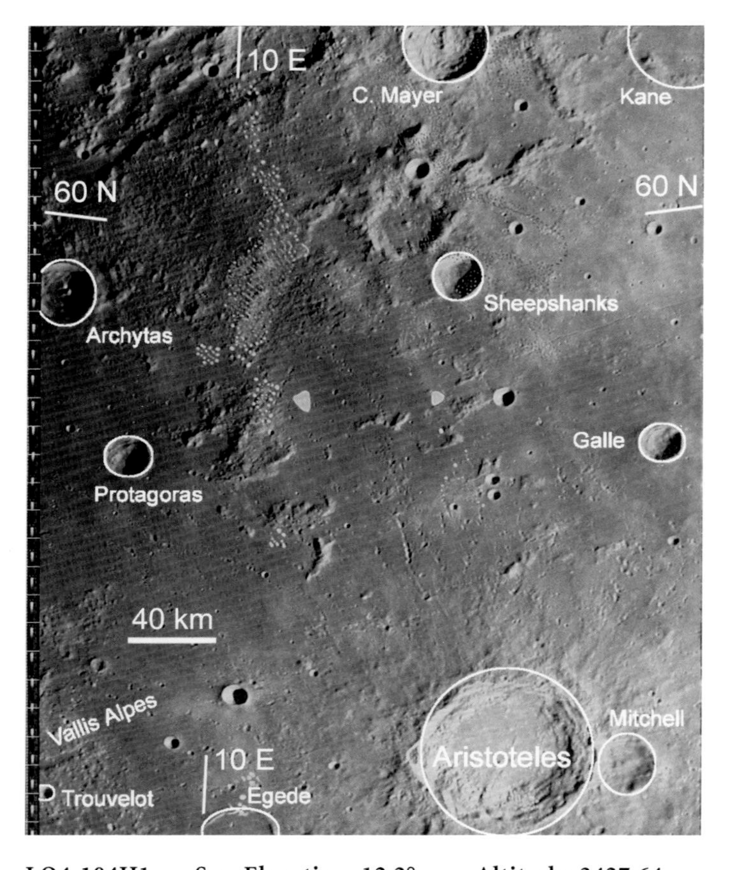

LO4-104H1 Sun Elevation: 12.2° Altitude: 3427.64

The ejecta from Eratosthenian crater Aristoteles have been laid down in two phases, separated by only minutes. The first phase consisted of finely divided material ejected at a very low angle. This formation has smooth radial ridges and troughs. Subsequently, chains and loops of small sharp secondary craters form from impactors that have been thrown high above the surface and land after the initial blanket has settled. The mare to the north, still part of Mare Frigoris, is no longer in the Imbrium trough that lies outside of its main ring. The depression that has been flooded to form the eastern part of Mare Frigoris may have been formed by the intersection of the second outer troughs of the Imbrium and Serenitatis Basins. Of course, these circles intersect at two points; the other intersection is at Sinus Medii, another flooded depression.

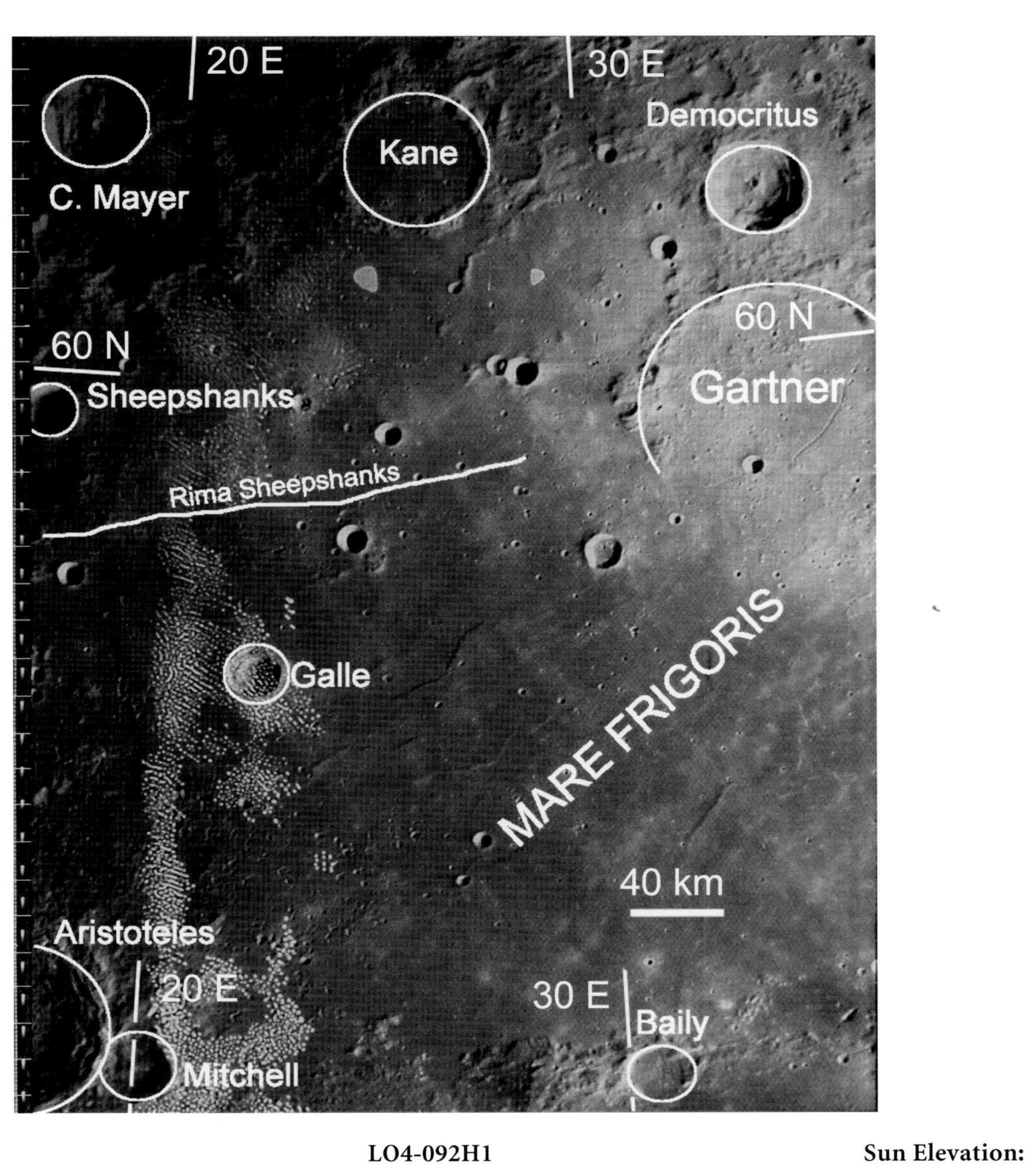

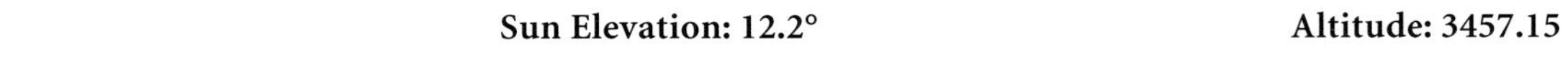

LO4-092H1 **Sun Elevation: 12.2°** **Altitude: 3457.15**

Several clues suggest that this area of Mare Frigoris is relatively shallow. These indications include the projecting rim of Gartner and its faint companion to the northwest, the presence of so many mare ridges, some with circular structure suggesting underlying crater rims, and the concentric pattern of Galle, suggesting that it has bottomed out into a layer beneath the mare material. The majority of flow scarps drop toward the northwest, suggesting that this area was flooded from the southeast.

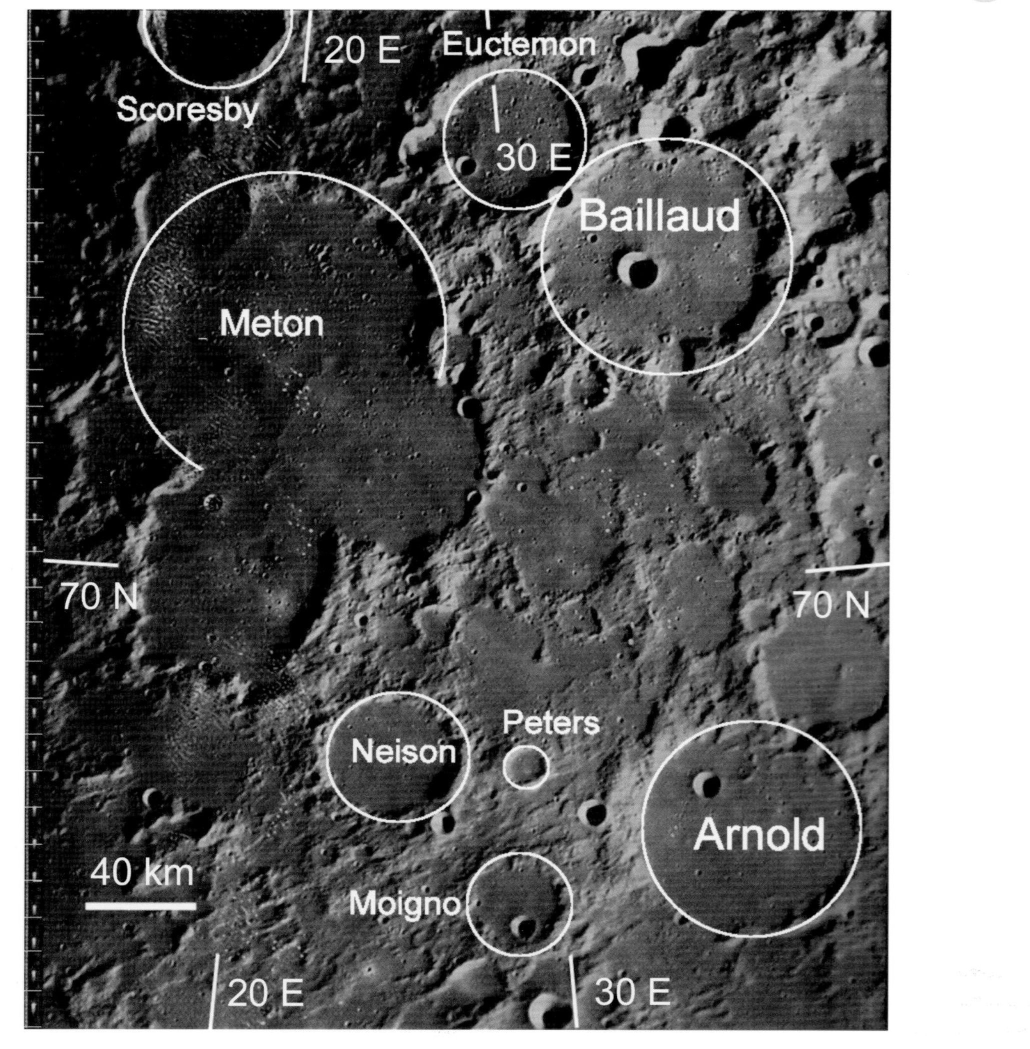

LO4-092H2 Sun Elevation: 12.2° Altitude: 3457.15

The floors of Meton and its similar companions are not inherently darker than the surrounding material (see Figure 1.3); they are just closer to the terminator.

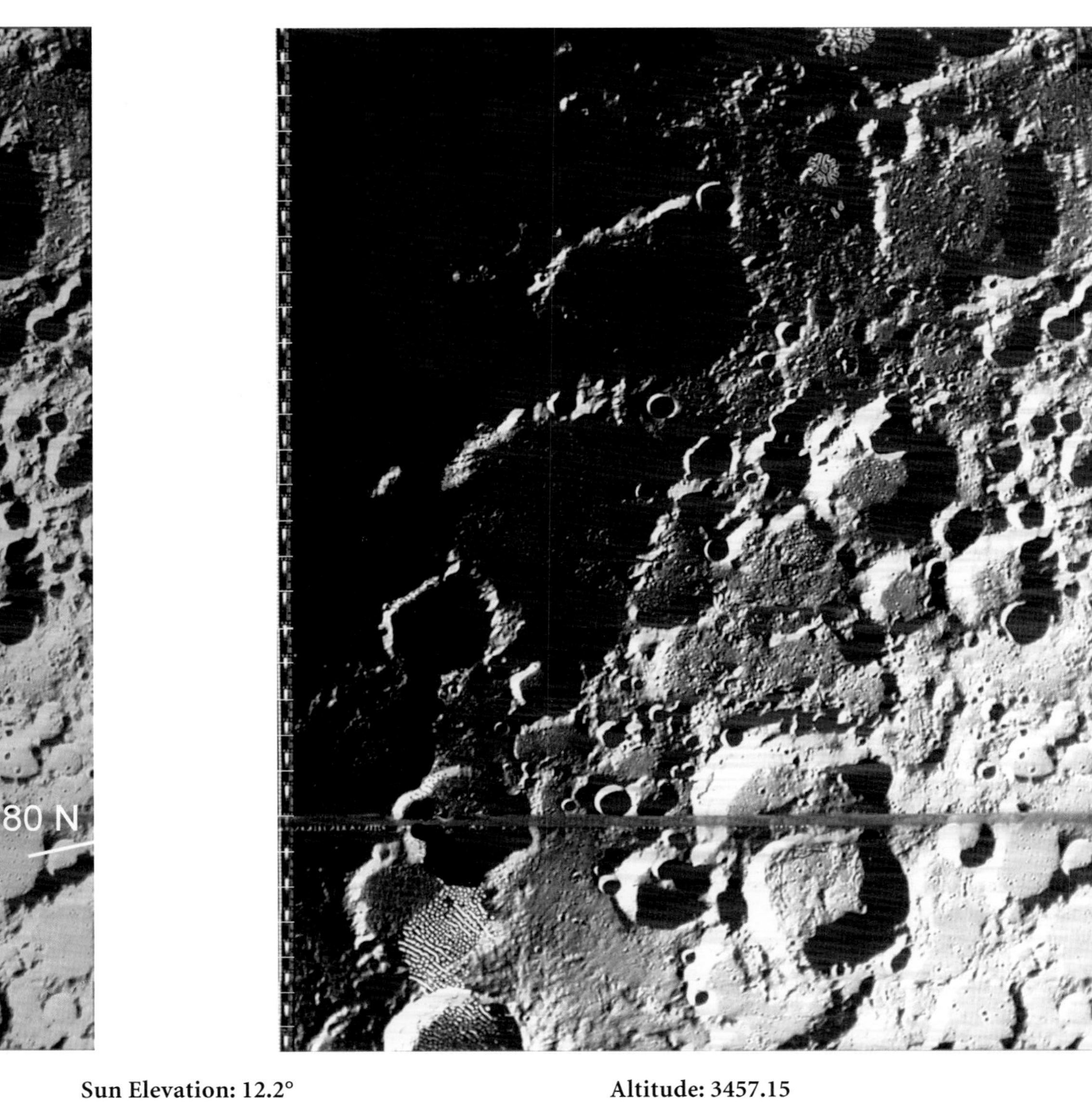

LO4-092H3 Sun Elevation: 12.2° Altitude: 3457.15

The cluster of secondary craters east of Byrd and Peary may include impactors from each of the Imbrium and Humboldtianum Basins; the pole is about equidistant from the center of each basin, relative to their sizes. The larger secondaries are likely to be from Imbrium and the smaller from Humboldtianum. Contributions may also have come over the pole from far side crater Rozhdestvenskiy.

40 E
Democritus
Thales
60 N
Gartner
Rima Gartner
60 N
De La Rue
MARE
FRIGORIS
40 km
40 E

LO4-080H1 Sun Elevation: 13.0° Altitude: 3479.03

The northeastern shore of Mare Frigoris is bounded by a highland plains unit, formed by a dusty covering of ejecta from the Humboldtianum Basin. De La Rue is just outside a ring of Humboldtianum, probably the main ring; that puts this area in an external trough of the basin. That impact may have essentially leveled the highland material in this area shortly before deposition of the pulverized ejecta, which further smoothed the area. Dark mantling material, often seen near the shallow edges of maria, not only covers this part of Mare Frigoris but also darkens the nearby highlands.

LO4-068H1 **Sun Elevation: 11.5°** **Altitude: 3487.74**

Endymion and De La Rue are just outside the 650 km main ring of the Humboldtianum Basin, in the first external trough (see LO4-165H1). Both are older than that basin, but De La Rue must be older than Endymion, judging by the relative degree of degradation. Endymion is classified as Nectarian. The brightness covering De La Rue, Strabo, and the surrounding area is the result of the fresh ejecta from Copernican crater Thales (LO4-165H1).

LO4-068H2 Sun Elevation: 11.5° Altitude: 3487.74

The main ring of the Humboldtianum Basin (LO4-165H1) crosses diagonally across the lower right corner of this photo, south of Cusanus and east of Schwabe. Radial ridges and valleys of ejecta pass between Baillaud and Petermann. There are clusters of small secondaries on the floor of Petermann and the very degraded group of three craters of similar size to the west of Petermann.

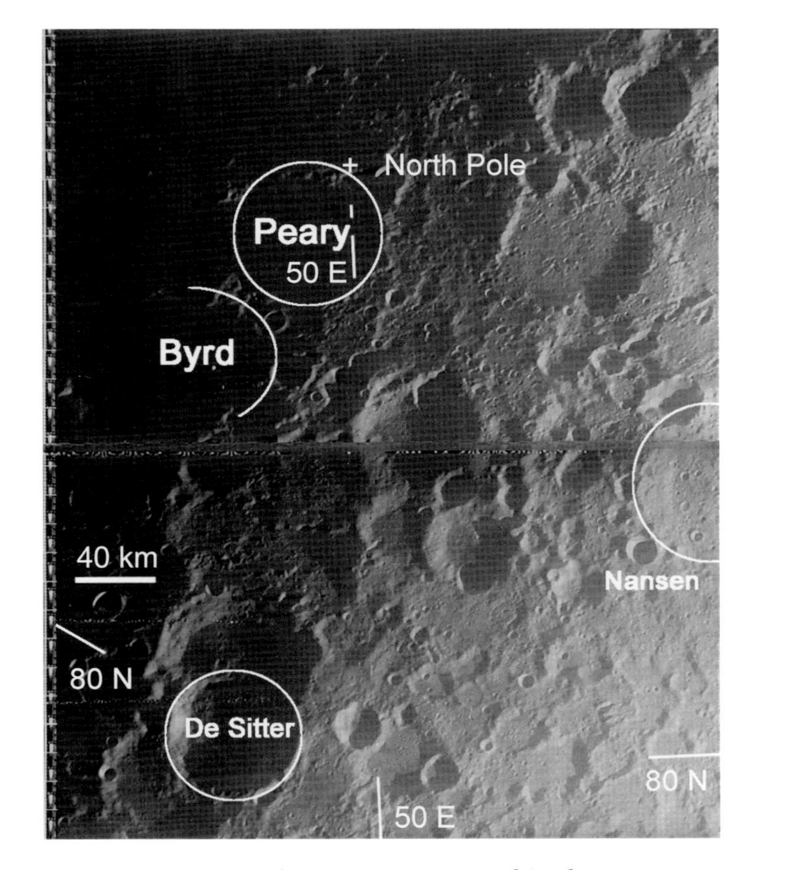

LO4-068H3 Sun Elevation: 11.5° Altitude: 3487.74

The area of the right side of this photo is dominated by ejecta from the Humboldtianum Basin. Some ridges, at right angles to those from Humboldtianum, are radial to the Imbrium Basin. Fridtjof Nansen allowed his specially constructed ship *Fram* to be frozen into polar ice near Siberia in 1993, intending to drift with the ice toward Greenland. As *Fram* made its closest approach to the pole, Nansen and a companion left the ship by dogsled and reached 86° 13' north latitude. Nansen's leadership in programs to repatriate prisoners of war after World War I and relief of famine resulted in his receiving the Nobel Peace Prize in 1922.

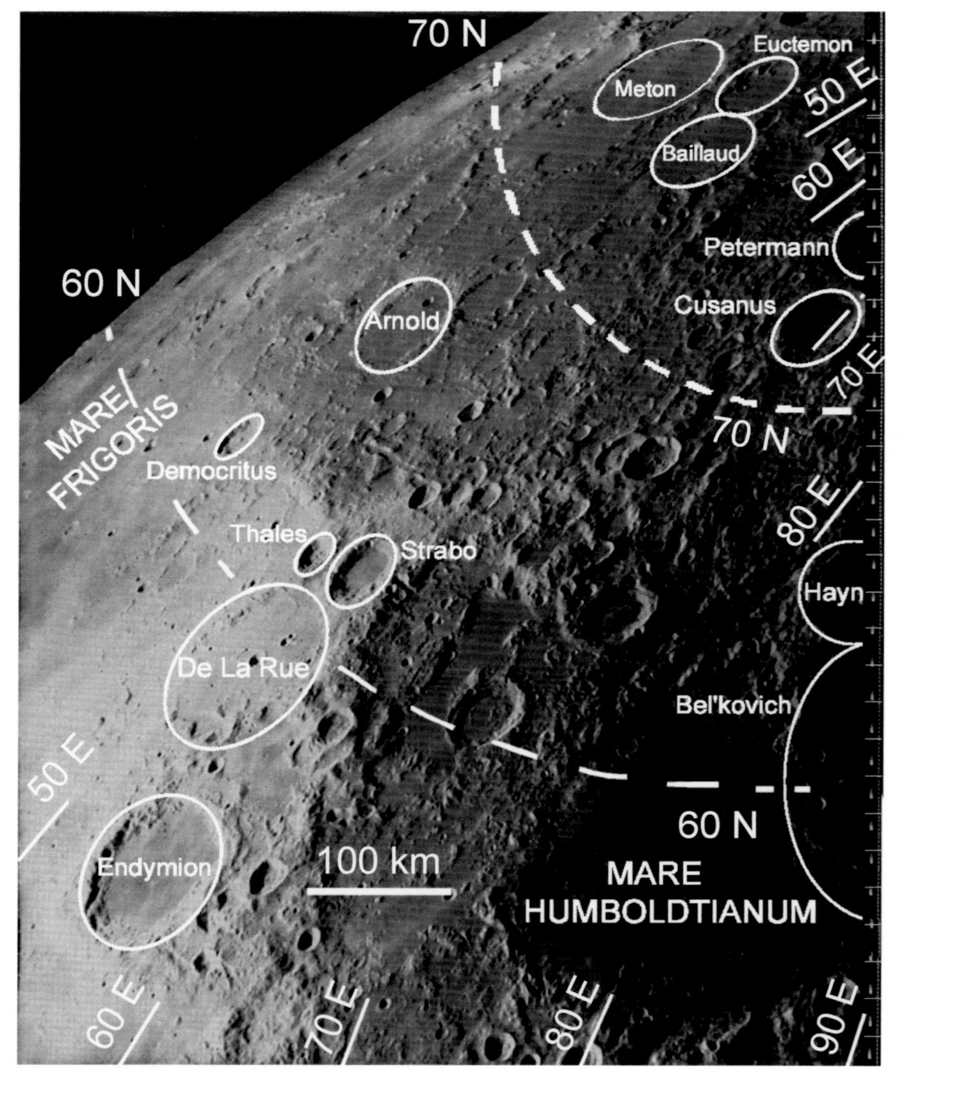

LO4-165H1 Sun Elevation: 13.9° Altitude: 5486.51

The ringed structure of the Humboldtianum Basin can clearly be seen in this apolune photo taken near the end of the mission. Mare Humboldtianum nearly fills a ring that may be an inner ring. The next ring is subdued in this sector, but the ring after that, running just east of the rims of De La Rue and Strabo, may be the main ring. Mare partly floods the area between the ring bounding the mare and the main ring. Radial ridges and valleys, such as those crossing the rim of Strabo and both north and south of Mare Humboldtianum, are radial to this ringed basin.

Chapter 12

South Polar Region

12.1. Overview

The near side South Polar Region covers the area between about 55° south latitude and the South Pole and between the western and eastern limbs.

The region is entirely highlands. On a large scale, its topography records the early over-cratered bombardment of the Moon after the crust formed. The entire South Polar Region is within one diameter of the main ring of the South Pole–Aiken Basin on the far side, so it has been essentially covered with ejecta from that basin. However, that ejecta has been so degraded by further bombardment that it is difficult to identify specific features such as ridges and secondary craters that might have been due to the South Pole–Aiken impact.

Western Sector of the South Polar Region

The Bailly Basin, unflooded by mare, is within this sector, toward the western limb (Figure 12.1). The lack of mare flooding is an indicator of a relatively thick crust in this area.

Central Sector of the South Polar Region

Figure 12.2 shows the central part of the near side South Polar Region. The highlands here are a continuation of the near side central highlands, a feature that reaches all the way to the lunar equator.

Eastern Sector of the South Polar Region

Figure 12.3 shows the eastern limb of the Moon, including the eastern edge of the near side South Polar region. The southern part of the Australe Basin is in this sector.

The South Pole

The best parts of the photos taken near the South Pole on each of the orbits of Clementine have been carefully assembled and are shown in Figure 12.4.

As can be seen from Figure 12.4, the sun is rarely seen at the floors of craters that are deep relative to their size because the lunar axis is inclined by only 1.5° relative to the plane of its travel around the sun. More of this permanently shadowed area is near the South Pole than near the North

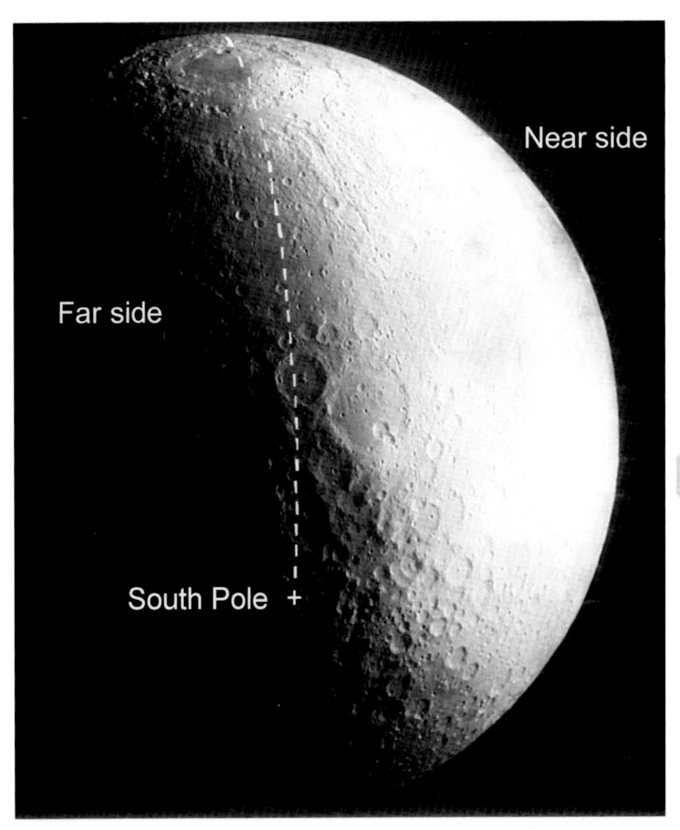

Figure 12.1. Western Sector of the South Polar Region (LO4-193M). The white dashed line marks the western limb, the meridian at 90° degrees west longitude. The Bailly ringed basin is in the center of this photo, in the northern part of the South Polar Region. The 90° meridian runs through the crater Hausen, adjacent to Bailly. The Orientale Basin is to the north, near the visible edge of the Moon in this photo. Ridges of Orientale ejecta can be seen reaching all the way to the edge of the South Polar Region.

Pole. As the surface is radiating heat to deep space and because the dusty surface (gardened by small meteorites) is a reasonably good insulator against thermal conductivity, the temperature at the floors of these craters is near absolute zero, as low as 40 degrees Kelvin. The surface dust at such cold ambient temperatures is a good trap for volatiles such as hydrogen that arrive there by chance. Hydrogen is available from the solar wind, in the form of ionized protons. Such protons that impact in sunlit areas on the Moon are likely to be neutralized and thermally excited into the space

Figure 12.2. Central South Polar Region (LO4-106M). The dashed line is the zero meridian. The small dark area in the upper left corner of this picture is Mare Nubium, in the floor of the Nubium Basin. The very sharp crater south of Mare Nubium is Tycho, which spreads its rays preferentially to the east, over the southern part of the near side central highlands. The large crater Clavius is directly to the south of Tycho, just within the South Polar Region. The bright area southeast of the center of the photo is due to the ejecta of Copernican crater Stevinus. The dark splotchy area in the lower right is the western portion of Mare Australe. The shadowed floor of the small far side Schrodinger Basin is below and to the right of the South Pole.

Figure 12.3. Eastern South Polar Region (LO4-008M). The dashed line is the eastern limb, the 90° east meridian. Mare Australe is in the upper left. The smaller Schrodinger Basin is below it, with deep valleys running between it and the Australe Basin.

around the Moon. If they do not escape, they "hop" from place to place (landing and being reemitted).

The Lunar Prospector Mission carried instruments to detect concentrations of hydrogen and found them near each of the poles. The hydrogen, perhaps combined with oxygen (plentiful in the minerals of the lunar surface) in the form of ice, is likely to have been trapped at or near the surface of these permanently shadowed areas.

The nature of these concentrations of hydrogen and perhaps of ice is of great interest. If sufficiently rich concentrations of suitable materials can be found in the low gravity of the Moon, they might provide a useful source of rocket fuel or propellant, rather than lifting supplies from Earth.

Another special characteristic of this region near the pole is its location near the South Pole–Aiken Basin. Rocks scattered here could be the deepest ever excavated from the crust (and possibly from the upper mantle of the Moon). An expedition to gather these rocks could also establish the age of the oldest lunar basin and help determine whether basin bombardment was continuous from the time of the solidification of the crust or whether there was a separate, later event of basin bombardment.

12.2. High-Resolution Images

The Lunar Orbiter photos of the South Polar Region (Table 12.1), like those of the North Polar Region, have a high degree of overlap because they were taken from a near polar orbit. The rectangular sets of three subframes in a high-resolution exposure taken at perilune have the pole in or near the bottom of the southern subframe (designated H1). The illumination angle in the more northern subframes (H2 and H3) is about 10°, but the angle approaches 0 toward the pole. Of course, the pole is always within 1.5° of the terminator. The actual South Pole was in shadow during this mission.

Although the polar photos were not taken on every orbit over the poles, there is still so much overlap in the middle and southern subframes that only every other photo is shown here.

After an early success (LO4-005H1, -H2, and -H3). the planned high-resolution photos of the eastern South Polar Region were spoiled by the failure of the thermal door. Near the end of the mission, apolune photos were taken of that area. One subframe of these (LO4-184H3) is presented near the end of this chapter to provide complete coverage of the near side South Polar Region. Table 12.1 shows the high-resolution images of the South Polar Region in schematic form.

The following pages show the high-resolution subframes from south to north and west to east. The photos are in the order LO4-193H1, LO4-193H2, LO4-193H3, LO4-179H3, LO4-186H1, LO4-186H2, ... After LO4-184H3, an apolune photo, are LO4-005H1, LO4-005H2, and LO4-005H3.

Figure 12.4. South Pole (USGS Astrogeology PIA00002, NASA). This mosaic covers the South Polar Region within 10° of the pole. The near side is in the upper half of the photo. Crater Shoemaker, near the pole, was in shadow throughout the Lunar Orbiter 4 and Clementine Missions.

Lattitude Range	Photo Number														
27 S–56 S N	194, 186	180, 172	167, 160	155, 148	142, 136, 131	124, 119	112, 107	100, 095	188, 083	076, 071	064, 059	076, 071	184H1, H2	178H1, H2	009
55S–65S	193H3	179H3	166H3	154H3	130H3	118H3	106H3	094H3	082H3	070H3	058H3	044H3	184H3	178H3	005H3
65 S–75 S	193H2	179H2	166H2	154H2	130H2	118H2	106H2	094H2	082H2	070H2	058H2	044H2			005H2
75 S–90 S	193H1	179H1	166H1	154H1	130H1	118H1	106H1	094H1	082H1	070H1	058H1	044H1			005H1

Longitude at 50 S	90 W, 80 W	70 W	60 W, 50 W	50 W, 40 W	40 W, 30 W, 20 W	10 W	0 E/W, 10 E	20 E	30 E	40 E	50 E	60 E	50 E, 60 E, 70 E		80 E, 90 E

Table 12.1. The cells shown in white represent the high-resolution photos of the South Polar Region that are printed in this chapter (LO4-XXX H1, -H2, and -H3, where XXX is the Photo Number). The shaded cells within the South Polar Region represent photos that are not shown in this chapter because of redundancy; however, they are in the included CD. North of this region are the Orientale Basin Region, the Humorum Basin Region, the Nectaris Basin Region, and the Eastern Basin Region. Exposure 184H3 was taken from apolune in afternoon sun.

LO4-193H1 **Sun Elevation: 9.3°** **Altitude: 3519.21 km**

Drygalski is a Pre-Nectarian crater (149 km diameter). It has a strong central peak and a well-defined rim, although it has been degraded by deposits from subsequent impacts. It has been covered with ejecta from the Bailly Basin (LO4-193H2, LO4-179H3), whose main ring passes just to the northeast of Le Gentil. Striations near Drygalski that come from the northeast are radial to Bailly. Drygalski has impacted broader ridges and valleys at right angles to those from Bailly; they are radial to the South Pole–Aiken Basin. Drygalski is on the topographic ring of that basin.

LO4-193H2 Sun Elevation: 9.3° Altitude: 3519.21 km

Hausen is a well-formed crater with a rugged central peak, a suggestion of an inner ring, and a well-preserved ejecta blanket. It is the largest crater (167 km) of either the Eratosthenian Period or the Copernican Period. Only Humboldt (207 km, Late Imbrian) is larger among impact events after Orientale. The ejecta of Hausen, which overlies the main ring and floor of the Bailly Basin, shows a herringbone pattern, especially to the north. This pattern results from lateral velocity imparted to the edges of a tongue of ejecta during landing. Outside of the Hausen ejecta, to the north of Bailly, is a heavy cluster of Orientale secondaries that help to establish the sequence as first Bailly, then Orientale, then Hausen.

LO4-193H3 Sun Elevation: 9.3 Altitude: 3519.21 km

This area is covered with the sculpted ridges and valleys (such as Vallis Baade) of the Inner Hevelius Formation, the heavy ejecta blanket from the Orientale Basin. Many of the craters such as Catalan are caused by objects thrown high from Orientale that subsequently fell on the fresh ejecta blanket of Orientale (the ejecta blanket followed a faster path at a low inclination to the surface). The flatness of the terrain in the lower left of this photo results from its formation as the floor of the ancient Mendel-Rydberg Basin, which continues west into the far side. The main ring of this basin is north of Guthnick and east of Yakovkin; the eastern sector of this ring can be seen in LO4-179H3.

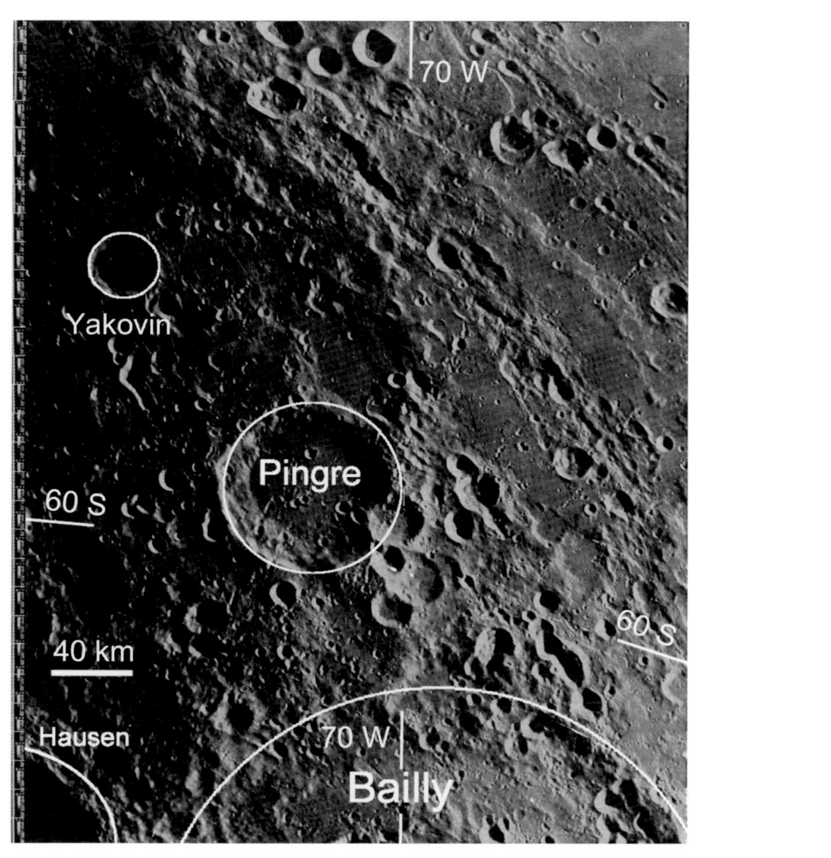

LO4-179H3 **Sun Elevation: 7.1°** **Altitude: 3591.83 km**

Heavy clusters of Orientale secondary craters dominate this area. The main ring of the Medel-Rydberg Basin can be seen entering this photo from the upper left corner, curling around Pingre, and intersecting the main ring of the Bailly Basin (the Bailly Basin has obliterated the older ring). Although the rings of the Mendel-Rydberg Basin are much degraded, the depression of its floor is very clear in the Clementine elevation data.

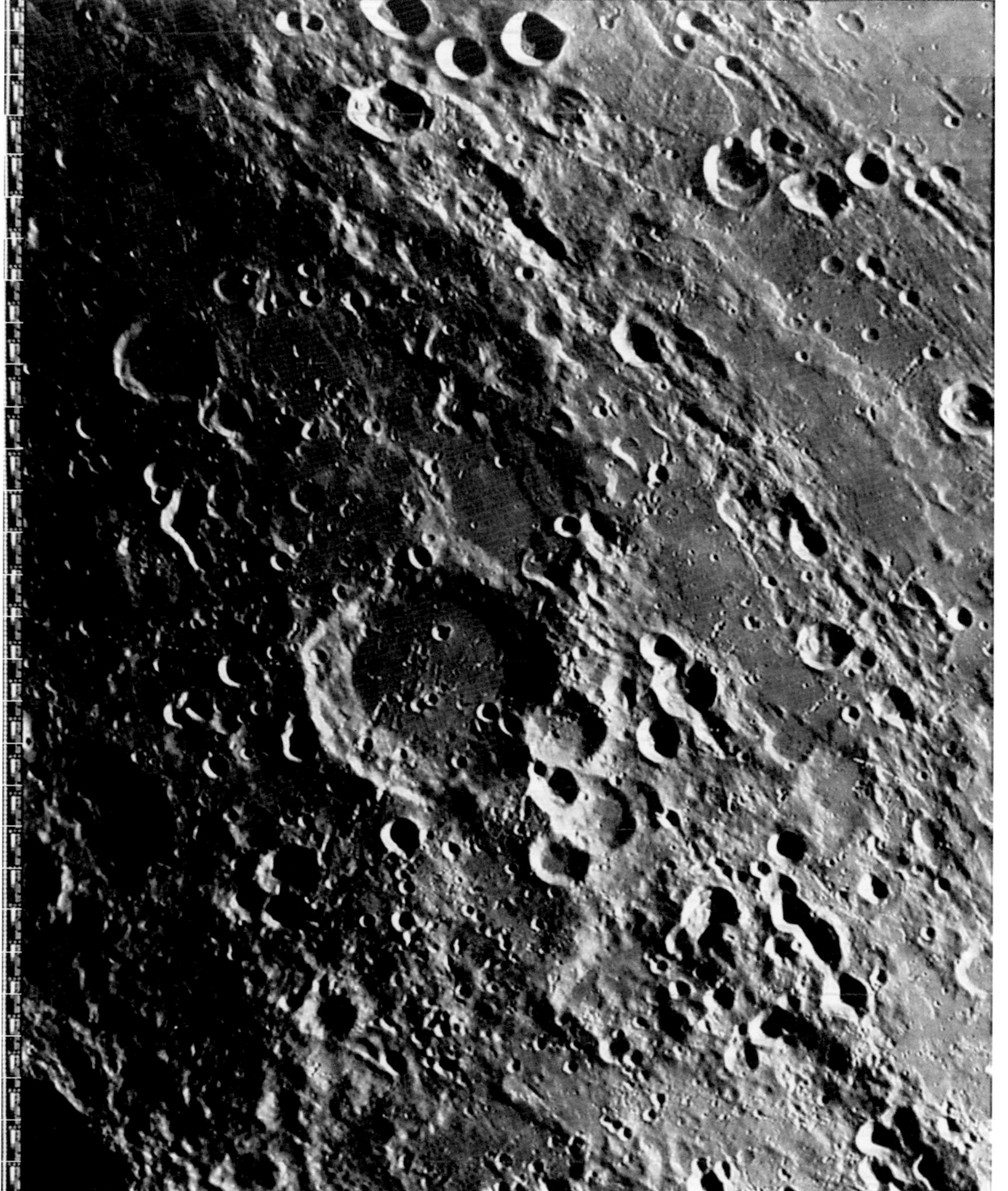

LO4-166H1 **Sun Elevation: 7.0°** **Altitude: 3592.60 km**

As is typical in the photos of this mission that cover the South Pole, not only the pole itself but much of the surrounding territory is in shadow, some of it in permanent shadow, exposed to the cold of deep space without heat from the sun. This is one of the better views of Malapert, showing it to be of a chaotic form. It is not clear whether it is a crater whose walls were pushed aside by other impacts or simply a chance arrangement of debris from multiple impacts.

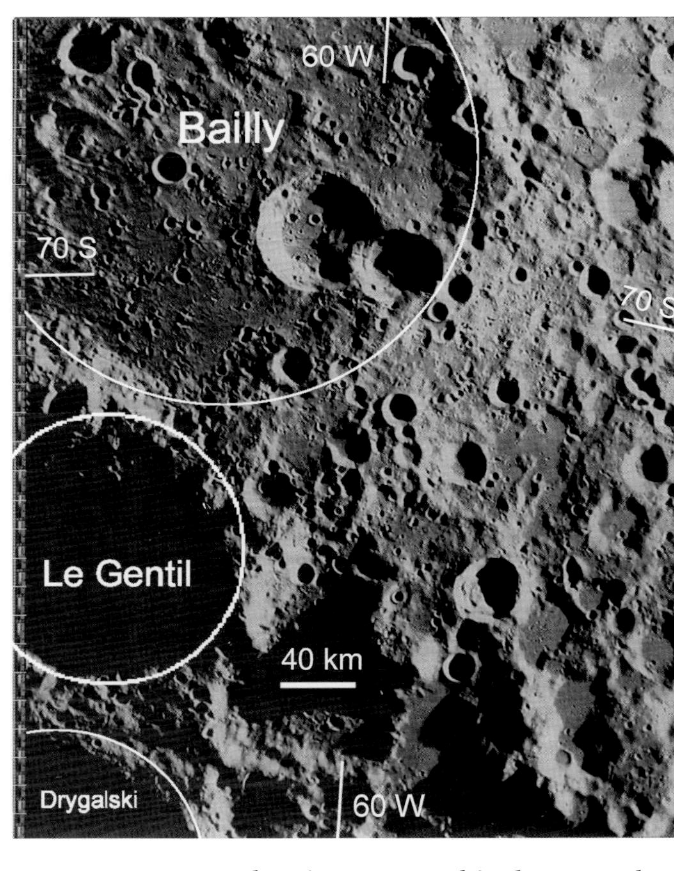

LO4-166H2 **Sun Elevation: 7.0°** **Altitude: 3592.60 km**

This is the most complete view of Bailly, just big enough at 300 km to form a multi-ringed basin. An outer ring runs near the right edge of this photo and into Le Gentil; sectors of an inner ring are also visible. Secondaries from Orientale dominate the region, but earlier craters have been covered with a more uniform pulverized material, probably also part of Orientale ejecta.

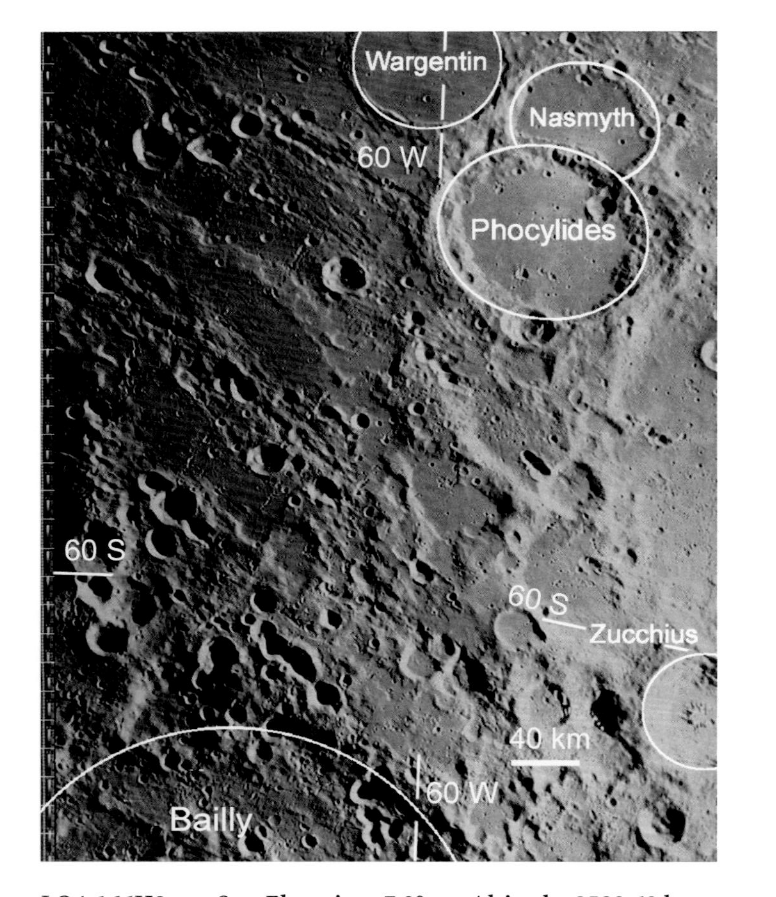

LO4-166H3 **Sun Elevation: 7.0°** **Altitude: 3592.60 km**

The chains of secondaries are radial to Orientale. The heavy ejecta blanket seems to end at a ridge that is running from the eastern edge of Phocylides down to the northern edge of Zucchius. This is a sector of the main ring of the Schiller-Zucchius Basin. Plains units have been formed by the powdery outer Hevelius Formation of Orientale covering flat crater floors such as those of Phocylides, Wargentin, and Nasmyth and the floor of the Schiller-Zucchius Basin (LO4-154H3).

LO4-154H1 Sun Elevation: 10.8° Altitude: 3613.43 km

The 80th southern parallel (the 80° S line of latitude) is shown here as a dashed arc. In three-dimensional geometry, a parallel is a circle, but in this photograph, taken when the spacecraft was at 72.07° S, the circle is foreshortened. Craters near the South Pole, like those at the North Pole, are often named after explorers of the poles of Earth. Robert Falcon Scott, whose crater is shown particularly well in this photo, led the second party to reach the South Pole of Earth in January 1912, 1 month after Amundsen had been there. Scott died along with two of his men on the return from the Pole.

LO4-154H2 Sun Elevation: 10.8° Altitude: 3613.43 km

This area has not only been bombarded with primary impactors during the formation of the Moon and after, but has been subjected to burial with ejecta from the South Pole–Aiken Basin, the Bailly Basin, and secondary craters probably from the Orientale Basin. Klaproth, young enough to have a reasonably intact rim in its northeastern sector, has had other sectors of its rim impacted in turn by Casatus and two other craters. However, its floor seems young, flat, and relatively smooth. Perhaps tectonic forces have leveled pulverized ejecta in that area. Whatever happened to smooth the floor happened long ago. Under high magnification, a heavy peppering of smaller (2 km and less) craters can be seen.

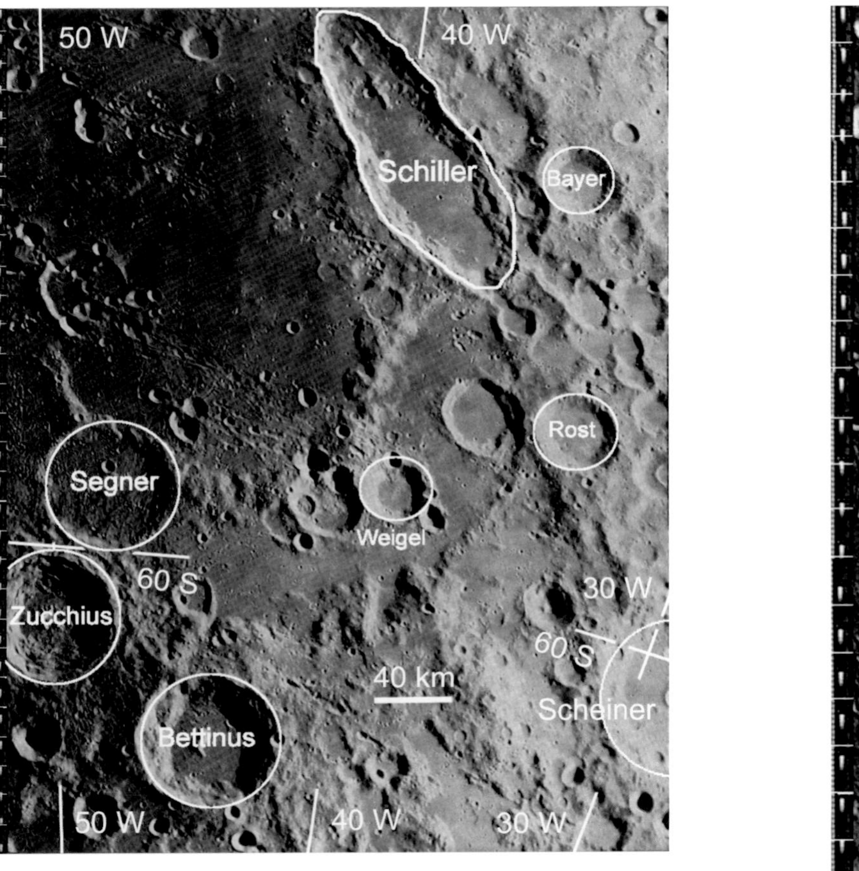

LO4-154H3 Sun Elevation: 10.8° Altitude: 3613.43 km

An outer ring of the Schiller-Zucchius Basin can be seen running just east of Schiller, south to the edge of Rost, and around beneath Zucchius. Within this outer ring is the main ring, incomplete in its northern sector. Secondary chains on the floor of this basin are from Orientale to the northwest. Schiller was probably formed by the simultaneous arrival of two or more primary impactors coming in at a very low angle to the surface.

LO4-130H1 Sun Elevation: 8.7° Altitude: 3574.83 km

Smaller craters, less than 30 km in size, are relatively deep. If they are near a pole their floors are likely to be in permanent shadow. This is especially true if they are on the northern part of the floors of larger craters whose rims then help to shade them.

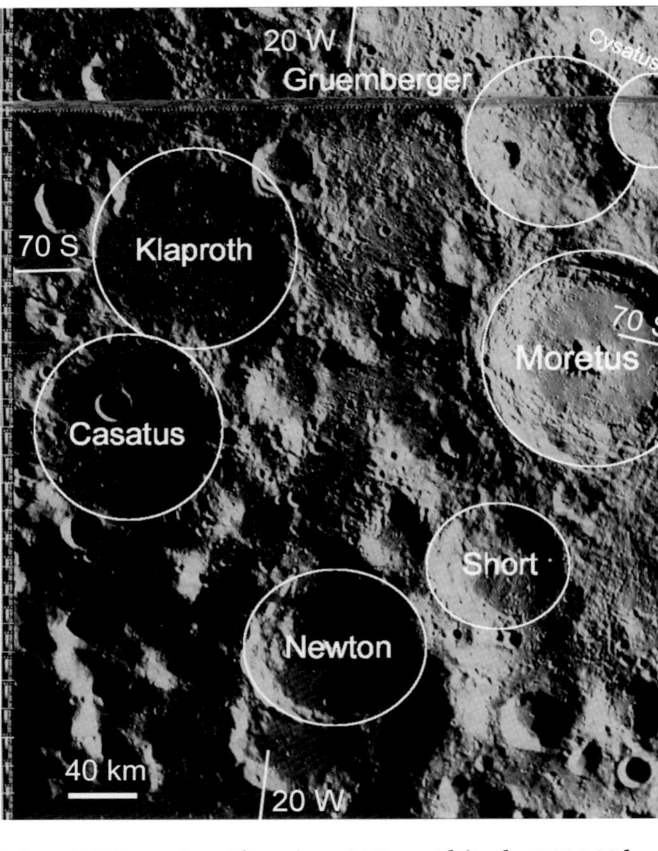

LO4-130H2 **Sun Elevation: 8.7°** **Altitude: 3574.83 km**

The ancient, over-cratered terrain of this area is interrupted by Eratosthenian Crater Moretus, with a well-formed central peak and an intact ejecta pattern.

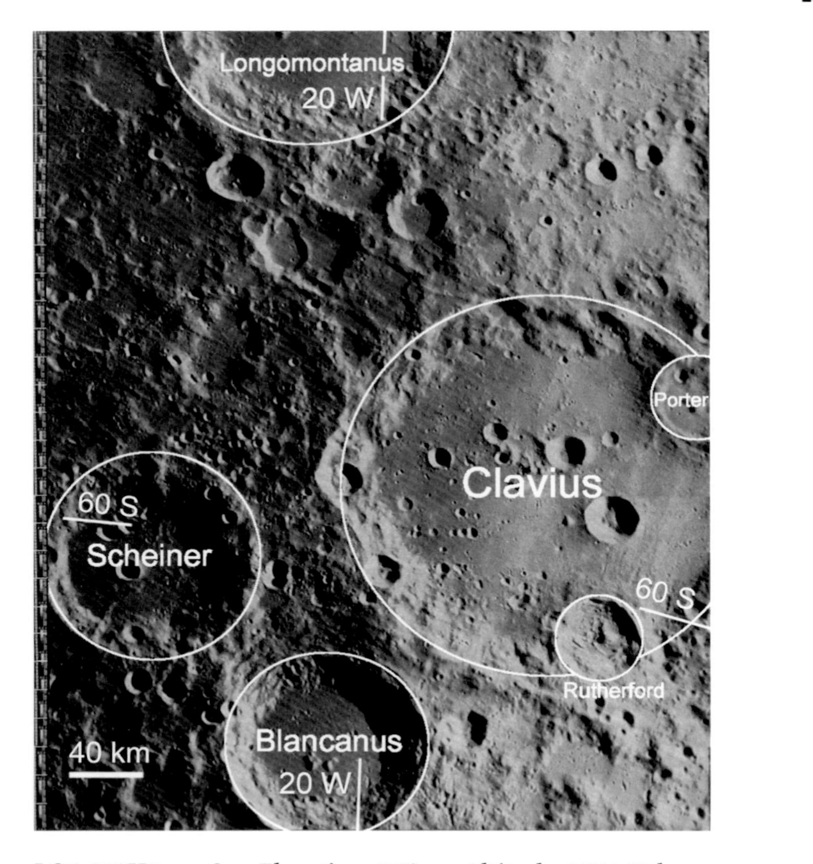

LO4-130H3 Sun Elevation: 8.7° Altitude: 3574.83 km

The smoothing of the terrain north and south of Clavius is probably caused by either its ejecta blanket or the shock of its impact on a preexisting accumulation of ejecta from other impacts. At 225 km, it is about as big as a crater can be; it is in the transition range between craters and basins. The original central peak may have been larger than it is now. It may have been largely buried by ejecta. Rays from Copernican crater Tycho to the north have also landed in this region. This area may contain secondaries from both the Orientale and Imbrium Basins.

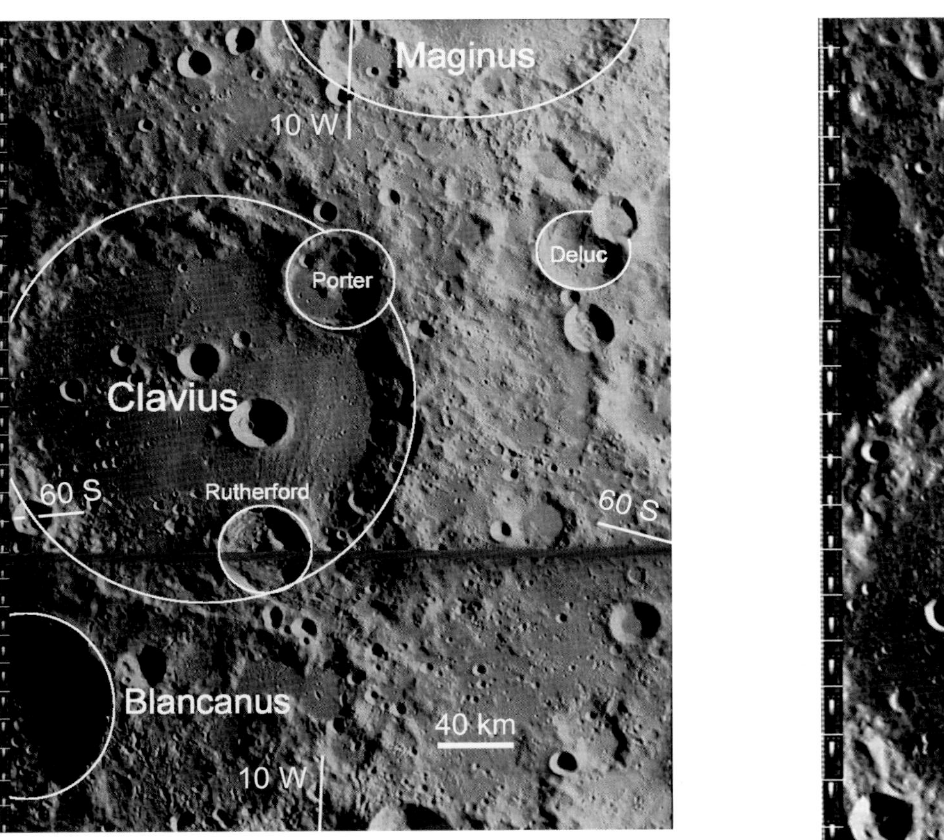

LO4-118H3 **Sun Elevation: 7.8°** **Altitude: 3554.96 km**

Slumps from the rim of Clavius to its floor are about 15 to 20 km in width. This size is typical of the slumps in crater walls, independent of the diameter of the crater. Smaller craters tend to have slumps in the same absolute size range.

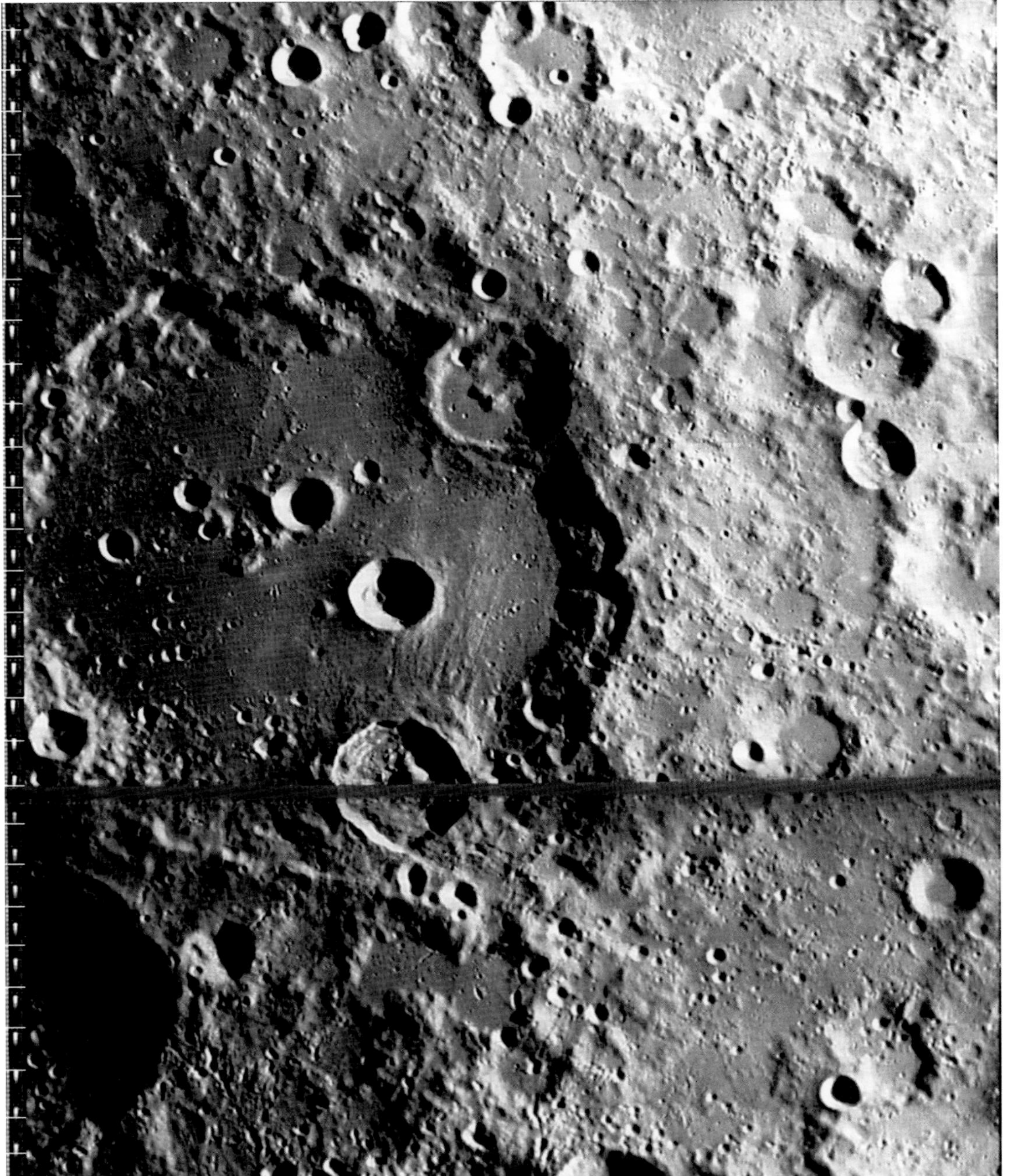

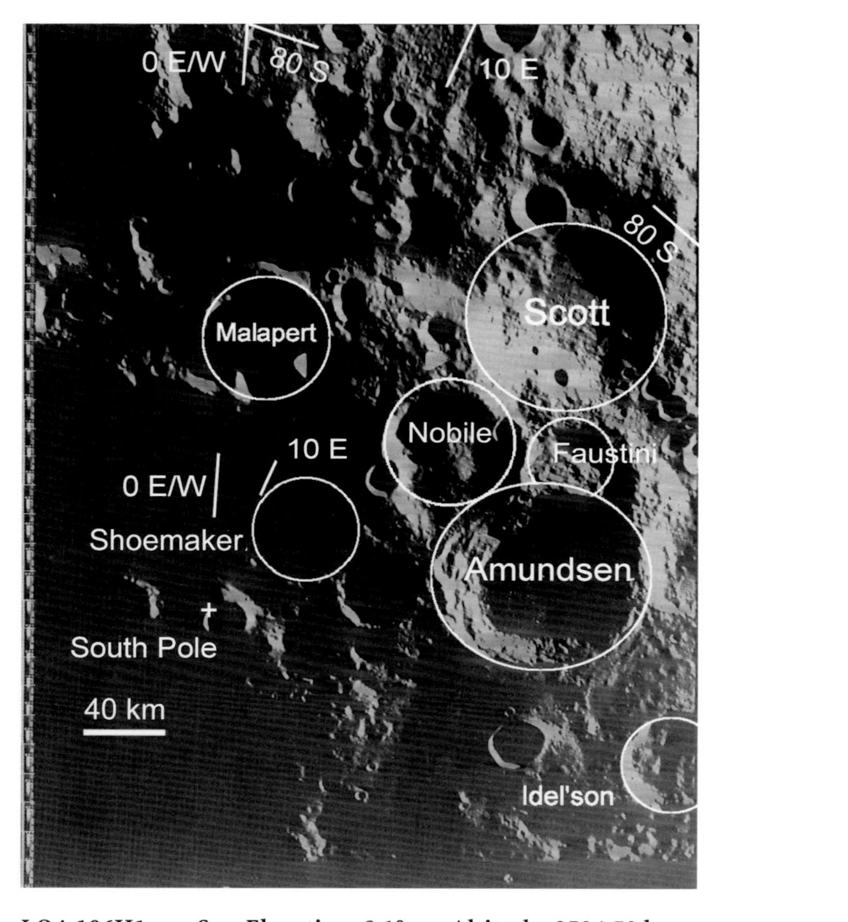

LO4-106H1 **Sun Elevation: 9.1°** **Altitude: 3534.50 km**

Eugene M. Shoemaker, the founder of the USGS Astrogeology Branch, has been a major leader in revealing the impact origin of most Moon craters and some on Earth. Gene's successful career was ended by a fatal automobile accident as he sought to investigate yet another crater in Australia. At the end of its successful mission, the spacecraft Lunar Prospector was crashed into this permanently cold crater floor (now named after Gene) in an attempt to detect water in the crash plume. The spacecraft carried a vial containing his ashes, and Gene became the first person to be interred beyond Earth.

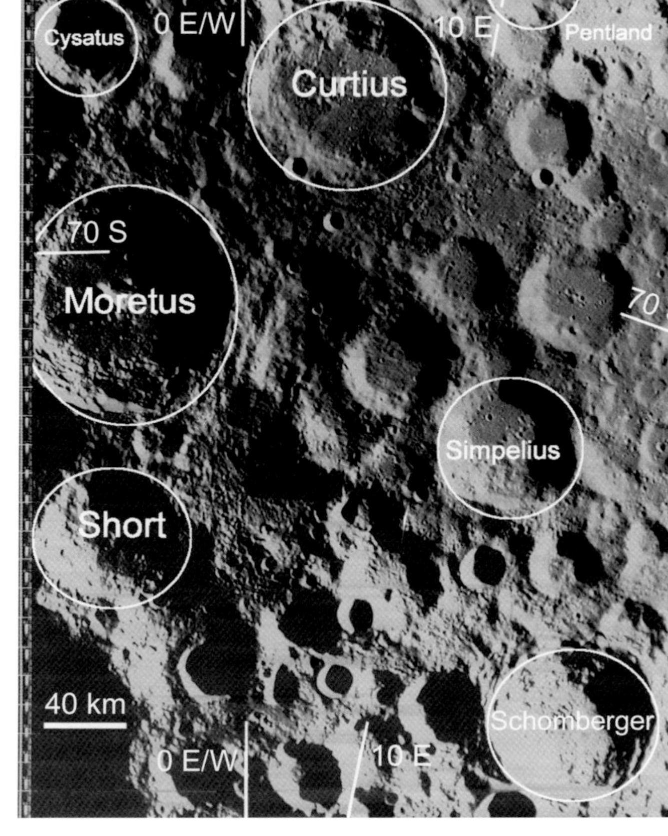

LO4-106H2 **Sun Elevation: 9.1°** **Altitude: 3534.50 km**

There is a transition in this area, especially to the east, between heavy cratering in chaotic debris and cratering in a relatively flat surface. This could mark the transition between the inner heavy ejecta from the South Pole–Aiken Basin and the outer plains-forming and secondary field of that basin. To the west, the equivalent area has been transformed by smaller basins from impacts that arrived long after the formation of the South Pole–Aiken Basin, but to the east there are fewer such basins.

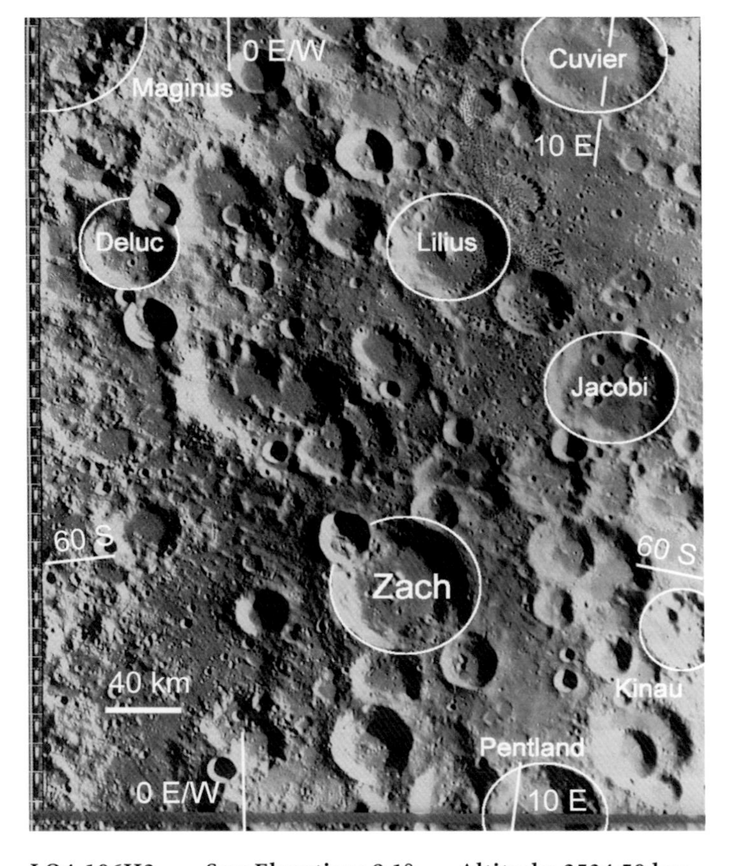

LO4-106H3 **Sun Elevation: 9.1°** **Altitude: 3534.50 km**

This area is about one basin diameter from Nubium and Nectaris, about 1.5 diameters from Imbrium, and about 0.5 diameter from the South Pole–Aiken Basin. This field of fresh 30-km secondary craters is radial to Imbrium. Some groups of small (2-km) secondaries are thought to come from the Nectaris Basin far to the northeast. Striations near the left edge of this photo are radial to Clavius.

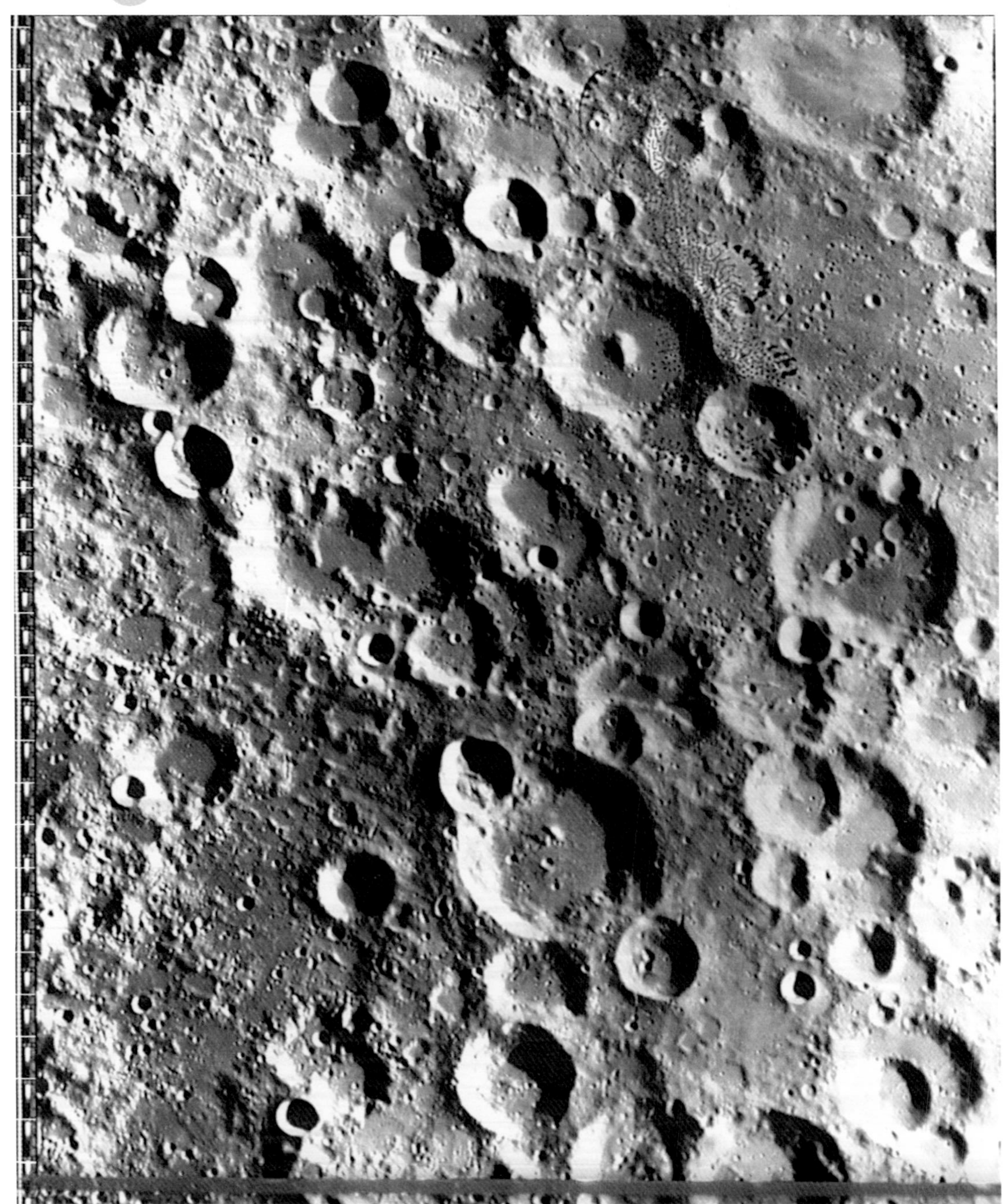

LO4-094H3 Sun Elevation: 9.8° Altitude: 3516.92 km

This flat area is about as far as one can get from basins and large craters on the near side. Still, there are many crater clusters of various sizes and degrees of degradation that are probably secondaries from remote basins. A basin has been proposed to be north of Mutus, but Clementine elevation data show that area to be relatively high. The area may be crust that has escaped major impacts by chance.

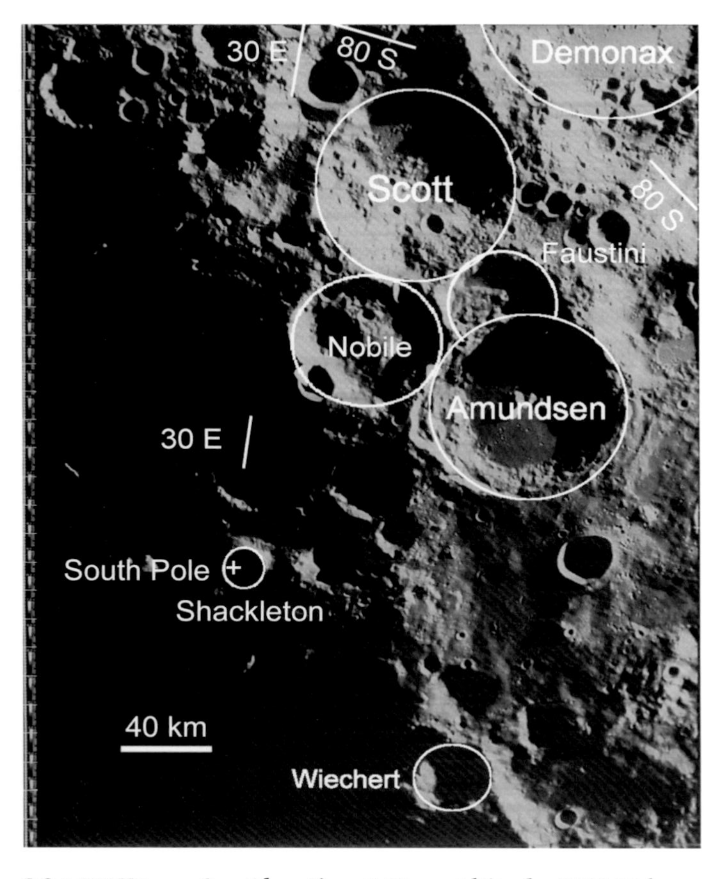

LO4-082H1 **Sun Elevation: 8.7°** **Altitude: 3502.94 km**

This is the sunniest view of crater Amundsen from Lunar Orbiter Mission 4. Roald Amundsen led the first expedition to reach the South Pole of Earth, succeeding in December 1910. He returned safely and continued explorations of both polar regions. The cluster of fresh-looking craters in the 10- to 30-km size range are probably secondaries from Imbrium, a full two diameters away from its main ring. This is further than secondaries are usually thrown, but because of the size of Imbrium, the curvature of the Moon may extend their travel. To put it another way, the ejection velocity was probably closer to escape velocity.

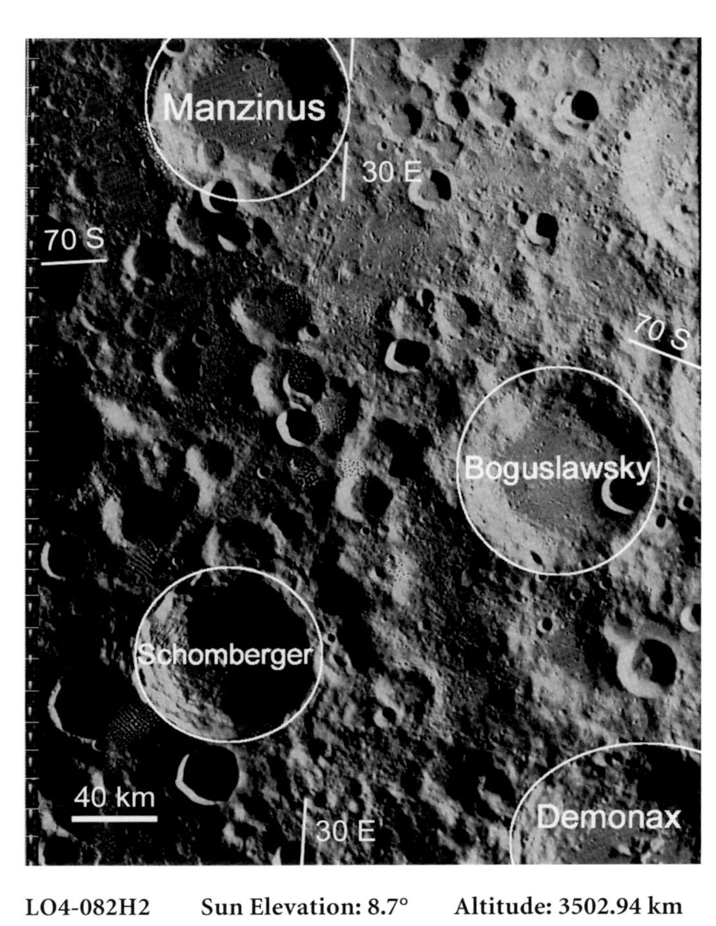

LO4-082H2 **Sun Elevation: 8.7°** **Altitude: 3502.94 km**

The prominent chains of craters here are radial to Imbrium. Schomberger has escaped such impacts; in fact, its ejecta has fallen on some secondary craters. It has been assigned to the Late Imbrian Epoch.

LO4-082H3 **Sun Elevation: 8.7°** **Altitude: 3502.94 km**

Vlacq, Nearch, and Mutus are believed to be Pre-Nectarian. A chain of secondary craters radial to the Nectaris Basin can be seen on the northwest rim of Nearch. Other crater chains between Tannerus and Nearch are radial to Orientale.

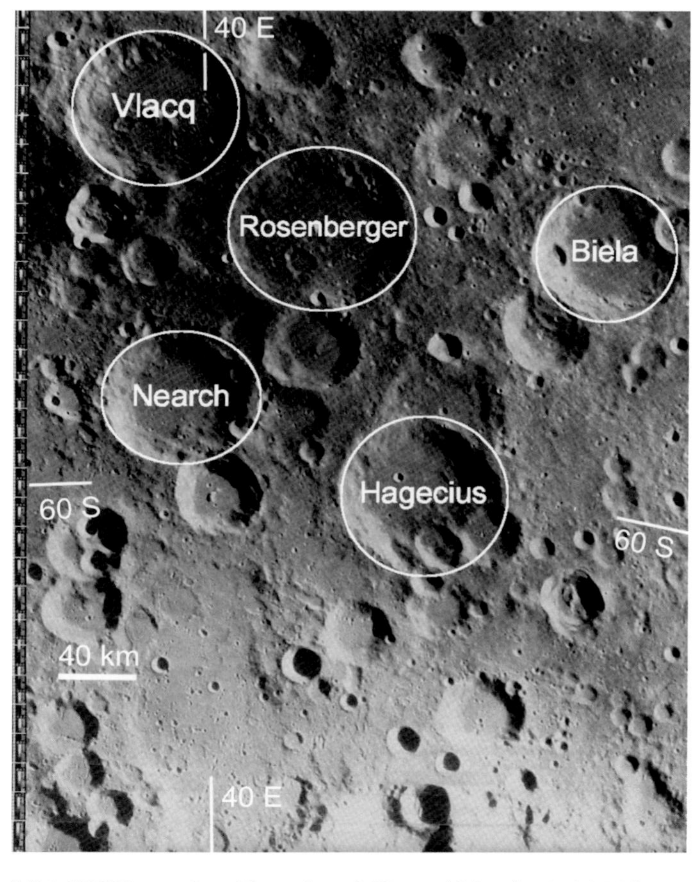

LO4-070H3 **Sun Elevation: 9.7°** **Altitude: 3496.53 km**

The long chain of small craters in the lower right of this photo is radial to the Nectaris Basin. Similar chains can be seen north of Biela.

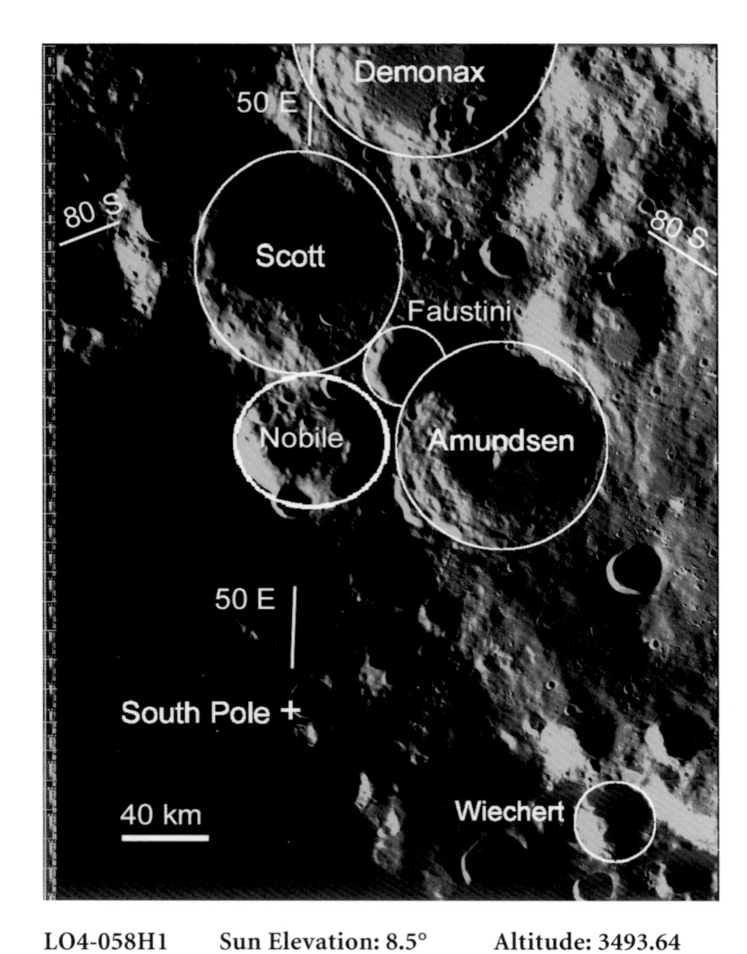

LO4-058H1 **Sun Elevation: 8.5°** **Altitude: 3493.64**

Nobile is well illuminated in this photo, but its floor remains in shadow. Umberto Nobile was an aeronautical engineer and an explorer of the Artic. He and Roald Amundsen were the first to fly over the North Pole of Earth (1926); they used a dirigible. Two years later, on another voyage of exploration, Nobile crashed in a dirigible on the ice near Spitsbergen. Amundsen died in an attempt to fly in and rescue Nobile, who survived.

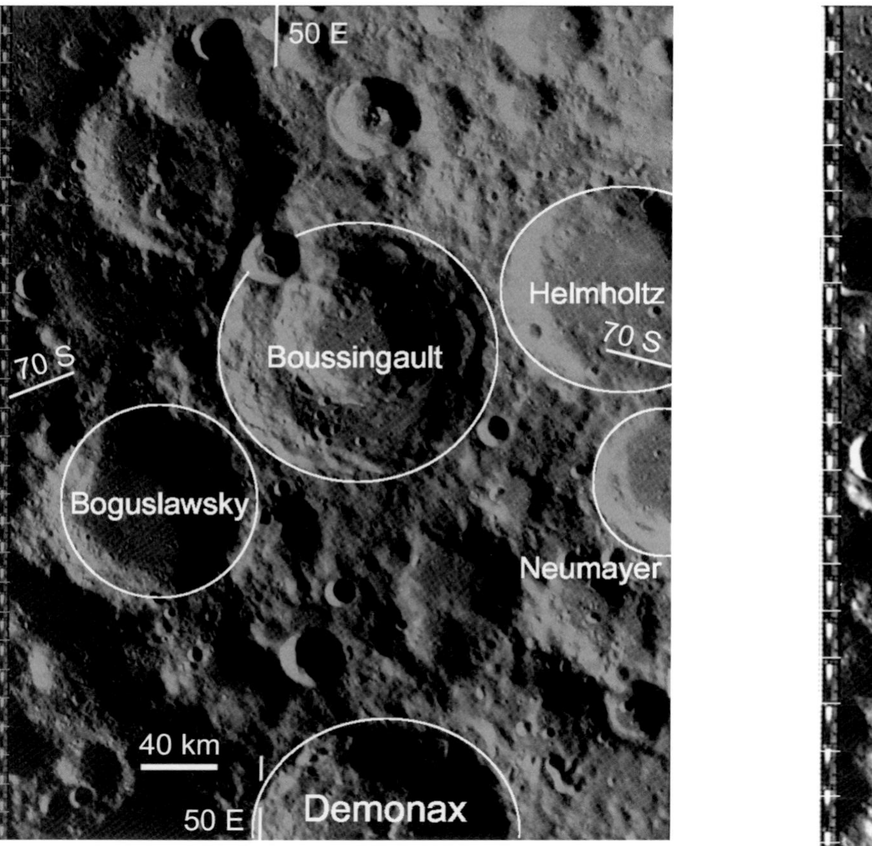

LO4-058H2 Sun Elevation: 8.5° Altitude: 3493.64 km

Boussingault is not a nested crater produced by impact into a layered target: a subsequent impact has hit its floor to create this unusual topography. The relative ages of craters in this area can be inferred from the sharpness of their features and degree of subsequent cratering. All the named craters in this photo are assigned to the Pre-Nectarian Period.

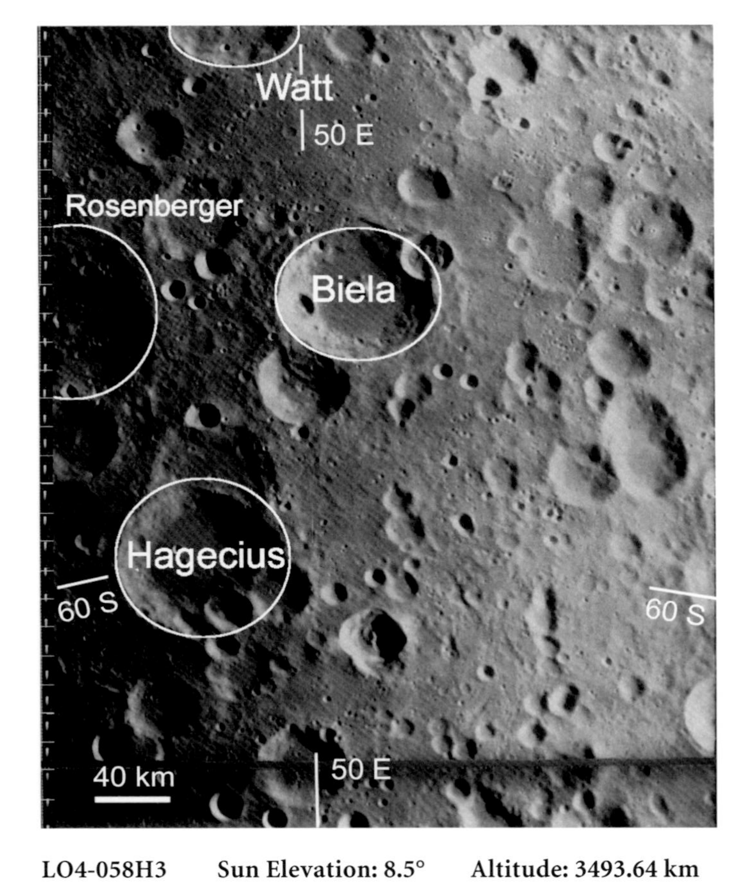

LO4-058H3 **Sun Elevation: 8.5°** **Altitude: 3493.64 km**

Near the top of this photo there are striations radial to Nectaris, to the northwest. Crater chains on the right side of the photo are radial to the Fecunditatis Basin to the north, but they seem too fresh to have come from a basin that old. Perhaps they came from the Crisium Basin, beyond Fecunditatis.

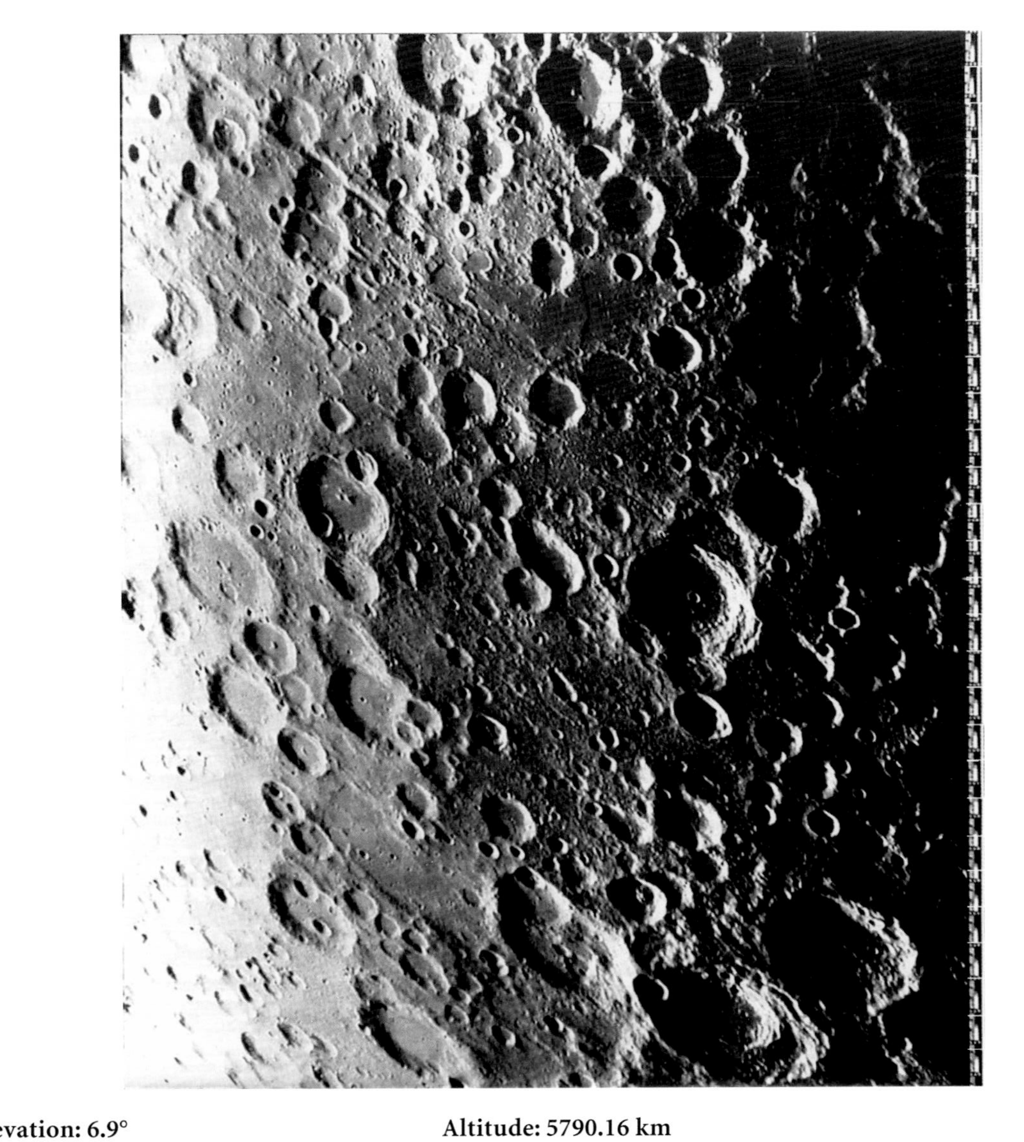

LO4-184H3 | Sun Elevation: 6.9° | Altitude: 5790.16 km

This is an apolune photo, taken at the end of the mission to fill in coverage that was missed at the beginning of the mission due to the thermal door problem. Consequently, the sunlight comes from the west. Crater chains that are aligned with Vallis Rheita come from Nectaris. Those that are more aligned with the north-south direction are radial to the Fecunditatis and Crisium Basins. The plains north of Biela are covered with Nectaris ejecta. Pontecoulant is of the Nectarian Period.

LO4-005H1 **Sun Elevation: 8.9°** **Altitude: 3497.58 km**

Photo LO4-005 was taken early in the mission, before the problem with the thermal door occurred. This photo spans the 90° east meridian, the boundary between the near side and far side. Traces of a very old, large (200-km) crater can be seen in the upper right of this photo, lying beneath Ganswindt and Idel'son. Its degraded rim is tangent to the north rim of Ganswindt and runs above the edge of the photo around to Hedervari.

LO4-005H2   **Sun Elevation: 8.9°**   **Altitude: 3497.58 km**

The heavy ridges and valleys coming from the northeast, passing west of Wexler and east of Neumayer, are radial to the Australe Basin. The ridges and valleys that cross them are radial to the Schrodinger Basin on the far side (see Figure 12.3). Neumayer is Nectarian. Hale appears to be much younger, probably Eratosthenian.

LO4-005H3 **Sun Elevation: 8.9°** **Altitude: 3497.58 km**

The center of the ancient Australe Basin is east of Lyot and north of Jeans. The floor of this basin nearly, but not quite, flooded with mare lava. Large craters such as Lyot and Lamb (LO4-009H2 in the Eastern Basin Region) seem to have stimulated the flow. The jagged dagger in the lower right corner of the photo is the tip of Vallis Schrodinger, a 310-km-long radial fracture from the far side Schrodinger Basin (see Figure 12.3).

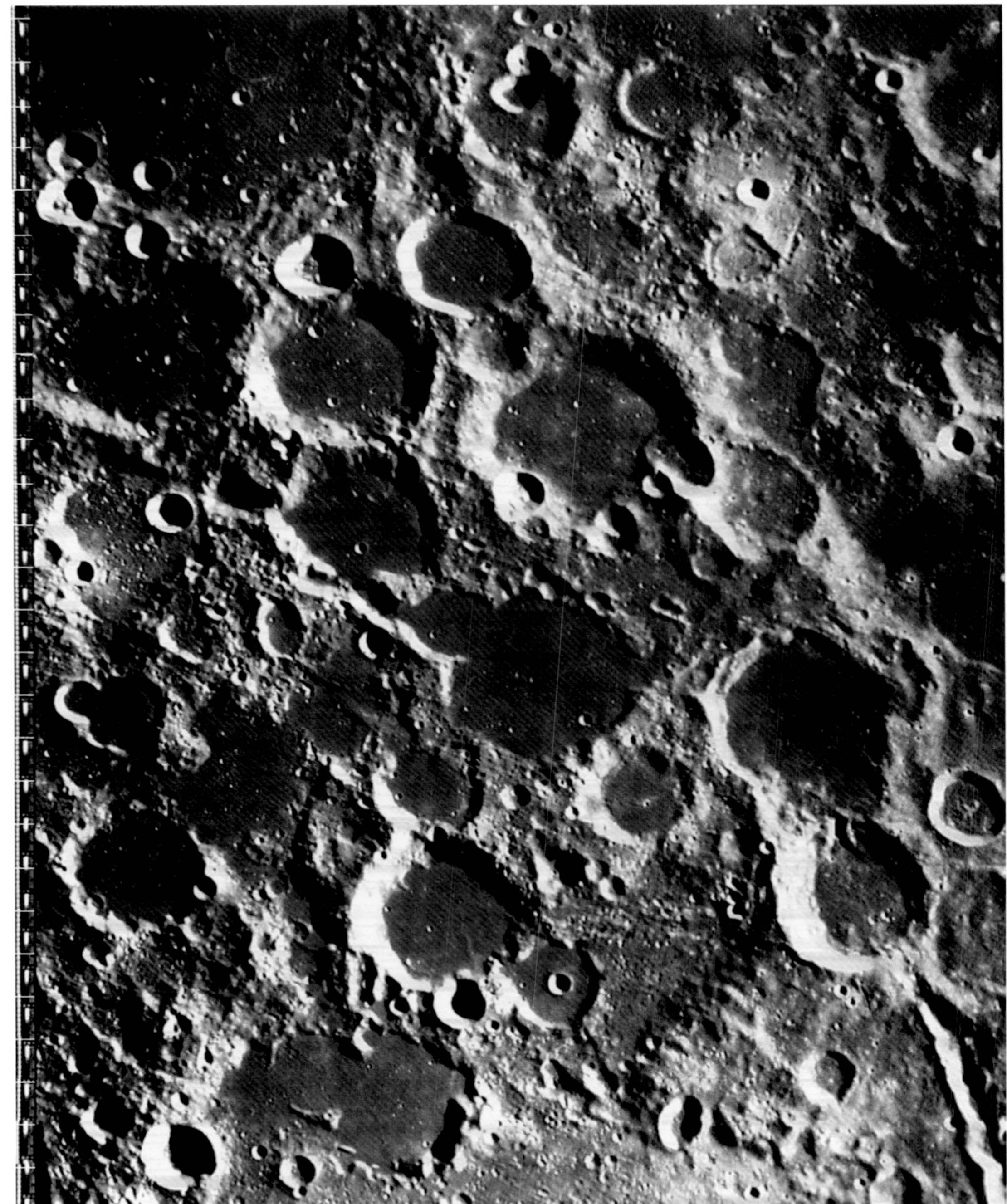

Glossary

Latin Words in Names of Features

Catena: *Bowl,* a chain of craters.
Dorsum: *Backbone,* a ridge, often wrinkled, in the surface of a mare.
Lacus: *Hollow,* a small depression filled with lava flows.
Mare: *Sea,* dark, smooth plains formed by lava flows in a large depression.
Mons, Montes: *Mount, mountains,* a peak or range of mountains.
Palus: *Swamp,* a small area with shallow lava flows.
Planitia: *Plain,* a low flat area.
Promontorium: *Promontory,* a highland peninsula extending into a mare.
Rima: *Crack, leak,* a narrow, long channel that may be sinuous (a sign of meandering lava flow), straight, or in an arc (signs of a fault).
Rupes: *Cliff,* a long abrupt change in elevation; a cliff along a vertical fault or an edge of a lava flow.
Sinus: *Bay,* an extension of a mare.
Vallis: *Valley,* a wide, long depression.

Age Ranges on the Moon (Oldest First)

Pre-Nectarian Period: The age range between the hardening of the crust and the impact event that formed the Nectaris Basin.
Nectarian Period: The age range between the formation of the Nectaris Basin and the Imbrium Basin.
Early Imbrian Epoch: The age range from the formation of the Imbrium Basin through that of the Orientale Basin.
Late Imbrian Epoch: The age range from the formation of the Orientale Basin through the most voluminous lava flows.
Eratosthenian Period: The age range of declining heavy lava flow and declining rate of impact. This period includes the formation of rayless post-mare craters like Eratosthenes.
Copernican Period: This age range extends from the time of formation of the oldest bright-rayed craters to the present. Copernicus, with very bright rays, is the typical young crater of this period. Since the rays of craters fade with time, craters with dimmer rays are assigned older ages in this period.

General Terms

Albedo: The inherent brightness of a surface.
Apolune: The point on a lunar orbit farthest from the surface of the Moon.
Basin: A large crater-like depression containing one or more rings in addition to a rim. Also "multi-ringed basin."
Crater: A compact depression. Nearly all craters on the Moon are caused by impact; when fresh, impact craters usually are nearly circular and have raised rims.
Crust: The upper layer of rock, composed of minerals that have separated from a melt and risen because of their low density.
Ejecta: Material thrown out from a transient crater as a result of the energy released by an impact.
Limb: An edge of a celestial body like the Moon, as viewed.
KREEP: An acronym representing potassium (chemical symbol K), Rare Earth Elements, and Phosphor. These elements are late to crystallize from a cooling melt.
Main ring: The highest ring of a basin. Also called the rim of the basin or the topographic ring.
Mantle: The layer of rock below the crust, composed of minerals that have separated from a melt and fallen because of their high density.
Mascon: A concentrated mass that affects the gravity field of the Moon. Mascons are associated with pipes of high-density mantle material that flooded maria.
Perilune: The point on a lunar orbit closest to the surface of the Moon.
Ring: A circular ridge that is formed along with a basin. Internal rings are inside the main ring and external rings are outside the main ring.
Secondary crater: A crater formed by the impact of material ejected from another (primary) crater or basin.
Theia: The small planet that impacted Earth to form the Moon, named for a Greek goddess whose daughter was said to be Selene, the goddess of the Moon.
Topographic rim: The highest ring around a crater. Also called the main ring if the crater exhibits multiple rings and troughs (a multi-ringed basin).
Topography: The shape of a surface.
Transient crater: The material that is fractured, melted, and vaporized by a hypervelocity impact before further dynamic processes redistribute the material as ejecta, secondary impactors, and so on.
Trough: A shallow linear or curved depression, especially a circular valley between rings of a basin.

References

Bowker, 1971: D.E. Bowker and J.K. Hughes, *Lunar Orbiter Photographic Atlas of the Moon,* NASA SP 206.

Bussey, 2004: Ben Bussey and Paul D. Spudis, *The Clementine Atlas of the Moon*, Cambridge University Press, Cambridge.

Byrne, 2002a: C.J. Byrne, *Automated Cosmetic Improvements of Mosaics from the Lunar Orbiter,* Lunar and Planetary Science Conference (LPSC) 33, #1099.

Byrne, 2002b: C.J. Byrne, *A New Moon: Improved Lunar Orbiter Mosaics,* The Moon Beyond 2002: Next Steps in Lunar Science and Exploration, p. 7, LPI Contribution No. 1128, Lunar and Planetary Institute, Houston.

Byrne, 2003: C.J. Byrne, *Proposed High-Level Regional Focal Points for Lunar Geography,* LPSC 34, #1517.

Byrne, 2004: C.J. Byrne, *Evidence for Three Basins beneath Oceanus Procellarum,* Lunar and Planetary Science Conference (LPSC) 35, #1103.

De Hon, 1979: R.E. De Hon, *Thickness of the Western Mare Basalts,* Lunar and Planetary Science Conference (LPSC) 10, pp. 2935–2955.

Gaddis, 1997: L. Gaddis, et al., *An Overview of the Integrated Software for Imaging Spectrometers (ISIS),* Lunar and Planetary Science Conference (LPSC) 28, #387.

Gaddis, 2001: L. Gaddis, et al., *Cartographic Processing of Digital Lunar Orbiter Data,* LPS 32, Abstract #1892, Lunar and Planetary Institute, Houston (CDROM).

Gasnault, 2002: O. Gasnault, W.C. Feldman, C. d'Uston, D.J. Lawrence, S. Maurice, S.D. Chevral, P.C. Pinet, R.C. Elphic, I. Genetay, and K.R. Moore, JGR 107 (E10).

Gault, 1978: D.E. Gault and J.A. Wedekind, *Experimental Studies of Oblique Impact*, Lunar and Planetary Science Conference (LPSC) 9, v. 3, pp. 3843–3875.

Gillis, 2002: J. Gillis (ed.), *Digital Lunar Orbiter Photographic Atlas of the Moon*, www.lpi.usra.edu/research/lunarorbiter, Lunar and Planetary Institute, Houston.

Gonzalez, 2002: Rafael C. Gonzalez and Richard E. Woods, *Digital Image Processing, Second Edition,* Addison Wesley, Reading.

Greeley, 1976: R. Greeley and M.H. Carr, Editors, *A Geological Basis for the Exploration of the Planets,* NASA Report SP-417.

Hawke, 2004: B.R. Hawke et al., *The Origin of Lunar Crater Rays,* LPSC 35, Abstract #1477, Lunar and Planetary Instituite (CDROM).

McEwen, 1994: A. McEwen, P. Davis, and A. Howington-Kraus, *Evidence for a Pre-Nectarian Impact Basin in Northwestern Oceanus Procellarum*, Lunar and Planetary Science Conference (LPSC) 25, Abstract # 869, Lunar and Planetary Institute, Houston.

Quaide, 1968: W.L. Quaide and V.R. Oberbeck, *Thickness determinations of the lunar surface layer from lunar impact craters,* Journal of Geophysics Research 73:5247–5270.

Schultz, 1972: P.H. Schultz, *Moon Morphology,* University of Texas Press.

Spudis, 1993: P.D. Spudis, *The Geology of Multi-Ring Impact Basins: The Moon and Other Planets,* Cambridge University Press, Cambridge.

Spudis, 1996: P.D. Spudis, *The Once and Future Moon,* Smithsonian Institution Press, Washington, DC.

Stevens, 1999: R. Stevens, *Visual Basic Graphics Programming, Second Edition,* Wiley, New York.

Whitaker, 1999: E.A. Whitaker, *Mapping and Naming the Moon,* Cambridge University Press, Cambridge.

Whitaker, 1970; E.A. Whitaker et al., *Atlas and Gazetteer of the Near Side of the Moon,* NASA SP-241.

Wilhelms, 1987: D.E. Wilhelms et al., *The Geologic History of the Moon,* USGS Professional Paper 1348, US Government Printing Office, Washington, DC.

Wilhelms, 1993: D.E. Wilhelms, *To a Rocky Moon,* The University of Arizona Press.

General Index

IAU Named Features

All of the annotated features are listed here, with the exception of the craters. These craters are only those that are mentioned at least once in the notes. For a full list of annotated craters, see the CD.